0-07-060362-6	McDysan/Spohn	*ATM: Theory and Applications*
0-07-042586-8	Minoli	*1st, 2nd, & Next Generation LANs*
0-07-042588-4	Minoli	*Imaging in Corporate Environments*
0-07-042591-4	Minoli/Vitella	*ATM & Cell Relay Service for Corporate Environments*
0-07-046461-8	Naugle	*Network Protocol Handbook*
0-07-911889-5	Nemzow	*Enterprise Network Performance Optimization*
0-07-046322-0	Nemzow	*FDDI Networking: Planning, Installation, and Management*
0-07-046321-2	Nemzow	*The Token-Ring Management Guide*
0-07-049309-X	Pelton	*Voice Processing*
0-07-707778-4	Perley	*Migrating to Open Systems: Taming the Tiger*
0-07-049663-3	Peterson	*TCP/IP Networking: A Guide to the IBM Environment*
0-07-051143-8	Ranade/Sackett	*Advanced SNA Networking: A Professional's Guide to VTAM/NCP*
0-07-051506-9	Ranade/Sackett	*Introduction to SNA Networking*, Second Edition
0-07-054991-5	Russell	*Signaling System #7*
0-07-054418-2	Sackett	*IBM's Token-Ring Networking Handbook*
0-07-057628-9	Simon	*Workgroup Computing: Workflow, Groupware, and Messaging*
0-07-057442-1	Simonds	*McGraw-Hill LAN Communications Handbook*
0-07-060360-X	Spohn	*Data Network Design*
0-07-063263-4	Taylor	*The McGraw-Hill Internetworking Handbook*
0-07-063638-9	Terplan	*Benchmarking for Effective Network Management*
0-07-063636-2	Terplan	*Effective Management of Local Area Networks: Functions, Instruments and People*
0-07-067375-6	Vaughn	*Client/Server System Design and Implementation*

To order or to receive additional information on these or any other McGraw-Hill titles, please call 1-800-822-8158 in the United States. In other countries, contact your local McGraw-Hill representative. **BC15XXA**

ISDN
Implementor's Guide

ISDN
Implementor's Guide

Standards, Protocols, & Services

Charles K. Summers

McGraw-Hill, Inc.

New York   San Francisco   Washington, D.C.   Auckland   Bogotá
Caracas   Lisbon   London   Madrid   Mexico City   Milan
Montreal   New Delhi   San Juan   Singapore
Sydney   Tokyo   Toronto

Library of Congress Cataloging-in-Publication Data

Summers, Charles K.
    ISDN implementor's guide : standards, protocols, & services /
Charles K. Summers.
        p.   cm.—(McGraw-Hill series on computer communications)
    Includes bibliographical references and index.
    ISBN 0-07-069416-8
    1. Integrated services digital networks.   I. Title.   II. Series.
TK5103.75.S93   1995
621.382—dc20   95-9069
                                                                          CIP

    2 3 4 5 6 7 8 9 0   DOC/DOC   9 0 0 9 8 7 6

ISBN 0-07-069416-8

*The sponsoring editor for this book was Jerry Papke, the editing
supervisor was Caroline R. Levine, and the production supervisor was
Donald F. Schmidt. This book was set in Century Schoolbook by Ron
Painter of McGraw-Hill's Professional Book Group composition unit.*

Printed and bound by R. R. Donnelley & Sons Company.

McGraw-Hill books are available at special quantity discounts to use
as premiums and sales promotions, or for use in corporate training programs. For more information, please write to the Director of Special
Sales, McGraw-Hill, Inc., 11 West 19th Street, New York, NY 10011. Or
contact your local bookstore.

 This book is printed on acid-free paper containing 10% post-
consumer waste.

To Betty, Nina, and Tony
and my fellow workers at TeleSoft International, Inc.

Contents

Preface

In the years since the first Integrated Services Digital Network (ISDN) standards were presented, the viability of ISDN has been considered with enthusiasm and gloom. Proponents have continued along the path of strengthening and expanding the standards—demonstrating new methods of using the concepts and standards to present a unified access to the networks. Opponents, or skeptics, have continued to point out the difficulty of completing the cycle of standards, availability, and product access. As is true with most issues, proponents and opponents serve as bookends which may serve to strengthen the final outcome. The final outcome of ISDN is yet unknown, but there is considerable growth of access and use in many parts of the world. ISDN is now a system that can be used.

In the last few years, books have appeared to explain the concepts, uses, and specific technical standards of ISDN. These books cover marketing issues, in-depth technical explanations, high-level technical overviews, standards processes, architectures, general services, and many other issues. The rapid growth in the availability of books, in itself, is an indication of the increased acceptance of ISDN and a reflection of the growing interest in, and use of, ISDN.

This book is oriented toward those who need to know how the various protocols work together as a system. How do the protocols interact? If there is a base standard for ISDN, why are there so many variants and in what ways do they differ? How do the layers of the protocol interact and what issues in common do they have? How are state machines implemented and what are the advantages, and disadvantages, of each approach?

Interested readers may be involved in the implementation of ISDN within a product, or they need to understand the specific methods of providing services that they may need to use. This book is for all such people.

Charles K. Summers

Acknowledgments

Many people deserve to be thanked for the completion of this book. First, great thanks to Gerald T. Papke, my editor at McGraw-Hill, for his patience and support while I juggled the tasks of writing and supporting the business needs at TeleSoft International, Inc. Thanks go next to Matt Wagner, of Waterside Productions, for acting as my agent to find a path to present this book to the publishing world and general public. Finally, thanks go to Gary C. Kessler for generously volunteering to read the final manuscript in preparation for publication. Any errors still remaining are solely my responsibility.

I would also like to acknowledge the various people in my life who made this book possible. Thanks go to Betty Cannon and Nina Friedman for the general support and guidance which allowed this book to come into existence. Anthony C. Graves and Shannon M. Hogan, of AT&T Bell Laboratories, supported me in my initial efforts to properly learn the basics of communications protocols, and ISDN in particular. Next, thanks go to Charles D. Crowe, my business partner and friend, and all the other employees of our company TeleSoft International, Inc.

I also thank Linda Melvin and Scott Loftesness for volunteering to read the manuscript as it was developed. Due to time pressures on the work, I was not able to take advantage of their generous offers— but that does not decrease my appreciation. Finally, I want to express my appreciation to Janet Ruhl and Raymond E. Feist for providing general information on publishing and agents.

Abbreviations and Acronyms

AAL — ATM Adaptation Layer (B-ISDN)

ABM — Asynchronous Balanced Mode (HDLC)

ADPCM — Adaptive Differential Pulse Code Modulation

AIS — Alarm Indication Signal

ANSI — American National Standards Institute

ASCII — American Standard Code for Information Interchange

ATM — Asynchronous Transfer Mode

BCD — Binary Coded Decimal

B-ISDN — Broadband ISDN

BECN — Backward Explicit Congestion Notification (Frame Relay)

BISYNC — Binary Synchronous Communications

BRI — Basic Rate Interface

C-plane — Control plane

CCITT — International Telegraph and Telephony Consultative Committee (old name of ITU-T)

CEI — Connection Endpoint Identifier (Q.931)

CES — Connection Endpoint Suffix

CIR — Committed Information Rate

CLLM — Consolidated Link Layer Management (Frame Relay)

CLP — Cell Loss Priority (ATM)

CODEC — COder-DECoder

CPE — Customer Premise Equipment

CRC — Cyclic Redundancy Check

CRV — Call Reference Value (Q.931)

CS — Convergence Sublayer (B-ISDN)

CUG — Closed User Group (X.25)

DCE — Data Communications Equipment or Data Circuit-terminating Equipment

DE — Discard Eligibility (Frame Relay)

DLCI — Data Link Connection Identifier

DMI — Digital Multiplexed Interface

DoD	Department of Defence	ITU	International Telecommunication Union
DTE	Data Terminal Equipment		
EEPROM	Electronically Erasable Programmable Read-Only Memory	ITU-T	ITU-Telecommunication Standardization Sector (formerly called CCITT)
EID	Endpoint IDentifier	LAN	Local Area Network
ETSI	European Telecommunications Standards Institute	LAP	Link Access Procedure
		LAPB	Link Access Procedure Balanced (X.25, layer 2)
FCS	Frame Check Sequence (HDLC)	LAPD	Link Access Procedure on the D-channel
FDM	Frequency Division Multiplexing	LAPF	Link Access Procedure Frame-mode
FECN	Forward Explicit Congestion Notification (Frame Relay)	LAPM	Link Access Procedure for Modem
FM	Frequency Modulation	LC	Local C-plane
GC	Global C-plane	LCI	Logical Channel Identifier (X.25)
GFI	General Format Identifier (X.25)	LED	Light Emitting Diode
HDLC	High-level Data Link Control	LIFO	Last In, First Out
		LLC	Low-Layer Compatibility
HDTV	High Definition TeleVision	LLD	Low-Level Driver
HEC	Header Error Control (ATM)	LLI	Logical Link Identifier
IA5	International Alphabet No. 5	LMI	Local Management Interface (Frame Relay)
ICE	In-Circuit Emulator	LT	Line Termination
IDN	Integrated Digital Network	MH	Modified Huffman (Facsimile standards endcoding)
IE	Information Element		
I/O	Input and Output	MLP	MultiLink Procedure (X.25)
IP	Internet Protocol		
IPE	In-band Parameter Exchange	MODEM	MOdulator-DEModulator
ISDN	Integrated Services Digital Network	MR	Modified READ (Relative Element Address Designate) (Facsimile standards encoding)
ISO	International Organization for Standardization		
		N-ISDN	Narrowband ISDN

NNI	Network-Network Interface	S-plane	Supervisory plane
NRZ	NonReturn to Zero	SAPI	Service Access Point Identifier (LAPD)
NT	Network Termination	SAR	Segmentation and Reassembly (B-ISDN)
NT1	Network Termination 1		
NT2	Network Termination 2	SCSI	Small Computer Interface System
OAM	Operation and Maintenance (ATM)	SDH	Synchronous Digital Hierarchy
OSI	Open Systems Interconnection	SDL	Specification Description Language
PABX	Public Access Branch Exchange	SLP	Single Link Procedure
PAD	Packet Assembly Disassembly	SONET	Synchronous Optical Network
PBX	Private Branch Exchange	SPID	Service Profile Identifier (Q.932)
PCM	Pulse Code Modulation	SS7	Signalling System 7
PDU	Protocol Data Unit	SSCOP	Service-Specific Connection-oriented Protocol (B-ISDN)
PH	Packet Handler		
PLP	Packet Layer Protocol	STM	Synchronous Transfer Mode
PM	Physical Medium (sub-layer)		
		STM-1	Synchronous Transfer Mode 1
POH	Path OverHead (ATM)		
POT	Plain Old Telephone System	SVC	Switched Virtual Circuit
PPP	Point-to-Point Protocol	TA	Terminal Adaptor
PRI	Primary Rate Interface	TC	Transmission Convergence (sublayer)
PRM	Protocol Reference Model		
		TCP	Transmission Control Protocol
PSDN	Public Switched Data Network		
		TE	Terminal Equipment
PSPDN	Packet-Switched Public Data Network	TE1	Terminal Equipment 1 (ISDN)
PSTN	Public Switched Telephone Network	TE2	Terminal Equipment 2 (non-ISDN)
PVC	Permanent Virtual Circuit	TEI	Terminal Endpoint Identifier
QOS	Quality of Service	TDM	Time Division Multiplexing
RAM	Random Access Memory		
		TID	Terminal IDentifier (Q.932)
ROM	Read-Only Memory		

U-plane	User plane	VCI	Virtual Channel Identifier (B-ISDN)
UART	Universal Asynchronous Receiver/Transmitter	VC	Virtual Circuit
		VPC	Virtual Path Connection (B-ISDN)
UNI	User-Network Interface	VPI	Virtual Path Identifier (B-ISDN)
USID	User Service IDentifier (Q.932)	XID	Exchange Identification
VCC	Virtual Channel Connection (B-ISDN)		

Introduction

Integrated Services Digital Network (ISDN) is an attempt to tie together various telecommunications services into a unified system.
There are many forms of data transmission in use. These include the
traditional forms of voice, fax, and binary file transfer. ISDN gives a
common method of identification of data form. This allows multiuse
access of services on common transmission lines. Initially, because of
lack of consensus on use of identification parameters, many services
will require specific Customer Premise Equipment (CPE). The continued evolution of standard use of ISDN provided information will
allow generic use of equipment on common lines.

ISDN directly allows cost savings by reducing the number of communication lines needed—depending on particular tariff structures in the
area. Other potential savings are realized by simultaneous multiple
uses of the transmission medium and by using internetworking capabilities of public national, or international, digital switching systems.

Knowledge of the mechanisms used within ISDN can be used to
evaluate equipment. It is also mandatory for implementation of new
products. This requires knowledge of all layers and the ways that
they interact.

ISDN as a Software System

Each protocol layer of ISDN provides separate, and distinct, functions
for the system. In this manner, each layer is totally independent and
modularized. However, layer 1 must be able to communicate information to layer 2. Layer 3 must be able to pass information to layer 2,
and so forth. Thus, in addition to knowledge of the specific protocols
within a layer, information is needed about how the interfaces work.
ISDN must be looked at as an integrated system of protocols.

The first, or "lowest," level of ISDN is the physical layer. The physical layer provides the basic digital electronic communication across
the transmission medium. Most of the functions are provided by semiconductor chips. These chips must be initialized according to the

needs of the hardware of which they are part. There must also be a way of communicating information, and commands, between the rest of the system (particularly layer 2) and the chip. This software entity is sometimes called a Low-Level Driver (LLD).

The next layer is the data link layer. It provides an error-detecting (and retransmitting, if necessary) data transfer function. It will receive information from layer 1 and make requests of the hardware.

The third layer is the network layer. For ISDN, this is usually a type of signalling protocol (based on ITU-T Recommendation Q.931). The network layer provides a connection-oriented call setup. It normally only directly communicates with layer 2 and the "upper layers." Sometimes the first three layers are referred to as *chained layers* because these layers are used between endpoints on a communication line. Other layers, if necessary, are passed transparently between the origination and termination equipment.

Finally, there are two classes of provided functions which connect all the layers into a working system. Sometimes these are referred to as the Supervisory-plane (S-plane) and User-plane (U-plane) by the standards. They are called *planes* because they act as a background for all of the module layers and may, depending on use, be considered a "lower" or "higher" layer. Together, all these modules create a functioning ISDN system.

Approach to Material

There is one main question approached in this book. How does it all work together? This is done by analyzing the protocols according to how they can be implemented and the impact of each layer upon other entities of the system. The chapters on the protocols first go into the philosophy behind the protocol. What purpose does it serve? What is its use and why are the specific features important?

The next step is to provide a companion analysis of the standard upon which the protocol is based. No attempt is made to be complete in coverage of all the details of the standard. One reason is that duplicating the material will not be of direct service to the reader. The other reason is that every specific implementation must be guided by the needs of the specification relevant to the certification requirements for the equipment in that area. What will be provided is an analysis of the relevant concepts, and methods of implementation, needed for implementation of any ISDN system—or for evaluation of ISDN equipment that may be used in the home or business environment.

Finally, the interface and management primitives will be examined. How does this protocol layer communicate with other layers? What services are needed from the software environment? What types of internal data are needed for efficient protocol handling?

The last part of the book goes into greater detail on specific implementation concerns. What alternatives are available for implementation of protocol state table systems? What are areas of concern for real-time programming in an ISDN system, and what are the different approaches to them? How do the S- and U-plane (referred to as coordination and management entities) interfaces work with the system?

How to Use This Book

This book is divided into three parts. Part 1 covers basic architecture issues. This is an overview of the history of ISDN and some of the evolution of the standards. Some coverage is given to isolating the differences among ISDN specifications around the world. What types of differences exist? What items are in common? Part 1 also starts the discussion about the protocols by reviewing the basic International Telecommunication Union-Telecommunication Standardization Sector (ITU-T) documents covering the basic architecture of the standards. As a side note to prevent later potential confusion, this book uses bytes and octets interchangeably. Octets are used within ITU-T documentation and bytes are primarily used by software people.

Part 2 covers the protocols. The first three chapters of this part discuss the *chained layers* of the physical, data link, and call management network layers. Broadband ISDN (B-ISDN) and Frame Relay are also discussed in this part. The remaining chapters are related to services that may be commonly used with ISDN. These include *bearer services* (such as ITU-T Recommendations X.31 and X.25 or voice or fax) and rate adaptation protocols used by terminals on the ISDN system.

The final part narrows in on implementation concerns, as mentioned above. Together, these three parts cover the areas of standards, protocols, and services. They may also be broken into the areas of history, use, and implementation. The direct purpose of this book may be divided into three main categories: an implementation guideline, a companion to the standards, and an aid to the analysis of ISDN features.

Implementation guideline

The second two parts are of greatest importance to the implementor of an ISDN system. Start with Chap. 10 to gain insight into different methods of implementing the state machine for the protocol. Proceed to Chaps. 11 and 12 for further architectural considerations and then review relevant protocols in Part 2. The overall architecture of the system needs to be firmly designed before implementing the actual protocols. Remember that, as a system, all of the modules must be able to work together. Also, by starting with the system architecture, many parts will be similar in form. This saves implementation time and can aid in maintenance of the software.

Standards companion

If there is need for analysis of a particular protocol or service, the place to start is the chapter devoted to that protocol. It may also be useful to first read through Chap. 2 to gain a better understanding of the general architecture. Note that most protocols do not operate in isolation. Reading the specific chapter, however, will help determine what other protocols may be of immediate relevancy.

Analysis of features

Chapters 5, 6, and 9 should be of immediate use. Read Chap. 5 for general mechanisms of conveying bearer service information. Chapter 6 goes into greater detail on some specific bearer services. Chapter 9 is relevant if Data Termination Equipment (DTE) is to be used in conjunction with ISDN.

If the specific need is to evaluate the way that software in a product has been implemented, general needs are similar to that mentioned above in reference to "Implementation Guideline." Implementing, or analysis of implementation, follows similar requirements.

Summary of Chapters

The chapters of this book are organized into three parts: basic architecture, the protocols, and implementation concerns. Chapters 1 and 2 are involved with the history of ISDN, its growth from the Open Systems Interconnection (OSI) model, and the general architecture documents concerning ISDN. Chapters 3 through 9 work with the specific protocols including the standards, interworking of layers, and specific features used within the protocols. Chapters 10 through 12 discuss particular architectural issues involved with implementation of a real-time system and with protocol state machines with an emphasis on the use of such with ISDN.

The OSI model and beginnings of ISDN

Chapter 1 is a general discussion of ISDN. It includes some of the specific history of the evolution of the standards. It also covers the general aspects of the OSI model that is used as a foundation for the architecture of ISDN.

N-ISDN, B-ISDN, and auxiliary data protocols

Narrowband-ISDN (N-ISDN) and B-ISDN refer to the general bandwidth (or data transmission rates) used within ISDN. Some of the specific basic architectural requirements for both are covered in Chap. 2. Additionally, some general architectural requirements for data services supported by ISDN are covered. Finally, areas of diversity between various national, and international, specifications are discussed.

Physical layer

The physical layer (layer 1 of the OSI model), discussed in Chap. 3, is responsible for the physical transmission of data. Within ISDN, this layer also is effectively responsible for the physical transmission requirements for a High-Level Data Link Control (HDLC) use of the communication line. The physical layer interacts with layer 2 and the management entity. Control information and data are communicated between modules by use of an LLD.

Data link layer

The data link layer (layer 2 OSI) provides data integrity to the higher-layer modules. For ISDN signalling, this standard is provided by ITU-T Q.921, which is commonly referred to as LAPD (or Link Access Procedure on the D-channel). General HDLC requirements are also covered in Chap. 4, with extended reference to two other commonly used layer 2 HDLC derivatives—ITU-T X.25, layer 2 (Link Access Procedure Balanced, or LAPB), and ITU-T V.120, layer 2. V.120, specifically, is based on the Q.921 standard.

Network layer, call management

Layer 3 of the OSI model is called the network layer. Generically, layer 3 provides independence from knowledge of specific connection protocols. As such, it is heavily involved in the setup of connections and may also be used for data transfer. The ITU-T Q.931 standard is used for connection service with ISDN. This standard, with the complementary ITU-T Q.932 standard used for *supplementary services,* is explored in Chap. 5.

Bearer services

ISDN is useless as a system unless it can provide necessary services to the user. Chapter 5 discusses the various parts of Q.931 that are used for conveying information about the use of the connection. Chapter 6 goes into greater detail about specific bearer capability and level compatibility fields, which can be used for service identification. Specific services, such as voice, fax, X.31, and rate adaptation protocols (ITU-T Recommendations V.110 and V.120) are covered.

Frame Relay

Frame Relay is used over communication lines that are considered to have low error rates. Two specific services are used. These are Permanent Virtual Circuits (PVCs), which require specific endpoints to be established by the switch, and Switched Virtual Circuits (SVCs), which use ISDN to allow flexible connections to be set up. Chapter 7 discusses these standards and those of management requirements that help to make Frame Relay a useful service.

Broadband-ISDN and Asynchronous Transfer Mode

One objection to widespread use of ISDN is that the speed is too slow for modern uses. B-ISDN, covered in Chap. 8, is part of the evolution of standards to allow the digital networks to be used for higher speeds in a standard fashion.

DTE protocols in relation to ISDN

DTE is often used on current analog switching systems. Such equipment must be usable on ISDN to make ISDN a useful replacement. Various methods (including V.110 and V.120 which are covered in Chap. 6) exist that allow use of DTE protocols over ISDN. Chapter 9 emphasizes the various issues involved as well as specific mechanisms that can be used.

State machine designs

Part 3 begins, in Chap. 10, with a discussion of different methods of implementation for state tables. A state table mechanism is normally ideally suited for data protocols. A state table consists of a *state* (which reflects past activity) and an *event* to be acted upon. The combination of current state and event determines a specific action to be taken.

Real-time programming and problems

Real-time events are items that must be serviced at the same time in which they occur. Reception of data falls into this category. Transmission of data also has critical regions. Certain problems tend to be common for all real-time systems and, thus, several are discussed in Chap. 11.

Coordination and management entities

Each protocol can exist by itself. However, a system of protocols needs to have methods to determine exactly where, and how, data must be sent. Error conditions arise. System resources are needed. Special needs are encountered for particular services. The job of the management entity (management plane) is to provide common services. The coordination entity (S-plane) is to provide supervisory capabilities. This last chapter, Chap. 12, discusses these needs and functions.

Standards and Specifications

A standard is a reference according to which other systems can be compared. A specification is a detailed explanation of how a particular system is to be used. The various ITU-T recommendations are standards. (These recommendations are still commonly referred to by the old name of the committee—the International Telegraph and Telephone Consultative Committee, or CCITT.) The actual, switch and national-specific, documents that are used in an implementation are specifications. This book covers the ITU-T recommendations rather than specifications—although some coverage is made of generic, and detailed, variances that exist in different implementation specifications.

Basic Architecture

1

The OSI Model
and Beginnings of ISDN

ISDN did not arise out of the ether. Nor was it devised to keep developers and systems integrators entertained. There were a number of problems that needed to be solved. ISDN is an approach to solve those problems.

What were these problems? The foremost of these was probably the multitude of services for which transmission lines are being used. Because there were so many different uses, there also had to be many different transmission lines in an office: One line for voice, one line for use of a modem, special lines to connect up to local networks, another line for connecting to television or other broadcast media, and so forth. The physical infrastructure necessary to support such a diverse set of communication capabilities was cumbersome and expensive.

Another purpose, however, does have to do with the imagination of people. The shift from analog to digital networks gave the capability to do many more diverse activities over transmission networks. A common structure would unify and standardize such a network. It was out of this mixture of dreams and practicality that ISDN developed.

Evolution of Transmission Systems

The first transmission systems were basically a set of cables that were strung from the house (or office) to the central office. At that point, a human operator would manually physically connect two lines together to provide a complete circuit. As the number of lines continued to increase, this method became more and more difficult to maintain. Therefore, automatic systems were developed to physically connect the lines into complete circuits.

Digital computers began to be developed in the 1950s, and it was observed that this type of machine could be used to create a specialized system that could logically connect together lines without having to physically shift electrical connections. This allowed a much larger number of lines to be handled by a single switching system—faster and without manual intervention. The lines were still using analog transmission techniques but were switched using digital communication logic. The next obvious step was to use digital transmission methods from the user's location to the switch. For larger companies, and in high-traffic situations such as between switching systems, this was done in increased amounts through the 1970s and 1980s. ISDN is a step to allow universal digital access to the switch.

Analog transmission

Human voices are an example of an analog form of transmission. As sound waves are created, the air molecules are compressed in "waves," with higher-pitched sounds causing faster, or closer, waves to be formed. Thus, it was natural to duplicate this effect in the electrical transmission of voice signals. Another way of looking at analog transmission is that of a continuously varying signal. The signal, as shown in Fig. 1.1, can fluctuate in speed and intensity, but it can be represented by a continuous pattern.

A communication line is useless for nondedicated use unless some method is available to allow control over the destination of the signal. The first method used, as mentioned above, was to speak to a tele-

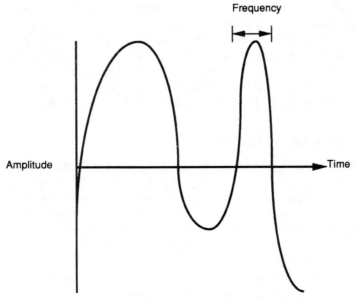

Figure 1.1 An example of analog transmission form.

phone operator and have the operator physically change the connection based upon the information given by the caller. The next step was to create a method of conveying the information in a precise pattern that could be interpreted by a machine. This method used the *tip and ring* signals available over an analog line. In brief, there were two major categories of information that could be carried over the analog line. One was voice signals. The other was whether the circuit was open or closed—*on-hook* or *off-hook*. By briefly interrupting the circuit in a regular pattern, numbers could be conveyed to the switching equipment. Normally, the number of "clicks" was equal to the number to be dialed. The number zero was generated with 10 clicks.

For many years, this type of signalling was the primary method of conveying signalling information. It had two disadvantages. The information had to be sent relatively slowly to distinguish between deliberate signalling information and that of electrical fluctuations inherent over a long-distance medium. The other problem was that of continuity. Transmission between high-traffic areas was handled over lines which carried a multitude of separate connections *multiplexed* together over a single physical line. With this type of a system, it was impractical to disrupt the entire connection for information pertinent to only one call of many.

The solution to this was to use the other form of information conveyed on analog lines—voice signals. Normal speech has a precise range of "notes," or frequencies, that are expected. By using signals that are higher, or lower, than this normal spectrum, information may be transmitted along with speech. This method of encoding is called tone generation and can be in-band (at the same frequencies as are used for speech) or out-of-band—which is at 3700 Hz to keep it out of the normal range for audible speech. The terms *in-band* and *out-of-band* are also used to distinguish between signalling methods carried on the same logical or physical communication link and those carried on separate links, as represented in Fig. 1.2.

Digital transmission

Analog transmission, due to its continuous form, has definite distance limitations because of distortion, and attenuation, of the signal. Various devices can be used to strengthen and recreate the signal, but the problem is inherent. Discrete digital signals are much easier to transmit because there are only two possible values (or four, in the case of certain transmission methods) for the digital signal. This means that the signal can be transmitted for longer distances without reaching the point of being uncertain about its content. Plus, exact recreation of the signal is easier if necessary.

Another category of advantages that digital transmission systems have is that of being able to give explicitly coded signalling informa-

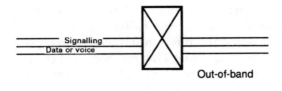

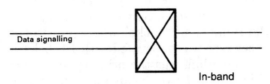

Figure 1.2 In-band versus out-of-band signalling.

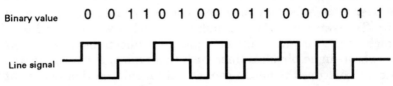

Figure 1.3 An example of digital transmission form.

tion. Breaking the circuit limits the speed of signalling. Tone information on an analog line is limited to what can be easily distinguished by machinery (most tone signalling methods are limited to 16 signals). Digital patterns, as indicated in Fig. 1.3, can be made indefinite in length and thus in complexity. A large number of possible signals, however, is useless without a standard that can be used by all. Otherwise equipment cannot communicate.

Analog and digital transmission are both means of conveying encoded information. Both can be part of a circuit—an electrical connection that is logically continuous from end to end. (Note that this no longer necessarily means physically connected—intermediary buffers may be used for the circuit to transfer the transmission from one physical line to another.) This is referred to as *circuit-switched*. Digital transmissions can also be "packetized"—broken up into discrete sections of a fixed maximum length. *Packet-switched* data transmission takes advantage of this to allow the information to be transmitted in bursts. This means that, when the line is not being actively used, some other connection can use the physical medium.

Data can be sent over circuit- or packet-switched lines. Data on a circuit-switched line can be in analog or digital format. Packet switches cannot be used for analog signals in their original form. However, data tend to come in bursts of information—then the line is unused until the next batch of data needs to be sent or received. During this period of time, the line is still available for data and is still being tariffed (or, at least, unavailable for any other use). Thus, a way of using lines for multiple purposes is very useful.

Multiplexing, circuit switching, and packet switching

One method of using the physical medium for multiple purposes is done through the process of multiplexing, for which there are various methods. Analog information can be multiplexed by allocating different frequency ranges to particular channels. This is called Frequency Division Multiplexing (FDM). Since digital information is in self-contained units, digital lines are often multiplexed according to time. That is, the physical line is used for one logical circuit for a period of time, and then, for the next period of time, it is allocated to another logical circuit. This method is known as Time Division Multiplexing (TDM) and is shown, compared to FDM, in Fig. 1.4. Neither of these methods, however, allows for the possibility of a logical circuit not being currently active. This leads to the topic of the advantages of packet switching for data.

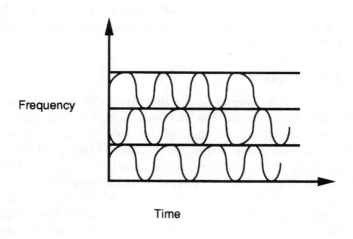

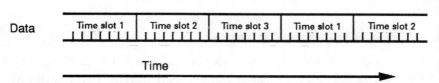

Figure 1.4 Time division multiplexing versus frequency division multiplexing.

Circuit-switched lines. Circuit-switched connections can also be referred to as dedicated connections. Although there may actually be physical breaks in the line within an electronic switch, once the connection is established, the line is available for use by both ends until the connection is released. Information about the destination is needed only during establishment since the line will stay in place at all times. Circuit switching is particularly useful for voice information. A relatively long time is needed to establish the link, but there is immediate feedback (silence on the line) to give users knowledge that they can disconnect the line.

Packet-switched lines. Packet switching is also known sometimes as *store and forward.* This is because a packet switch will receive a packet and then send it on when a time slot on a physical line becomes available. Also, because the line is not continuously available, information about the connection that it is to be forwarded on must be retained. This information is usually contained as part of a packet header and, thus, is part of overhead not necessary on a circuit-switched connection. Larger packets have less overhead and are generally more efficient. The network is also responsible for making sure that packets are received at the other end in the correct order.

Packet switching offers two advantages. One is that less bandwidth is necessary since all packets share the same physical lines—allowing for idle time. The other potential advantage is to the user. Tariffing structures may be such that there will be a cost advantage—particularly if data is sporadic in nature.

Integrated digital networks

Currently, it is rare to be able to send a message in its digital form all the way through the network. Various switches along the path are likely to be analog switches. At each analog switch, the TDM channels must be demultiplexed and then remultiplexed using FDM methods. At the point that the analog switch forwards to a digital switch, the same thing has to be done in reverse. The current structure creates bottlenecks and also causes signalling problems for end-to-end information. Integrated Digital Networks (IDNs) have fully digital paths available. Creation of a global IDN needs a common structure and signalling method. This method is called ISDN.

Open Systems Interconnection Model

A standard structure for a system requires a global architecture. The OSI model is a reference model developed by the International Organization for Standardization (ISO). The OSI model consists of

TABLE 1.1   The OSI Model as Related to ISDN

Layer	Name	ISDN signalling significance
7	Application layer	Host-to-host, application specific
6	Presentation layer	Host-to-host, application specific
5	Session layer	Host-to-host, application specific
4	Transport layer	Host-to-host, application specific
3	Network layer	I.451/Q.931, I.452/Q.932, I.453/Q.933
2	Data link layer	I.441/Q.921, Q.922
1	Physical layer	I.430, I.431, I.432

seven layers. Each layer is responsible for particular requirements of a system. See Table 1.1 for a description of the layers as they relate to ISDN.

Physical layer (layer 1)

Layer 1 handles the electrical and mechanical requirements necessary to transfer data between two adjacent devices in a network. Examples of these include EIA-232-E, EIA-530, V.24, V.28, and V.35. ISDN uses I.430 as a standard for basic rate ISDN lines and I.431 (with G.703 and G.704 for electrical and frame formats) for primary rate ISDN.

Data link layer (layer 2)

Layer 2 handles the protocol necessary for error-free communication between two nodes in a network. This implies the potential for error detection and retransmission. Some protocols require that the error detection and retransmission be instigated at higher levels. Examples of this layer include IBM's Binary Synchronous Communications (BI-SYNC) and ISO's HDLC. Narrowband ISDN (N-ISDN, basic rate, and primary rate) use LAPD (ITU-T Q.921) for signalling.

Network layer (layer 3)

Layer 3 specifies the protocols necessary for network control functions. Such functions include call setup and termination, accounting, and routing. Examples include the Department of Defense (DoD) Internet Protocol (IP). N-ISDN uses ITU-T Q.931 for signalling call control procedures.

Upper layers (layers 4–7)

The lower three layers are sometimes called the *chained layers* because these are the layers which are utilized from node to node. These lower layers are also used at the endpoints, or hosts, but at this point there must be a reason for all this data being shifted around. Layers 4 through 7 specify those protocols that give the data a purpose for existing. However, because there are so many different uses for data and ways in which they are manipulated, it becomes difficult to mandate the layers. Some layers are "ignored" by some applications, and some are split into multiple sublayers. This occurs at the lower layers also but not for nodes that want to communicate.

Layer 4, the transport layer, exists to coordinate various protocols and classes of service. A well-known example of this layer is the DoD's Transmission Control Protocol (TCP), which is part of the Internet protocol suite. Layer 5, the session layer, coordinates interprocess activity including required synchronization, communication, and, possibly, further error detection and retransmission.

Layers 6 and 7 are both concerned with application guidelines. Layer 6, the presentation layer, gives guidelines for general services that many diverse applications may want to utilize such as encryption or data compression. The official application layer, layer 7, deals with types of application services—electronic mail, file transfer services, automatic directory services, and so forth.

The OSI model allows a conceptual picture of systems working with each other. The interfaces allow for the possibility of *open* systems that can work with each other. This model is a necessary foundation block for the creation of an encompassing service model such as ISDN.

ISDN History and Development

ISDN began its life as an approach to an escalating problem. This problem was how to integrate existing services into one general structure and, at the same time, provide the means for future uses and applications that were not currently in existence. William Stalling's book, *ISDN and Broadband ISDN* (2d ed., Macmillan, 1992), does an excellent job in covering much of the political, economic, and technical histories of the development of ISDN standards.

What were the general problems to be addressed by the various "working groups" of the CCITT? (This committee is now known as the ITU-T, and documents produced by this group are referred to as ITU-T recommendations.) The largest problem was to provide a structure that would allow for the growth of technology and consumer needs. A global communications network does not come into being overnight, and thus, the structure must allow for future growth. ISDN has taken over two decades to advance to its current feasible state. It will take many more years to truly become an international facility.

The ITU-T Recommendation I.120 goes into detail about the general principles and hoped-for evolution of ISDN. The principles are generally in the categories of:

- Support of a wide range of services encompassing voice and data requirements
- Support of switched and nonswitched communication lines
- Standardization of connections into the network
- Emphasis on digital services with allowance for coexistence with existing analog services and networks
- Use of layered protocols to allow interworking
- Acknowledgment that national needs will need to be allowed for within the international network environment

Most of the ISDN recommendations, released by the ITU-T, are contained in the I-series of recommendations which have been produced by an internal study group now called Study Group XVIII. Many of the primary I-series documents are covered by additional series designations. In general, this book will refer to the more specific series name rather than the general I-series name, but both names will initially be given.

Evolution of standards

The first documents relating to ISDN came out of the Study Group in 1980. These documents were related to general principles for ISDN. The first documents that started addressing technical requirements were issued in 1984 (Plenary Assemblies of the CCITT took place every 4 years). There were many gaps (labeled "for further study") in the standards, but the standards were sufficient for manufacturers and service providers to start demonstration of general feasibility of ISDN. Some implementations were done in advance of the 1988 recommendations—such as Germany's 1TR6 and AT&T's Digital Multiplexed Interface (DMI). These forerunner implementations were useful in the determination of additional features or requirements but were not directly compatible with the final recommendations from CCITT. The 1988 recommendations were sufficiently well defined to allow compatible ISDN implementations.

Note, however, that the ITU-T recommendations are just that—recommendations. Each country, area, or even manufacturer is free to implement in the manner that they see fit. However, the successful implementations will be able to work together—which means that they will follow the basic recommendations as presented. Gary C. Kessler's book, *ISDN: Concepts, Facilities, and Services* (2d ed., McGraw-Hill, 1993), presents many of the various international specifications, in addition to the basic ITU-T recommendations.

The various I-series documents were broken into separate categories:

I.100 Series—general concepts.   Terminology, concepts, and glossaries allow for a general introduction to ISDN. These documents are a core for the philosophy of the standards and are useful in projecting possible future directions.

I.200 Series—service capabilities.   Services, according to the ITU-T, entail aspects of compatibility, equipment design for use with the services, testing and maintenance procedures, financial and accounting rules, and the ability to track users of services. In brief, these documents are meant to cover not only what services are to be provided but, also, how they will be used.

I.300 Series—overall network aspects and functions.   These are recommendations meant to cover the architectural requirements for specific ISDN data forms—such as N-ISDN (basic rate and primary rate), B-ISDN, and Frame Relay and will encompass other data forms in the future.

I.400 Series—user-network interface aspects.   These recommendations cover many of the actual protocols used over the networks. Layer 1, 2, and 3 protocols are covered in addition to multiplexing, rate adaptation, and specific interface requirements. These recommendations will form the core of the protocol section of this book.

I.500 Series—internetwork interfaces.   These are recommendations covering aspects of having various networks coexist with ISDN.

I.600 Series—maintenance principles.   The networks, themselves, are the focus of these recommendations. They also include methods of testing how the network is interacting with specific types of nodes.

Features and benefits of ISDN

There has already been some general discussions of the benefits of ISDN. Customers benefit from an open system because it fosters competition and diversity of equipment. An international system provides even greater value. Currently, there are many implementations of ISDN which are not directly compatible with each other. However, continued work is being done on consolidation and interworking of the systems.

Imagine a single piece of equipment that can provide file transfer, electronic mail, directory services, facsimile production, video, voice, games, and more. Service providers have a great overhead in having to support multiple standards. Single standards provide ways to cut costs, which can benefit users and service providers. Global markets can be open to common equipment. ISDN provides a mechanism for worldwide access to the data infrastructure.

Many modern services require fairly expensive equipment. However, what is really provided is data. Medical equipment can be split into data-gathering and data-processing equipment. The data-gathering equipment can be dispersed throughout an area, with the more expensive data-processing part located centrally. This can significantly lower health costs and, at the same time, provide greater coverage. Service garages often have diagnostic equipment for analysis of many automobile problems. Why not connect your car up to your phone line and have the problems diagnosed remotely? Review all of your grocery needs at home with direct access to your current inventory, dial up the grocery store, browse through their selections, order, and have the groceries ready to pick up or be delivered.

Data can always go in both directions. The previous paragraph was concerned primarily with data going away from the user—but it works in both directions. Emergency crews could have immediate access to recent x-rays or other medical data. Uncertain as to whether you turned off the lights in the house? Turn them off from a pay phone. Go on vacation with the house in "vacant" mode—lights off or rotating in a random fashion, appliances on cycled reduced power and heat (or air conditioning) at a maintenance level. Call the house from the airport before you head home and have everything ready when you arrive.

So, benefits range from cost savings to global interconnectivity. Features are limited only by imaginations. Still, these are only possibilities. It will depend on the networks, manufacturers, and consumers to decide what is most desired.

Protocols

As mentioned above, this book will focus on the I-400 Series of recommendations with some additional coverage of general architecture. ISDN was initially designed around the utilization of 64-kbps services—basic voice and data needs. Ongoing committee work is consolidating this work and expanding it to include greater speeds. The 1992 ITU-T recommendations provide a solid base for N-ISDN protocols and, therefore, these will be covered in the greatest detail. B-ISDN is at a much less evolved stage, but architectural details are now known, and standard protocols will continue to be developed.

Chapter 2 will begin the discussion of the general protocol architecture of the various forms of ISDN. It will also present many of the protocols of the I-400 Series and the ways that they interact with one another.

2

N-ISDN, B-ISDN, and Auxiliary Data Protocols

ISDN is designed to evolve. This is required because it can not be fully implemented at once; it must be compatible with existing services until they are able to be upgraded to make full use of the new networks. Because of this requirement, many architectural recommendations are necessary. For the network designer, or the person involved with standardizing new features into the networks, these recommendations are vital. An implementor who is working on an existing standard within existing networks is more directly concerned about the specific needs for the equipment or service.

All of the specific recommendations are based upon the general architecture and service recommendations. Therefore, a short review of the general principles of ISDN architecture may be useful to an implementor or someone who needs to evaluate services or equipment. However, this book will not go into detail on the general principles because the primary purpose is to focus on specific implementation issues and protocols. In keeping with this purpose, this chapter will end with a discussion of the international variants upon the ITU-T recommendations, why they exist, and what forms they take.

ISDN Reference Models

An important reference model for ISDN is given in ITU-T I.411. This is a model of connectivity to the ISDN. As can be seen in Fig. 2.1, the model allows for non-ISDN (designated Terminal Equipment 2, or TE2) equipment to be connected via a terminal adaptor (TA). At this point, Reference point S, non-ISDN equipment appears the same as ISDN-specific equipment (Terminal Equipment 1, or TE1). Reference point T allows the possibility of intermediary switching functions such

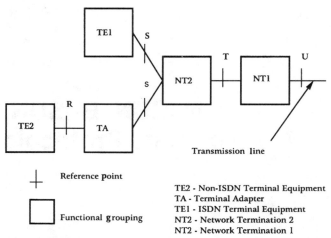

Figure 2.1   Reference points for an ISDN. (*Adapted from ITU-T Recommendation I.411.*)

as local area networks (LANs), Public Access Branch Exchanges (PABXs), or terminal controllers (called Network Termination 2s, or NT2s) before connecting into the Network Termination 1 (NT1) which allows entry to the network. For ISDN equipment being used without an intermediary switch, this is sometimes known as the S/T reference point. Note that the original diagram from I.411 does not explicitly include the U reference point. In North America, the U reference point is useful for end-user equipment. In most other places the NT1 is considered part of the network rather than on-premise customer equipment.

ITU-T Recommendation I.320 describes the ISDN Protocol Reference Model (ISDN PRM) in detail. A list of specific ISDN needs is given, which allows for the various information flows and interworking in an ISDN. An important part is reference to the Control plane (C-plane) and (U-plane). The Management plane is also important for various functions and can incorporate supervisory functions (for which use it is called an S-plane). They are called planes because they have parallel access to all of the layered protocols.

Control refers to controlling information for the network resources. This is information such as call reference values (CRVs), protocol state information, call establishment and clearing, providing supplementary services, and controlling the use of an established network connection. These are services which are supplied to the user application but which the user application does not directly access. The C-plane can be divided into local and global significance. If it has only adjacent (to the next node) significance, it is considered local. Control information used by a remote entity has global significance to the node.

The U-plane, on the other hand, is concerned with the transfer of data between user applications. The C-plane works to set up a call; the U-plane acts to use the connection that has been established. Such user data may be completely transparent to the switch, or the switch may provide intermediary services (such as A/μ-law conversion for digitized voice).

Architecture

ITU-T Recommendation I.325 summarizes the basic architectural model of an ISDN, as shown in Fig. 2.2. Note that the primary aspect of this model is to show how various capabilities can be used within an ISDN. I.325 continues in its explanation of how the various access elements connect up to a national ISDN, which then interconnects through international transit connection elements to allow end-to-end access throughout the international ISDN.

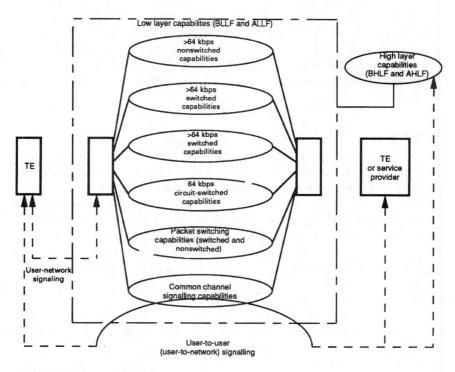

BLLF Basic Low Layer Functions
ALLF Additional Low Layer Functions
BHLF Basic High Layer Functions
AHLF Additional High Layer Functions

Figure 2.2 Basic architectural model of an ISDN. (*From ITU-T Recommendation I.325.*)

It is important that each type of ISDN access use the same basic architecture. In the ITU-T Recommendations I.120, I.121, and I.122 the basic principles of N-ISDN, B-ISDN, and Frame Relay are covered. I.325 and I.327 cover additional network aspects of N-ISDN and B-ISDN. Finally, it is important to remember that various data services must be able to be supported by the various access methods of ISDN.

N-ISDN

N-ISDN was the original ISDN, and, thus, the N prefix is often left off of the term unless broadband is to be discussed in the same context. The first iteration of recommendations was designed to support existing voice requirements and data services. For this purpose, a bit rate of 64 kbps was considered sufficient (with improvements in coding methods, digitized voice now normally only occupies 32 kbps of the data stream). So, much of the original documentation, though universal in nature, had N-ISDN in mind.

ITU-T I.120 was mentioned briefly in Chap. 1. It lays down the philosophical foundation for ISDN. In particular, it mentions that it is necessary to work with national variants that are still in the process of converting over to comply with the I-series recommendations. It also highlights the fact that high bit rates must be allowed. I.325 was discussed at the beginning of this section; it covers general architectural requirements.

B-ISDN

The process of creating standards is a slow one. ISDN was conceived of, as an idea, in the late 1960s. The first documentation was released in 1980. The first recommendations, sufficient for preliminary implementation, were released in 1984. By 1984, it was becoming obvious that 64 kbps was not going to be sufficient for many uses of the ISDN. So, in conjunction with even more complete documentation for N-ISDN, the first recommendations for B-ISDN appeared in the 1988 *Blue Books* (named for the color of the covers of the printed manuals). Considerable effort is being concentrated on defining the needs of B-ISDN as quickly as possible.

ITU-T I.121 appeared in the Blue Books, but it wasn't until 1990 that Study Group XVIII of CCITT approved its first set of recommendations on B-ISDN. These consisted of 13 I-series recommendations, giving first foundations to the architectural and functional principles of B-ISDN. The general structure and service capabilities are found in ITU-T I.121, and overall network aspects and functions are found in ITU-T I.327. The B-ISDN functional recommendations (I.361, I.362, I.363) are covered in Chap. 8.

I.121 is a philosophical architecture document (similar to that of I.120 for N-ISDN). Four of the characteristics mentioned for Asynchronous Transfer Mode (ATM) of the B-ISDN are

- High flexibility of network access due to the cell transport concept and specific cell transfer principles.
- Dynamic bandwidth allocation on demand with a fine degree of granularity.
- Flexible bearer capability allocation and easy provision of semipermanent connections due to the virtual path concept.
- Independence from the means of transport at the physical layer.

ITU-T I.327 first emphasizes its adherence to the basic architectural principles established for N-ISDN. It then expands the picture seen in Fig. 2.2 to include the use of broadband capabilities between the local function capabilities. Appendix 1 of ITU-T I.327 then proceeds to give three functional model examples for implementation of B-ISDN. These models include a star structure where customers have direct individual links to the local exchange available. Two other examples allow for multistar structures with intermediary distribution facilities.

Frame Relay

The general concept of Frame Relay was concerned with being able to take advantage of a reliable digital network. Existing data protocols (such as ITU-T X.25) had redundant error detection and sequencing protocols at different layers. This was useful in the situation where the line transmission was considered very unreliable. However, with improved transmission reliability and speed, the overhead necessary for these additional protocol layers was difficult to justify. ITU-T I.233 discusses the general structure and service capabilities of frame-mode bearer services. Frame Relay is discussed in greater depth in Chap. 7.

Terminal adaption

There is currently a lot of DTE in use that is not directly usable with ISDN. In Fig. 2.1, we saw that the architecture of ISDN allows for this. Using a TA, a piece of non-ISDN equipment can have equal access to the reference point S of the ISDN. What does that mean? How can a non-ISDN piece of equipment make use of ISDN?

Consider two types of commonly used terminals. One is a terminal using a serial port to connect to a modem. A data stream passes between the terminal and the analog line. There are a number of tasks needed in order to map such services onto ISDN. First, there must be a mapping between the signalling commands used and those of ISDN. Second, the data must be packetized into digital form. Third, a Universal Asynchronous Receiver/Transceiver (UART; RS-232-C, E, or other V-series interface) device may need rate adaption (packetizing may eliminate this particular requirement). The TA will handle physical requirements for the ISDN physical layer.

There are two popular methods used for terminal adaption of serial DTE. ITU-T I.463 or V.110 is popular in Europe and Japan. ITU-T I.465 or V.120 is popular in North America. These recommendations are examined in greater detail in Chap. 6.

ITU-T X.25 is also a popular method for call setup and data transfer. X.25 has been used for data transport for a long time, and the X.25 network cannot be ignored. ITU-T I.462/X.31 allows for synchronization of ISDN bearer service setup with X.25 Packet Layer Protocol (PLP) and in-band call control services. ITU-T X.30 can be used for terminals using the X.21 protocol. These are both discussed further in Chap. 9.

Transmission Structure

Some mention has been made of the bearer channel (B-channel) within the context of ISDN. It is important, particularly as B-ISDN is enhanced, to understand how digital transmission channels are divided. There are three basic classifications of TDM channels that operate in the digital network. These are:

- *D-channel.*   16 or 64 kbps (depending on basic versus primary interface)
- *B-channel.*   64 kbps
- *H-channel.*   384 (H0), 1536 (H11), or 1920 (H12) kbps and others

D-channels

The D-channel is the common channel signalling link. It will use LAPD as the protocol for the data link layer. This protocol is defined in ITU-T I.440/Q.920 and ITU-T I.441/Q.921. This is covered in greater detail in Chap. 4. For current discussion, the important part of this protocol is its addressing capabilities. A D-channel can be multiplexed into separate logical channels over the same physical channel.

The logical links available on the D-channel allow separation of functions at the ISDN. One logical link (two logical links in some specifications) will be used for the common channel signalling. Another link will be used for maintenance and management functions. This leaves other logical addresses available for additional uses. One such use is often to allow a low-speed data channel for user traffic. Q.921 gives guidelines as to how addressing information should be allocated.

B-channels

The bearer channels are the basic divisions used for user data traffic. Although B-channels may support a higher-level multiplexing scheme (such as used within ITU-T X.25 at layer 3), the ISDN will route the channel as a single entity. This does not really decrease the flexibility of use. A B-channel can be switched to an X.25 router (for example), and the X.25 router can then act as a network addressing subsystem for any in-band connections desired. However, the ISDN is not directly involved in this additional routing. The ability to handle in-band call control information is enhanced by the use of ISDN.

B-channels can be set up in three ways. The first is called *nailed-up* or semipermanent. This is done by arrangement, or subscription, with the local exchange provider. It is very similar in effect to a leased line. The other two methods are really access to various switching methods.

Nonpermanent B-channels are set up using the common channel (out-of-band) signalling of the D-channel. Within the signalling protocol, particular bearer services can be requested. A B-channel may be requested to be connected via a circuit switch or a packet switch. In the case of a circuit switch, either voice or digital information may be sent. A packet switch is restricted in the form of the data it may carry.

H-channels

H-channels, or high-bit-rate channels, are aggregates of the basic B-channels. The H0 channel is 384 kbps, or the equivalent of six B-channels. An H11 channel is 1536 kbps, or the equivalent of 24 B-channels, and so forth. The purpose of H-channels is to allow an end-user application to divide up the digital transmission line according to individual needs. These needs may include video, high-quality audio, or other high-density data requirements. H-channels, because of their speed, are available over Primary Rate Interfaces (PRIs) or in B-ISDN configurations.

N-ISDN

N-ISDN is broken into two general types of interfaces. These are called Basic Rate Interface (BRI) and PRI. As shown in Fig. 2.3, a BRI consists of one 16-kbps D-channel for signalling and other low-speed data transfer and two B-channels, each at 64 kbps. Some methods (such as MultiLink Protocol within X.25) allow the two B-channels to be used as a combined unit providing 128 kbps.

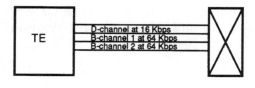

Figure 2.3 BRI channel division.

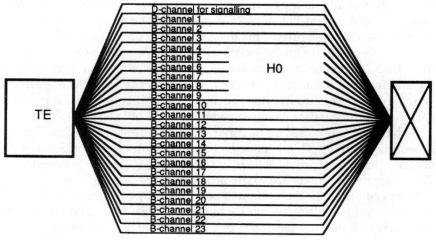

Figure 2.4 PRI channel division.

A PRI makes use of an entire digital trunk line. In the United States, Canada, and Japan this is based on a rate of 1.544 Mbps, which corresponds to the capacity of a T1 transmission facility provided by AT&T. In Europe, the standard reference is called E1 and has a rate of 2.048 Mbps. A PRI is divided into bearer channels and one D-channel (which is usually at 64 kbps for PRI). A standard PRI for North America and Japan would have the capacity for 23 B-channels and one D-channel (as seen in Fig. 2.4). In Europe, it is possible to have 30 B-channels and one D-channel on a PRI. It is also possible to configure the line using H-channels as long as at least one D-channel (this D-channel may actually be used to control multiple PRIs) is available for signalling information.

B-ISDN

B-ISDN, because of its data cell architecture, allows for many different, varying, effective data speed rates. The current capacity depends on the requirements of the data stream and the facilities available. Some of the details for full interworking with N-ISDN are still under study, but different implementations (such as Synchronous Optical Network, or SONET) allow for continued research and architectural improvements. B-ISDN is considered in greater detail in Chap. 8.

Bearer Services

An ISDN is worthless unless it can provide service to the users at the endpoints. The ITU-T I.200 Series goes into these services. There are three general types of services: bearer, teleservice, and supplementary services. Bearer services provide for communication between hosts in real-time and without additional intervention from the exchange once the connection is established. Teleservices may include some additional intervention by the network after the connection has been made, and supplementary services enhance the capabilities of both bearer services and teleservices—often on the basis of switch and subscription capabilities. Bearer services will be focused on in this book along with acknowledgment and discussion of methods to integrate further services into ISDN equipment. Supplementary services are discussed in Chaps. 5 and 6.

There are eight categories of bearer service defined for circuit-switch mode. These are discussed in ITU-T I.230 and detailed in ITU-T Recommendations I.231.1 through I.231.8. Packet-mode services have their own three categories and are defined in ITU-T I.232.1 through I.232.3. These recommendations define basic service types and data groupings as seen in Table 2.1. B-ISDN incorporates the

TABLE 2.1   Bearer Service Categories for Circuit-Mode and Packet-Mode

ITU-T reference	Description
	Circuit-mode bearer service categories
I.231.1	64 kbps unrestricted, 8 kHz structured
I.231.2	64 kbps, 8 kHz structured, usable for speech information transfer
I.231.3	64 kbps, 8 kHz structured, usable for 3.1-kHz audio information transfer
I.231.4	Alternate speech, 64 kbps unrestricted, 8 kHz structured
I.231.5	2 × 64 kbps unrestricted, 8 kHz structured
I.231.6	384 kbps unrestricted, 8 kHz structured
I.231.7	1536 kbps unrestricted, 8 kHz structured
I.231.8	1920 kbps unrestricted, 8 kHz structured
	Packet-mode bearer service categories
I.232.1	Virtual call and permanent virtual circuit
I.232.2	Connectionless
I.232.3	User signalling

SOURCE: Adapted from ITU-T Recommendation I.230.

bearer services discussed in ITU-T I.230 and expands them with various service classes for high-bandwidth categories.

ISDN provides for various services by means of Information Elements (IEs) that are contained within the signalling protocol messages. This is described in greater detail in Chaps. 5 and 6. However, there are three IEs that are of great significance to this function. They are the Bearer Capability, the Low Layer Compatibility, and the High Layer Compatibility IEs. These IEs, and their possible values, are found in ITU-T Q.931 (in addition to general descriptions in the I.200 Series).

I.400-Series Recommendations

The I.400-Series Recommendations are concerned with the User-Network Interfaces (UNIs) into the ISDN. Many of these documents are published in duplicate in other ITU-T series, particularly those of the Q-series, X-series and V-series. A UNI is concerned with how a user makes use of the network. Thus, the recommendations include documents about the chained layers to be used from node to node for common signalling. They also include methods of multiplexing and rate adaption. Finally, they provide mechanisms for connecting up existing types of equipment to the ISDN.

Physical layer

The physical layer recommendations are discussed in Section 3 of the I.400 Series and include ITU-T I.430 and I.431. These are covered in Chap. 3. I.430 and I.431 are concerned with the transmission and physical protocols necessary to maintain the transmission link into the ISDN. I.430 is oriented toward BRIs, and I.431 is for PRIs.

Data link layer

Data link layer recommendations are in Section 4 of the I.400 Series and currently include ITU-T I.440 (Q.920) and I.441 (Q.921). Cross-referencing of series is being discouraged by the ITU-T; therefore, further recommendations are likely to be issued only in the Q-Series. One such document is ITU-T Q.922, which specifies modifications to Q.921 for use with frame-mode bearer services (or Frame Relay). Q.920 and Q.921 specify architectural issues for the data link and LAPD. These are covered in Chap. 4, with Frame Relay specifications covered in Chap. 7.

Network layer

The network layer recommendations are concerned with call control procedures. Thus, Section 5 of the I.400 Series covers basic call control and supplementary services. ITU-T I.450, I.451, and I.452 are replicated in the Q Series as Q.930, Q.931, and Q.932. The Q Series is further extended with Q.933, which is concerned with signalling specification for frame mode basic call control (Frame Relay). Frame Relay specifications are discussed in Chap. 7 with basic call control and supplementary services in Chap. 5.

Multiplexing, rate adaption, and support of existing interfaces

These recommendations in Section 6 of the I.400 Series are concerned with the support of existing terminal interfaces. The supported DTE types include those that support an X.21, X.21 bis, and X.20 bis interface (ITU-T I.461, also replicated in X.30). ITU-T X.25 packet mode DTEs are supported in I.462 (X.31). Finally, serial interfaces are indicated in I.463 (V.110) and I.465 (V.120). The X-series supportive recommendations are discussed in Chap. 9, with the serial interfaces covered in Chap. 6, since they make heavy use of bearer service negotiation within the signalling channel.

Variances between Recommendations and Specifications

The ITU-T recommendations are just that—recommendations. There is considerable benefit to following them as closely as possible. These benefits include access to a global market, easier interconnectivity, and standardization of equipment. However, the various switch manufacturers, and regional and national specifications do not completely agree with one another. There is a movement (with the U.S. national ISDN specifications and the European Telecommunication Standards Institute, or ETSI, Euro-ISDN specifications) to consolidate ISDN specifications. As mentioned in the principles of the evolution of ISDN, however, this does take time.

Most current ISDN specifications follow the same procedures for basic call control, and all are standardized on the physical layer. The ITU-T ISDN recommendations allow considerable variance, with many optional features and uses, in order to facilitate a slow integration of existing networks into a global ISDN. The following subsections discuss these variations in greater detail.

Optional features

The recommendations are broken into two broad categories—mandatory and optional. Mandatory features must be implemented by a particular specification to be considered an ISDN. Optional features may be used, or not, as decided by the network designers. In some instances, optional features are required by a specification. In other instances, they are forbidden. In yet other specifications, they may be truly optional. In the case of completely optional features, usage may depend on the testing capabilities needed for certification of equipment. Some examples are given in the following subsections that refer to items covered in greater detail in later chapters.

Overlap versus en-bloc signalling.   One major optional feature is support of overlap signalling (en-bloc is mandatory). En-bloc signalling is the situation where all necessary addressing information is contained in a single packet sent to the closest exchange. All information is contained within a *block* of data. In the case of overlap signalling, a sequence of packets can be sent to the exchange, with an acknowledging packet sent back when sufficient information has been received to process the call. Overlap signalling requires more switch resources and is, therefore, not always supported.

Support of D-channel X.25.   In the discussion of D-channels, it was mentioned that one logical link of the D-channel could be used for X.25 packet services. This DX.25 link uses LAPD at the data link layer but with X.25 PLP used as a data transport mechanism at the network layer. Support of this logical link is optional. Additionally, the switch may require out-of-band signalling call setup to use this logical link. Other switches may offer the service only on a permanent basis or may offer in-band call control in addition.

Support of optional frames and packets.   ITU-T Q.921 provides data link services. One such frame defined in Q.921 is called an EXchange IDentification (XID) frame. Support of the XID frame, and associated procedures, is optional. In ITU-T Q.931, various packet message types are optional to the use of ISDN. For example, the return of a CONNECT ACKNOWLEDGE (CONN_ACK) message-type packet by a user endpoint is considered optional. A SETUP ACKNOWLEDGE (SETUP_ACK) message-type packet is not expected if overlap sending is not supported. In some specifications, optional frames and packets are ignored. In others, they are considered error conditions.

Support of optional states.   The use of Terminal Endpoint Identifiers (TEIs) was discussed briefly as part of the addressing method for Q.921. A TEI can be *fixed* (permanently assigned by agreement with the local exchange) or *automatic* (negotiated for with the local exchange). The various protocol states of Q.921 allow for fixed or automatic TEIs. If only fixed TEIs are supported, part of the protocol states become unnecessary.

In Q.931, state transitions are dependent on packet message types supported. Therefore, if overlap sending is not supported, associated protocol states are not needed (see Chap. 5). Other states are associated with activities that may be part of optional supplementary services and, thus, not needed if not supported.

Support of optional timers. ITU-T Q.931 states that network layer timers designated as T305, T308, and T313 are optional on the user side of a connection. Other network layer timers are needed, or not, depending on support of services.

Handling of error conditions. Sometimes the ITU-T recommendations allow a choice from a set of actions. The choice is left to the requirements of the local specification. An example is that of data link protocol errors. A management response can be to issue a message to verify the current status of a link or to remove the link and start again. Handling of received nonsupported optional message types and IEs may entail active error handling or ignoring the errant message.

Use of different optional fields. An ITU-T Q.931 message type includes a set of mandatory fields. It also usually includes a number of optional fields. Sometimes these fields overlap in meaning. An example of this is the Keypad IE. A Keypad IE is designed to carry numeric information between the equipment and the switch. This could be information such as a dialed number on a keypad—or it could be credit card information for a collect (or bill-to) call. Another optional IE is called the CalleD Number (CDN), which contains the address of the terminating equipment being addressed. Each of these optional IEs could be used in a call origination, and each is used within different specifications.

National features

Evolution of ISDN includes the need to allow for national requirements. These requirements are necessary to allow for migration to ISDN, and many will be necessary for features not supported in general. Some features are truly national in scope, and the ITU-T recommendations allow for this by having the concept of *codesets* available within the channel signalling elements. A codeset of 0 indicates international standard use of message types and IEs. A codeset of 5 indicates use for national purposes. The various codesets are discussed more fully in Chap. 5.

Supplementary services. These extensions to basic bearer services and teleservices are always exchange specific, although ITU-T Q.932 gives architectural and functional structure to implementation. Supplementary services include voice functions such as *hold, transfer,* and *conference,* which allow switch intervention for new channel routing.

Optional features. This category of features was discussed above. Many such features will be implemented on a national or regional level. As such features are used and tested in various ISDNs, they may become common in multiple specifications.

Network requirements

A network has limited, and specific, facilities available to it. Based on these limitations, various requirements for use must be imposed. The following subsections focus on some of these categories of requirements.

Line activation. The physical interface can be active, with electrical signals being transmitted, or it can be dormant. In Europe, the line should remain inactive until it is needed (incoming call or request to place an outgoing call). In North America, the physical line can remain active even when not in active use.

TEI management. TEI management procedures are given in ITU-T Q.921. However, certain usage is left to the discretion of the network specifications. For example, a BRI can use a single TEI for call control on both B-channels. It can also allocate separate TEIs for each B-channel. A local switch may support only fixed TEIs. If a TEI is removed, due to temporary or permanent line problems, the TEI may be renegotiated immediately upon line activation, or it may wait for a call to be established. These various procedures affect switch resources.

Packet sizes. Packet switches require buffers for each packet passing through. Some switches may have smaller, or larger, buffers available. Small buffers save switch resources. Large buffers enable greater throughput with less overhead. Such tradeoffs play a part in the requirements for use of the switch.

Timer values. It was mentioned above that certain timers are optional, and others are needed only if optional features are supported. Timers also have lengths. They may also have *counts*—the number of times the event can happen without change in state. These timers, both at the data link layer and at the network layer, affect the traffic level on the transmission lines.

Flag insertion. The HDLC protocol includes information about what to do between frames. This is called interframe filling. Some networks require continuous flag insertion on the line. Others may have other needs. This is discussed further in the next chapter.

Physical requirements. There are two major methods of digitizing voice into a form compatible with digital networks. These two sets of rules are called the A-law and the μ-law. The A-law is used in Europe, while the μ-law is used elsewhere. This is a type of physical requirement necessary for use in internetworking. (If both ends used the same method without requiring switch intervention, the specific coding law would be unimportant.)

Test Suites and Standards Evolution

Equipment designed to work with a particular specification must be tested and, possibly, certified in order to be allowed to be marketed to work with an ISDN. These test suites are designed according to the specifications. Test suites, like the specifications themselves, are always in the process of change. These changes may include revisions to test for new features, changes in interpretations of the specifications, or additional tests for completeness.

Testing usually includes functionality and error handling. Error handling makes sure that the equipment is able to handle both correct and incorrect frames and packets. Sometimes this simulates loss of packets or frames on the transmission line. Sometimes this is done so that, in the case of interworking with other networks, new network equipment can be integrated without fear of causing problems with other nodes.

The ITU-T recommendations are also in a continual state of evolution. Items labeled "for further study" are completed. New features and services are added. Each of these make ISDN implementation, testing, and manufacturing part of an ongoing evolution of the network.

The Protocols

3

Physical Layer

The physical layer of ISDN is concerned with the ability to transmit and receive information over a transmission medium. There are requirements for all parts of the interface—electrical (or other medium, such as light manipulation), formation, and protocols. The physical layer has the distinction of being at the lowest layer of the OSI model and is, indeed, the foundation for all telecommunication standards.

The exact protocols and requirements of the physical layer depend on use and speed. A slower use of the medium allows for less protocol overhead but also allows for more direct software control. Broadband use of the physical layer, on the other hand, requires a more involved physical protocol so that hardware can be designed to support it at the necessary speeds. This chapter will focus on the software and protocol needs relative to the physical layer but will discuss, at a higher level, the various tasks required of hardware needed to support physical layer integrity.

Physical Line Requirements

The physical layer is involved with manipulating a physical medium (such as electricity, light, microwaves, or whatever). The ITU-T I.4xx recommendations are primarily concerned with the method of use, with the G.xxx recommendations dealing more with the physical media. Most of the examples given in the IUT-T I.43x recommendations use electrical signals as examples.

Although Chap. 1 presented analog and digital signals as completely separate types of signals, a digital signal is really a special class of analog signal. All electrical signals are continuous. A digital signal, however, is created or interpreted by taking samples at discrete time periods. The value of the signal at this sampling period is the value used. This sampling method means that a method of synchronization is needed between the two endpoints. This is usually done via a framing pattern and a clock signal (internally or externally generated).

The various ITU-T I.4xx, ITU-T G.9xx, and the ITU-T G.7xx recommendations discuss the specific signal requirements. In this book, we will discuss these signals only in terms of their logical operation.

Synchronization and signal balancing

When a series of signals is transmitted or received, it is necessary to know just when, and how, the signals are to be interpreted. This process is known as synchronization. There are two basic requirements for this. The first is that there be some method of timing to sample the signal. The second is that there is some pattern that can be recognized as the beginning of a frame. There are various methods of doing this. There will be a pattern that is sent during "idle" time which gives the ability to keep the sampling synchronized. There will also be a pattern that enables the receiver to know when data is now being transmitted. Various specific methods of doing this will be discussed in this and later chapters.

Another requirement for electrical signals is that the signal be balanced for direct current use. This balancing is associated with the voltage variations used for conveying the signal information. For example, if a 0 was conveyed only by a high voltage, a series of such 0s would mean a long period of high voltage—causing both extra power demands as well as making it more difficult to maintain sampling integrity. Suppose that the signal is to be sampled one million times a second. A series of 1000 0s would take a 1ms to transmit or be received. If this signal did not have variation, and the sampling clocks were not precisely the same, it would be easy to be in error as to whether there were meant to be 999 zeros, 1000, or 1001. Based on these needs, a number of encoding schemes have been devised.

Digital encoding

As can be seen in Fig. 3.1, a number of digital encoding schemes can be used to provide both synchronization methods and a method of balancing the signal. Most of these rely upon alternations of at least one of the signal values. The simplest method is to use two different voltage levels for the binary values. This method is called NonReturn to Zero (NRZ). This method is simple and may provide reliable values in the case of recorded media, but it causes problems for synchronization and balancing, as mentioned in the previous subsection.

Another category of encoding is called multilevel binary. In this case, more than two voltage levels are used. The pseudoternary method (which was used in the example of a digital signal in Chap. 1)

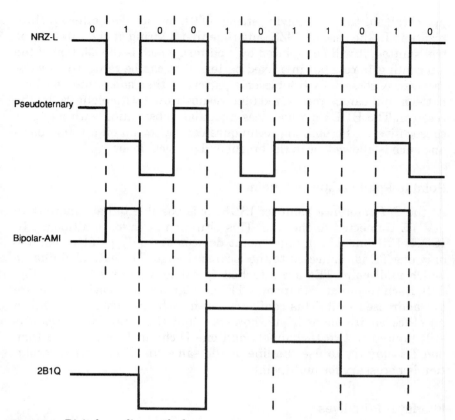

Figure 3.1 Digital encoding methods.

is used for the *S* interface for BRI. It sets a value of *1* if no line signal is present and alternates between high and low voltage for successive 0s. The opposite approach is used for the method known as bipolar-AMI. Both of these methods still have the possibility of long sequences of one value (1 for pseudoternary; 0 for bipolar-AMI).

It is also possible to use more than three voltage levels to encode the signal. One such approach, used for ANSI T1.601 *U* interface standards, is called 2B1Q. Every 2 bits are encoded by one of four voltage levels. This allows greater use of available bandwidth and increases effective speed.

Scrambling

The third category of encoding bases itself on a simpler coding standard but replaces part of the bit stream in order to remove instances that may cause confusion. The data encoding methods called B8ZS and HDB3 are both based upon bipolar-AMI but perform special actions, upon transmitting or receiving strings of 0s, to allow for better synchronization. In the case of B8ZS, any time that a string of eight 0s is to be transmitted, it will be replaced by a pattern which is in violation of the alternating 1s rule but in a fixed pattern that can be recognized by the receiver. A potentially troublesome pattern is thus substituted for by a pattern that can be recognized (and resubstituted with eight 0s) by the receiver. The HDB3 method does a similar substitution with any four consecutive 0s. Scrambling techniques can be set up on a frame basis and such is allowed for in the broadband physical layer.

Point to Point versus Multipoint

At the *S/T* reference point for ISDN, it is possible to have more than one TE connected to the line. This situation is called multipoint. B-ISDN, PRI and *U* reference point devices are only point to point— only one TE is connected to the local exchange. The extended line, to which multipoint TEs are attached, is called a passive bus (see Fig. 3.2). Each terminal gets its own TEI and ignores any packets that are not addressed to it. This configuration may be particularly useful in an office environment. Note, however, that the total data stream is still limited to two B-channels and one D-channel. Since more than one TE may try to use the line at the same time, contention resolution is necessary for multipoint.

Interlayer Primitives

The various ITU-T recommendations list specific primitives for specific layers and entities. These primitives fall into four major categories: requests, indications, responses, and confirmations. A request is a

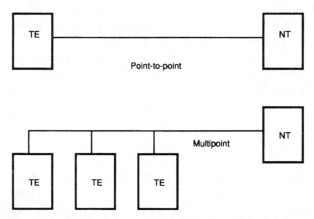

Figure 3.2 Multipoint versus point-to-point. (*Adapted from ITU-T Recommendation I.430.*)

primitive from a higher level asking for services from a lower level. (The management functions are always considered to be a higher level.) A confirmation is the acknowledgment by the lower level to a request. An indication is notification from a lower layer to a higher layer of an event or need for resources. The response primitive is a method of supplying acknowledgment or resources in accordance with the lower layer's indication primitive.

The next category for primitives is the layer involved. Primitives assigned to the physical layer will have *PH* as part of their label. Similarly, the data link layer will have *DL* as part of the primitive label. An *M* used as prefix indicates that the primitive is used between the layer and the management controlling functions. Thus, a MPH_DEACTIVATE_IN is a primitive from the physical layer to the management entity indicating a particular event (line deactivation in this case). A PH_DATA_RQ is a request from layer 2 to the physical layer for data transmission. The ITU-T recommendations list primitives with a hyphen (such as PH-AR). In this book, we will use the underscore (such as PH_AR) in primitive names because such naming conventions are compatible with most programming language conventions. Table 3.1 gives the specific primitives associated with the physical layer.

The naming convention is useful because it can be (and often needs to be) extended past the ITU-T recommended minimum primitives. For example, a request for timer services can be known as a MPH_TIMER_IN. A buffer resource request can be called a MPH_BUFFER_IN. Semiconductor-specific requests can be added. Most such added primitives will fall into the management class.

The nonmanagement primitives are normally expected to be implemented as given. There are exceptions. ITU-T Recommendation I.430 gives, as a note, the information that a PH_DATA_RQ primitive implies underlying negotiation between layers 1 and 2. It is possible to link the PH_ACTIVATE_RQ primitive to the PH_DATA_RQ primitive since the latter implies the former. It may be useful to separate them for particular implementation needs.

Semiconductor Device Support

Events at the physical layer happen at high speeds. Thus, hardware support is often desired and may be necessary. A semiconductor device will often support some layer 2 requirements in addition to the physical layer. A LLD is usually used as a software interface between other entities and the physical layer hardware. Design of LLDs is covered in Chap. 11. This LLD handles the primitives needed to use the physical layer. As such, it may be considered a layer 1 entity, or it can be considered to be between layers 1 and 2.

TABLE 3.1   Physical Layer Primitives

Generic name	Specific name		Parameter		PRI use?	Message unit contents
	RE-QUEST	INDICA-TION	Priority indicator	Message unit		
L1 ←→ L2						
PH_DATA	x	x	REQUEST primitive only	x	Yes	Layer 2 peer-to-peer message
PH_ACTIVATE	x	x			No (RQ) Yes (IN)	
PH_DEACTIVATE		x	-	-	Yes	
M ←→ L1						
MPH_ERROR		x		x	Yes	Type of error or recovery from a previously reported error
MPH_ACTIVATE		x			Yes	
MPH_DEAC-TIVATE	x	x			No	
MPH_INFOR-MATION		x			No	Connected or disconnected

Source: ITU-T Recommendations I.430, I.431.

There are two phases of support provided by a semiconductor device—setup and active stage support. During setup, the semiconductor device is programmed to work correctly with the existing hardware for the equipment. In the active stage, several functions are provided. One function is to enable the line interface. This allows activation from the other endpoint. Another function may be to request activation locally (PH_ACTIVATE_RQ). (This function is not needed in higher-speed interfaces such as PRI or B-ISDN since the interface should always try to achieve activation.) There will also be run-time requests such as data transmission requests.

One side effect of using semiconductor devices for the physical layer is that many of the physical layer effects are not directly under the control of the software. For example, ITU-T Recommendation I.430 details the state table of sending and receiving particular framing patterns on the line. The state of the line changes depending upon the patterns received or transmitted. Most semiconductor devices do not allow direct access to the "info" patterns. Instead, the device allows the software access to the current framing-state (F-state). Some state transitions may

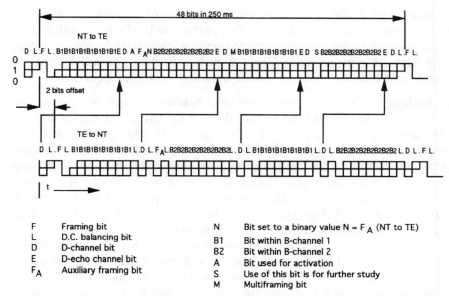

Figure 3.3 Frame structure at reference points S and T. (*From ITU-T Recommendation I.430.*)

happen so quickly that the LLD does not receive indication of the transitory state. Implementations must allow for these possibilities.

Basic Rate Interface

ITU-T Recommendation I.430 contains the user-network interface requirements for BRI at the S or T interface reference points. The BRI provides access to one 16-kbps D-channel for signalling (and maintenance and other optional logical link services) and two 64-kbps B-channels. The aggregate bit rate of the link is 192 kbps in both directions. This allows for the D-channel, both B-channels, and 48 kbps used for framing overhead (see Fig. 3.3). This interface may be in point-to-point or multipoint configuration.

Physical requirements

BRI at the S/T reference points are coded using pseudoternary encoding. The first 0 bit is of the same polarity (positive or negative voltage) as the balance bit of the framing bit (see Fig. 3.3). The TE uses the signal received from the local exchange for its synchronization timing.

Interframe time fill. Since the channels of the BRI are allocated in a TDM fashion, it is important to know when valid data is being transmitted or received. For BRI, the D-channel will contain 1s when the channel is idle. This is sometimes called *mark idle*. When data is to be transmitted on the D-channel, one flag pattern of the form 0 1 1 1 1 1 1 0 will precede the data. Another flag pattern will follow the valid data. It is possible to use the same flag to terminate one data frame and begin the next. As is explained in Chap. 4, such patterns require the ability to alter valid data so as not to conflict with flag (or abort) sequences.

D-channel echo, monitoring, and prioritization. There is 1 bit in the BRI bit sequence which is filled only by the local exchange (or Network Termination, or NT). This is the D-echo bit. The NT will copy a received D-channel bit into this bit, and the TE will monitor this bit. During data transmission, an identical bit from what was transmitted indicates that the data was received correctly. A different bit indicates that some other terminal on the passive bus is currently trying to send data on the D-channel. This is called a collision, and any current transmission will be terminated.

Monitoring of the D-channel echo bit serves another purpose. While waiting for data transmission, the TE will count the number of 1s received back. (As mentioned above, the TE must send 1s while idle.) This allows a prioritization method to be used for transmissions—particularly useful in multipoint cases.

Framing. Frames can be a bit confusing in the various ITU-T recommendations. A layer 2 group of data can be called a frame (layer 3 is often called a packet). A frame is also a physical layer grouping of information consisting of layer 2 data and special codes and overhead necessary for synchronization and balancing. A frame is really similar to that of a picture frame—it surrounds a group of data to make it more easily recognized. The physical layer frame requirements may be different between protocols, but the general purpose is the same.

S reference point versus U reference point. ITU-T I.430 only describes the interface at the S or T reference points. The ITU-T does not specify the U reference point interface—considering it to be part of the local exchange equipment. For the United States, the American National Standards Institute (ANSI) has designated T1.601 as the standard for a BRI UNI at the U reference point. This interface uses a 160-kbps line using 2B1Q data encoding. Each frame consists of a synchronization word of 18 bits, 12 groups of two octets of B-channel data, 2 bits of D-channel data, and 6 bits of overhead. The frames are further grouped into sets of eight frames, with the first frame having an inverted synchronization word, called a superframe.

Protocol states and software requirements

A state machine consists of a current state, which is the reflection of past events, and an event. In the case of BRI S UNI, there are three categories of events. A primitive can arrive to be handled, as discussed above. A timer can expire. Finally, a physical line signal can be received. There is also an optional set of actions associated with power sensing at the UNI. Table 3.2 gives this version of state table; if it is not applicable to the equipment, the power sensing sections can be ignored.

Timers. ITU-T I.430 lists one timer associated with the BRI. This T3 timer is initiated upon receipt of a PH_ACTIVATE_RQ (PH-AR, in the recommendation) primitive. The T3 timer is usually around 30 s in duration. If line activation is achieved within this period, the timer is canceled, and all is well. If line activation has not yet been achieved (framing states 4 through 6), attempts to activate are discarded, and the line returns to a deactivated state.

Another timer, T4, is used in some implementations. This timer is not directly mentioned in ITU-T I.430. It is useful based on the reality of physical links. There will occasionally be problems in the line—brief instants of static, power surges, and so forth. I.430 indicates that most types of problem will cause the link to be considered deactivated. This will cause repercussions throughout all the ISDN layers and, if possible, is to be avoided. The T4 timer, with a value ranging between $\frac{1}{2}$ and 1 second, is useful for this. If line activation is lost, the T4 timer is initiated. If line activation is reachieved before the timer expires, the physical layer continues without notification of problems to the upper layers. If the timer expires without having reachieved line activation, the normal primitive exchange listed in I.430 takes place.

Primitives. The above subsection on interlayer primitives was a fairly complete discussion of possible primitives for the physical layer. How are they actually used? The PH_DATA_RQ and PH_DATA_IN primitives convey data to and from the physical layer. These actions require that the line activation has been achieved and is in proper state. The other primitives are involved with activation, deactivation, and line status. It can be noted from Table 3.2 that MPH_DEACTIVATE_IN and PH_DEACTIVATE_IN are always located as pairs in the state table. This redundancy allows separate notification of layer 2 and the management entity—they may not both be necessary in a specific implementation. The MPH_DEACTIVATE_RQ allows the equipment to disconnect from the local exchange in order to conserve power and may not be necessary for a particular specification. Finally, the MPH_IN-FORMATION_IN primitive is useful in power sensing applications.

Physical line information. The data in Table 3.2 indicates what happens when a particular information frame (as described in Fig. 3.4) is received from the local exchange (NT). As was discussed above, a semiconductor device will usually only allow notification of change of

TABLE 3.2 Locally Powered TE Activation/Deactivation BRI State Table

State name	Inactive — Power off	Inactive — Power on	Sensing	Deactivated	Awaiting signal	Identifying input	Synchronized	Activated	Lost framing
State number	F1.0	F1.1	F2	F3	F4	F5	F6	F7	F8
INFO sent	INFO 0	INFO 0	INFO 0	INFO 0	INFO 1	INFO 0	INFO 3	INFO 3	INFO 0
Event									
Loss of power	/	F1.0	F1.0	MPH-II(d) F1.0	MPH-II(d) MPH-DI PH-DI F1.0	MPH-II(d) MPH-DI PH-DI F1.0	MPH-II(d) MPH-DI PH-DI F1.0	MPH-II(d) MPH-DI PH-DI F1.0	MPH-II(d) MPH-DI PH-DI F1.0
Application of power	F1.1	/	/	/	/	/	/	/	/
Detect power S	/	F2	/	/	/	/	/	/	/
Disappearance of power S	/	/	F1.1	MPH-II(d) F1.1	MPH-II(d) MPH-DI PH-DI F1.1	MPH-II(d) MPH-DI PH-DI F1.1	MPH-II(d) MPH-DI PH-DI F1.1	MPH-II(d) MPH-DI PH-DI F1.1	MPH-II(d) MPH-DI PH-DI F1.1
PH_AR	/	–	–	Start T3 F4	–	–	–	–	–
Expiry of T3	/	–	–	–	MPH-DI PH-DI F3	MPH-DI PH-DI F3	MPH-DI PH-DI F3	–	–
Receiving INFO 0	/	/	MPH-II(c) F3	–	–	–	MPH-DI PH-DI F3	MPH-DI PH-DI F3	MPH-DI PH-DI MPH-EI2 F3

TABLE 3.2 Locally Powered TE Activation/Deactivation BRI State Table (*Continued*)

	F1	F2	F3	F4	F5	F6	F7	F8
Receiving signal	/	-	-	-	-	-	MPH-EI1 F6	MPH-EI2 F6
Receiving INFO 2	/	MPH-II(c) F6	F6	F6	F6	-	MPH-EI1 F6	MPH-EI1 F6
Receiving INFO 4	/	MPH-II(c) PH-AI MPH-AI F7	PH-AI MPH-AI F7	PH-AI MPH-AI F7	PH-AI MPH-AI F7	PH-AI MPH-AI MPH-EI2 F7	PH-AI MPH-AI MPH-EI2 F7	PH-AI MPH-AI MPH-EI2 F7
Lost framing	/	/	/	/	/	MPH-EI2 F8	MPH-EI2 F8	-

- = No change, no action
| = Impossible by L1 definition
/ = Impossible
SOURCE: ITU-T Recommendation I.430.

51

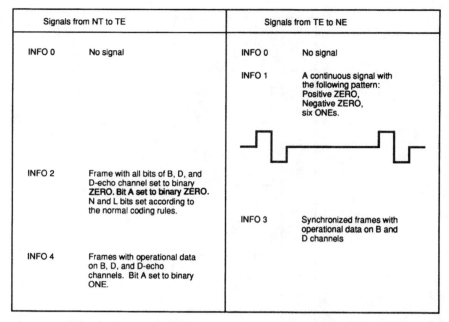

Signals from NT to TE		Signals from TE to NE	
INFO 0	No signal	INFO 0	No signal
		INFO 1	A continuous signal with the following pattern: Positive ZERO, Negative ZERO, six ONEs.
INFO 2	Frame with all bits of B, D, and D-echo channel set to binary ZERO. Bit A set to binary ZERO. N and L bits set according to the normal coding rules.		
		INFO 3	Synchronized frames with operational data on B and D channels
INFO 4	Frames with operational data on B, D, and D-echo channels. Bit A set to binary ONE.		

Figure 3.4 *S/T* reference point INFO signals. (*Adapted from ITU-T Recommendation I.430.*)

framing state. For example, if framing is lost within the activated state (F7), the semiconductor device will notify the LLD that the state has changed to the lost framing state (F8). By keeping track of the past state, the implementor can determine that this transition must have been done because of lost framing and knows that the software is to send a MPH_ERROR_IN primitive to the management entity.

Other information can be received from the semiconductor device—such as collisions (D-echo bit not equal to last transmitted D-channel bit) or underrun conditions (transmitted frame lost integrity because data was not available). Some of these conditions occur sporadically during a properly running system. Some situations (such as collisions) should happen only in multipoint environments. Other situations may indicate problems in software or hardware.

***U* interface support.**   Various semiconductor devices exist in support of ANSI T1.601 specifications. The main advantage of these is to reduce equipment cost. A separate NT1 device is not necessary to connect to the switching network. A LLD making use of a *U* interface will probably want to map states into the familiar I.430 F-states. However, various information is not available from the *U* interface. Only point to point is supported, and, therefore, collisions will not happen and D-bit monitoring (along with prioritization) is unnecessary. An implementor should use the manufacturer's documentation for guidelines to pseudo-I.430 state handling.

Primary Rate Interface

PRI use of ISDN can be broken into two categories. A 1.544-Mbps interface is currently used for North America, Japan, South Korea, and Taiwan (note that specific country lists are always changing). A 2.048-Mbps interface is used for Europe. These are roughly based upon the T1 and E1 digital transmission standards. PRI is used for medium-speed network access. This is achieved by making use of the H-channels discussed previously. The H-channel groupings can consist of fixed B-channels or can be assigned in an irregular fashion. ITU-T Recommendation I.431 is concerned with the physical layer of PRI.

Configuration requirements

There are a number of substantial differences in the use of PRI as opposed to BRI. PRI supports only point to point (as is true at the U reference point with BRI). The digital channel is, however, configured as separate 64-kbps B-channels (which can be grouped into higher-bandwidth H-channels). For Europe, 31 channels are supported. In the T1-based PRI, 24 channels are supported. If signalling is to be carried on the same PRI, one of the channels will be designated a D-channel but at a speed of 64 kbps.

A PRI should always remain active (if possible). This eliminates various primitives that may be used for BRI. Specifically, the PH_ACTIVATE_RQ, MPH_DEACTIVATE_RQ, MPH_DEACTIVATE_IN, and MPH_INFORMATION_IN primitives are eliminated. It is still useful for other layers to know the current status of the interface. Thus, indication primitives are still used for PRI, as may be seen in Table 3.3.

Data encoding.   PRI uses bipolar-AMI data encoding. In order to keep the line balanced, the European standard recommends use of HDB3. For the 1.544-mbps North American PRI, use of B8ZS is recommended.

Frame structure.   As can be seen in Fig. 3.5, PRI structure is very different between the 1.544- and the 2.048-mbps versions. Each frame of the North American version is composed of 1 framing bit followed by an octet (or byte) for each of the 24 possible channels. This amounts to a total of 193 bits per frame. Since 1 bit is not sufficient for synchronization and error-checking purposes, groups of 24 frames are used as a single multiframe. The F-bit thus gets expanded into a set of 24 bits that can be used for synchronization and error checking.

TABLE 3.3 PRI User-Side Physical Layer State Table

Initial state	F0	F1	F2*	F3	F4	F5*	F6
			Definition of the states				
Operational condition or failure condition	Power off at user side	Operational	FCI	FC2	FC3	FC4	Power on at user side
Signal transmitted toward interface	No signal	Normal operational frames	Normal operational frames	Frames with RAI†	Frames with RAI†	Normal operational frames	No signal
			New event, detected at the receiving side				
Loss of TE power	/	PH-DI MPH-EI0 F0	MPH-EI0 F0	MPH-EI0 F0	MPH-EI0 F0	MPH-EI0 F0	MPH-EI0 F0
Return of TE power	F6	/	/	/	/	/	/
Normal operational frames from network side	/	—	PH-AI MPH-AI F1	PH-AI MPH-AI F1	PH-AI MPH-AI F1	PH-AI MPH-AI F1	/
Reception of RAI*†	/	PH-DI MPH-EI1 F2	—	MPH-EI1 F2	MPH-EI1 F2	MPH-EI1 F2	MPH-EI1 F2
Loss of signal or frame alignment	/	PH-DI MPH-EI1 F2	MPH-EI2 F3	—	MPH-EI2 F3	MPH-EI2 F3	MPH-EI2 F3

TABLE 3.3 PRI User-Side Physical Layer State Table (*Continued*)

Reception of AIS‡	/	PH-DI MPH-EI3 F4	MPH-EI3 F4	MPH-EI3 F4	—	MPH-EI3 F4	MPH-EI3 F4
Reception of RAI and continuous CRC§ error report*	/	PH-DI MPH-EI4 F5	MPH-EI4 F5	MPH-EI4 F5	MPH-EI4 F5	—	MPH-EI4 F5

*For 2.048-Mbps ISDN PRI only, a network-option allows separation of RAI only from the error condition of RAI with continuous CRC error. In such cases, the above table is used as listed. For 1.544-Mbps ISDN PRI or 2.048-Mbps systems that do *not* support this optional procedure, state F5 (and the last row of the above table) is unused and can be merged into state F2. (State F2 varies from F5 only in the MPH_ERROR_IN parameter.)

†RAI = Remote alarm indication
‡AIS = Alarm indication signal
§CRC = Cyclic redundancy check
SOURCE: ITU-T Recommendation I.431.

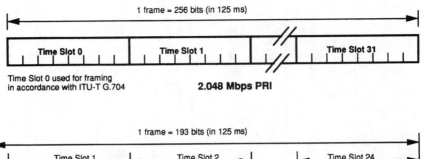

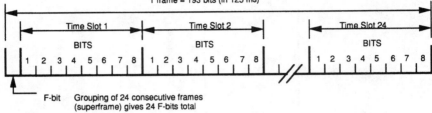

Figure 3.5 Frame structure for 2.048- and 1.544-Mbps PRI. (*Adapted from ITU-T Recommendation I.431.*)

The 2.048-mbps interface has a simpler structure. Each 256-bit frame is broken into 32 time slots of 8 bits each. The first time slot is called the framing channel and allows for synchronization, error checking, and maintenance needs. The remaining 31 time slots are allocated to the data channels.

Protocol states and software requirements

Since the PRI is to remain active at all times, the state table (shown in Table 3.3) is simpler than that for BRI. State F0 is a nonpowered state. State F1 is the operational state (on the TE, or user, side). F6 is a transitional state for when power returns. Finally, states F2 through F5 are temporary states indicating nonoperational status. Thus, a transition from operational state (F1) to nonoperational state allows for notification to layer 2 of a deactivated interface. Transitions from nonoperational states (F2 through F5) to the operational state have activation indication involved. Other transitions allow for error indications. The network-side state table is similar in form—although the exact definitions of events is different.

Broadband ISDN Interface

B-ISDN is the subject of intensive study within ITU-T. Many of the specific requirements for B-ISDN are still "for further study." This is true in ITU-T I.432, which defines the B-ISDN UNI for the physical layer. The physical layer for B-ISDN is based upon the ITU-T's Synchronous Digital Hierarchy (SDH). Because of the high speeds involved, much of the physical layer details have to be handled by hardware support and are not directly under the control of an ISDN systems implementor.

Two different bit rate interfaces are defined. One such is a bidirectional 155.520-Mbps interface. This is known as Synchronous Transfer Mode 1 (STM-1). The other one (STM-4) is that of 622.080 Mbps in one direction (or, optionally, both). If it is not fully bidirectional, the other direction will support 155.520 Mbps. In general, it will probably be necessary to use fiber optics to get these types of speeds—but ITU-T I.432 does not specify the physical medium. Most of ITU-T I.432 is concerned with the Transmission Convergence (TC) sublayer.

Cell transmission structure

The cell is the basic transmission unit of B-ISDN. For proper physical transmission and reception, cells are packaged into other groupings to provide error detection, framing, and other maintenance functions. Figure 3.6 shows the general hierarchy for STM-1. The lowest level is an ATM cell. This consists of a 5-byte header and 48 bytes of data (provided by layer 3—the ATM Adaptation Layer, or AAL).

Each ATM cell, consisting of 53 bytes, is then mapped into a C-4 frame. The VC-4 (C-4 with overhead) consists of 1 byte of header information followed by 260 bytes of ATM cells, then another header byte, another 260 bytes of ATM cells, and so forth. A total of 9 Path OverHead (POH) bytes are defined, giving a total capacity of 2349 bytes for a VC-4 frame package. Note that 260 is not an even multiple of 53, and, therefore, some ATM cells may be interrupted by POH bytes. Finally, a VC-4 cell package will be bundled into a 2430-byte STM-1 package. The format for STM-4 is not yet defined but will probably follow the SDH STM-4ce format.

Header error control and detection

The 5 bytes of header in the ATM cell include 1 byte called the Header Error Control (HEC) byte. This HEC byte allows for single-bit error correction and multibit error detection. B-ISDN can operate in two modes on reception. An error can cause automatic discarding of the cell or an attempt can be made to correct the error. In general, the HEC gives enough information to correct a single bit error—but, under certain transmission conditions, it may be faster to discard an error rather than go through the correction algorithm. Error correction may slow down overall throughput.

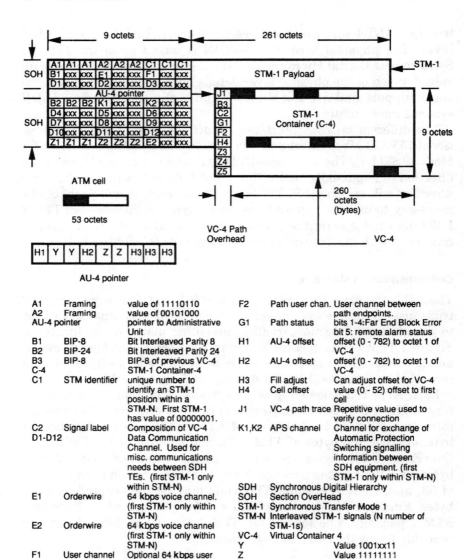

A1	Framing	value of 11110110
A2	Framing	value of 00101000
AU-4 pointer		pointer to Administrative Unit
B1	BIP-8	Bit Interleaved Parity 8
B2	BIP-24	Bit Interleaved Parity 24
B3	BIP-8	BIP-8 of previous VC-4
C-4		STM-1 Container-4
C1	STM identifier	unique number to identify an STM-1 position within a STM-N. First STM-1 has value of 00000001.
C2	Signal label	Composition of VC-4
D1-D12		Data Communication Channel. Used for misc. communications needs between SDH TEs. (first STM-1 only within STM-N)
E1	Orderwire	64 kbps voice channel. (first STM-1 only within STM-N)
E2	Orderwire	64 kbps voice channel (first STM-1 only within STM-N)
F1	User channel	Optional 64 kbps user channel (first STM-1 only within STM-N).

F2	Path user chan.	User channel between path endpoints.
G1	Path status	bits 1-4:Far End Block Error bit 5: remote alarm status
H1	AU-4 offset	offset (0 - 782) to octet 1 of VC-4
H2	AU-4 offset	offset (0 - 782) to octet 1 of VC-4
H3	Fill adjust	Can adjust offset for VC-4
H4	Cell offset	value (0 - 52) offset to first cell
J1	VC-4 path trace	Repetitive value used to verify connection
K1,K2	APS channel	Channel for exchange of Automatic Protection Switching signalling information between SDH equipment. (first STM-1 only within STM-N)
SDH		Synchronous Digital Hierarchy
SOH		Section OverHead
STM-1		Synchronous Transfer Mode 1
STM-N		Interleaved STM-1 signals (N number of STM-1s)
VC-4		Virtual Container 4
Y		Value 1001xx11
Z		Value 11111111
Z1,Z2		Reserved for future use
Z3-Z5		Reserved for future use

Figure 3.6 Mapping for ATM cells into the STM-1 signal. (*Adapted from ITU-T Recommendation I.432.*)

Scrambling requirements

One of the criteria for B-ISDN is that it be independent of the actual transmission method used. This implies that a specific digital encoding method may not be universal. In order to allow for balancing needs across a multitude of physical transmission methods, ITU-T I.432 specifies a general scrambling scheme. This applies only to the information field (the latter 48 bytes) of the ATM cell.

Future standardization

Many aspects of B-ISDN are yet to be defined. The general philosophy is to provide a standardized structure, based on the SDH, that can be independent of physical transmission media. Some implementations currently exist, such as ANSI's SONET, which are compatible with B-ISDN requirements. It can be expected that this implementation, and others to come, will be used as experimental vehicles to determine classes of functions and services necessary to define B-ISDN.

Summary of Chapter

The physical layer of ISDN is the foundation of ISDN. N-ISDN physical layer requirements are broken into two types of interfaces. These are BRI and PRI. BRI can further be broken into categories of use and location. BRI, at the S/T reference point, can be used in multipoint or point-to-point configurations. The physical layer interface at the U reference point is not generally defined in the various ITU-T recommendations, but T1.601 is used for such within North America.

PRI is distinctive from BRI in the ability to group B-channels into H-channels. Additionally, the D-channel for signalling can be optional if more than one PRI is handled by the same exchange and physical endpoint.

All of the N-ISDN physical layer definitions include notification to layer 2 of the state of the line activity. I.430 expands this to allow specific requests for line activation or deactivation. Such requests are not appropriate for PRI, which is expected to remain active if possible.

The philosophy of B-ISDN separates it from the physical transmission medium. Nevertheless, certain structural requirements—similar to bit framing in N-ISDN—exist for B-ISDN and are defined in ITU-T I.432. B-ISDN is oriented around the cell as a unit of transmission.

The physical layer interacts with the data link layer (layer 2) and the management entity. The next chapter discusses the basics of the data link layer. Management interactions are covered in Chap. 12, and various aspects of specific LLD design are handled in Chap. 11.

Chapter

4

Data Link Layer

Layer 2 of the OSI model is called the data link layer. It is primarily a means of providing flow and error control for data transfer over the physical link. As such, it must deal with the physical devices which handle the electrical transmission of data. Often the layer 2 code will use a defined set of primitives to the physical devices, particularly in the case of HDLC links. A LLD will be used to interpret these primitives into the specific register controls necessary to activate the semiconductor devices used for physical link access. LLDs are discussed in greater detail in Chap. 11. They must also work with higher layers to perform specific functions (data transfer is a primary use).

Architectural Position within ISO Model

The data link layer is positioned, within the ISO model, between the physical and network layers. As mentioned above, the interactions with the physical layer are often done via an intermediary in the form of an LLD. The network layer may not be needed if the link is currently dedicated and the network interactions are null.

In the case of ISDN, the data link layer protocol for signalling is defined by Q.921 (ANSI 602). The Q.921 protocol belongs to the family of HDLC protocols (as defined in ISO 4335 and other ISO documents). For BRI, it will interact with I.430 at the physical layer and Q.931 at the network layer. (PRI will interact with I.431 at the physical layer.) We will first cover the general aspects of HDLC protocols, followed by an in-depth coverage of Q.921 and then a discussion of related data link layer protocols that are often used on the ISDN bearer channels.

Use of the Data Link Layer within ISDN

ISDN does incorporate more than just signalling information. The D-channel can support multiple logical links due to the method of ad-

dressing used within Q.921. These channels can be used for signalling, maintenance, and special data services. Additionally, the B-
channels (or groups thereof, such as H0 channels) may also have
HDLC protocols on them. Such protocols include LAPB for the data
link layer of ITU-T X.25 or a variant of Q.921 such as ITU-T's V.120
data link layer. If a general knowledge of HDLC protocols is obtained,
it is possible to reutilize significant parts of any HDLC protocol module for multiple protocol use.

HDLC

The HDLC protocol can be used for many different types of configurations. The protocol allows for point-to-point (one station connected to
one other) and multipoint (each station having access to multiple stations) links. It can operate over various distances. Either half-duplex
(two-way but not at the same time) or full-duplex (simultaneous two-
way) operation can be supported. Finally, the protocol can be used for
either a host-to-terminal type of operation or between two peers (for
example, two PCs). The great variability of the HDLC protocol lends
itself to the creation of protocol subsets for specialized purposes—
such as LAPB, which is used for link-level X.25 communication.

Framing requirements

HDLC protocols have a number of features in common. The first item
is that frames of data are delimited by flags. A flag consists of a series
of logical bits of the form

```
01111110
```

These six 1s surrounded by 0s precede and follow a frame. (Some systems allow a flag to be "shared"—the same flag is used as an ending
flag for one frame and the beginning flag for the following frame.) In
order to allow this pattern to be part of the actual data, a procedure
known as *bit stuffing* is used. When data is transmitted, a 0 is inserted into the data stream after each set of five sequential 1s. Upon receipt, the 0s will be taken out—any 0 following five 1s is removed.
Thus the data pattern

```
10111111100011100111111111
```

will be transformed into

```
10111110100011100111111010111
```

upon transmission and will be changed back into the original form
upon receipt.

The transmission line must have some type of data pattern superimposed on it during idle times. Two common options are a FLAG IDLE situation (where flag characters are transmitted continuously while idle) or MARK IDLE (where a stream of fixed bits are transmitted). Another common feature is the *abort* sequence, which can be used to stop receipt of data during transmission. The abort sequence consists of seven consecutive 1s (a pattern illegal in all other circumstances).

HDLC fields

The HDLC frame is delimited by flags, but the frame itself also has a particular form. It consists of the following:

- The beginning flag
- An address field (which tells destination)
- A control field (which indicates the type of frame)
- Information data (if applicable)
- A frame check sequence (FCS), which allows for error detection

The beginning flag, address, and control fields are often referred to as the *header*. The FCS and ending flag(s) are referred to as the *trailer*.

The FCS is often encoded as a cyclic redundancy check (CRC) (either 2 bytes, referred to as CRC-16, or 4 bytes, referred to as CRC-32). A more complete description of the CRC calculation can be found in William Stallings book, *Handbook of Computer-Communications Standards* (Macmillan, 1987), Vol. 1.

As a brief description of the control field, let us say that there are three classes of frame types: Information, Supervisory, and Unnumbered. A more complete description of a possible set of control fields is given in the following section concerning Q.921, an ISDN-specific HDLC type of protocol.

The interframe flags, delimiting flags, and FCS are often handled directly by an HDLC-handling semiconductor chip without access or manipulation by layer 2 protocol software. The manipulation of the rest of the frame is sometimes also implemented in hardware but rarely as a full layer 2 protocol set.

Balanced versus unbalanced data links

A balanced data link allows both ends to send and receive both commands and responses. An unbalanced data link allows commands from only one end, sometimes called the *primary*. The *secondary* receives commands and sends back responses. The primary is in control of the session and maintains integrity of the link. Most HDLC protocols used within ISDN are of a balanced mode, although B-channels are often unrestricted as to actual usage. These protocols are said to be in Asynchronous Balanced Mode (ABM) and, thus, a common command is to Set Asynchronous Balanced Mode (SABM).

Commands and responses

Commands and responses are encoded within the control field for HDLC frames. A common set of such codes would include Information (I) frames used for transmission of data. Receiver Ready (RR) frames are used to indicate readiness to receive data and may also be used to acknowledge previously received data. A Receiver Not Ready (RNR) frame may be used to acknowledge received data but indicates that it is no longer able to accept data.

The Set (S) command is used in conjunction with an Unnumbered Acknowledgment (UA) response. The S command is usually noted as the type of mode that is to be entered. Thus, the SABM sets a link into asynchronous balanced mode, and a SNRM is used to set a link into normal response mode—with the sender of the command acting as the primary. A SABME command sets a link into ABM extended. The *extended* part refers to the use of extended fields allowing larger modulo frame sequencing.

The DISConnect (DISC) places the other end into a disconnected mode—often used for switched lines. A UA frame is the expected response. If an end finds itself in disconnected mode (without an explicit command to do so), a Disconnect Mode (DM) response can be sent autonomously. Finally, the FRaMe Reject (FRMR) response frame can act to reject a received frame.

Other commands and responses may be added within the HDLC protocol schema. These additional control field codes can be used for rejection schemes, or polling, or for unacknowledged data transfer.

Data transfer use of HDLC

Data transfer is the phrase used for transmitting data from one high-level (layer 3 or above) peer to another. These types of frames fall into the categories of acknowledged and unacknowledged data. The set of commands/responses dealing with acknowledged data include the I-frame, RR, and RNR. Special frames such as a FRMR or reject (REJ) may also work with acknowledged data, although these frames are not supported by all HDLC protocols. An unacknowledged frame can be lost without explicit knowledge of the protocol. Unacknowledged data transfer is done for speedy transfer when the information can be recreated or is of transient value.

Acknowledgment of data requires a method of identification of a particular frame. This is done via sequence numbers. Each frame is numbered according to a modulo scheme. Thus, the first frame will be frame number 0, the next frame number 1, and so forth up to the modulo number minus 1 (7, for example, in the case of modulo 8). At this point frame numbering starts over at frame 0. This identification allows specific acknowledgment and the ability to transmit more than one frame at a time before acknowledgment. The transmission of multiple frames before acknowledgment is known as windowing and is discussed in greater detail in the following Q.921 section.

Family of protocols

As mentioned in the introductory section on HDLC, the HDLC *schema* is considered to allow for a family of protocols that meet the general criteria. Link Access Procedure for Modem (LAPM) (V.42) is used for modem protocols. LAPD (Q.921) is used for ISDN applications. ITU-T Recommendation V.120 is used for rate adaption. Point-to-Point Protocol (PPP) can be used for multiplexing various high-level protocols over the same transmission media. Although they are each a specific protocol, with specific additions and requirements, the common subset of HDLC structure allows them to each be used over similar digital transmission facilities.

LAPD

LAPD is defined by ITU-T Recommendation Q.921. Its primary use is for transport of signalling information. It can also be used for maintenance and general data transport due to the format of its addressing field. LAPD is part of the HDLC family and, thus, uses the same flag delimiting method (with attendant bit stuffing to avoid confusion of data with control information). The general field format for LAPD is shown in Fig. 4.1.

Address field

The address field of the LAPD frame is made of 2 bytes (octets), as shown in Fig. 4.2. These 2 bytes include information concerning the Service Access Point Identifier (SAPI), the TEI, and a Command/Response (C/R) field bit.

The EA bit allows for versatility in the length of the address field. A 0 in the EA field bit indicates that additional bytes of information are included to be considered as part of the address field. A 1 indicates that this byte is the last one of the current information field. This method is used a lot in more complex command streams such as in Q.931. For LAPD, the first byte's EA bit is 0, and the second byte's EA bit is 1. Use of the EA bit allows the protocol module to distinguish between LAPD and LAPB (X.25, layer 2).

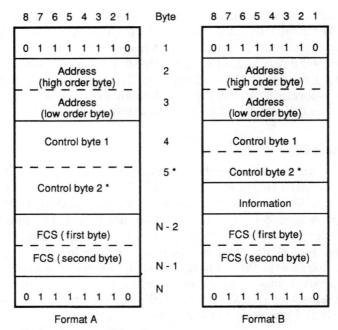

Format A is used for non-Information frames
Format B is used for Information frames (I, UI, FRMR, XID)

Control byte 2 is needed for frames with sequence information (I, RR, RNR, REJ).

Figure 4.1 LAPD general field format. (*Adapted from ITU-T Recommendation Q.921.*)

8	7	6	5	4	3	2	1	Octet (byte)
SAPI					C/R	EA 0		2
TEI						EA 1		3

EA   = Address field extension bit
C/R = Command/Response field bit
SAPI = Service Access Point Identifier
TEI = Terminal Endpoint Identifier

Figure 4.2 LAPD address field format. (*From ITU-T Recommendation Q.921.*)

The use of the C/R field bit is useful in cases where the peer protocols are not completely symmetric. For LAPD, there is a user side which is located at the user's equipment and a network side which is located at the network access point. Using the C/R bit in an asymmetric fashion allows each endpoint to be identified. For LAPD, the network side will always mark the C/R bit to be 1 for commands (and 0 for responses). The user side will mark the C/R bit as 0 for commands and 1 for responses.

TABLE 4.1   LAPD SAPI Values

SAPI value	Related layer 3 or management entity
0	Call control procedures
1–15	Reserved for future standardization
16	Packet communication conforming to X.25 level 3 procedures
17–31	Reserved for future standardization
63	Layer 2 management procedures
All others	Not available for Recommendation Q.921 procedures

SOURCE: ITU-T Recommendation Q.921.

Service access point identifier. The SAPI is used to determine the function of the data link. Table 4.1 indicates current values. Two values are particularly important for LAPD. These are the values of 0 (for call control—signalling) and 63 (for layer 2 management procedures). The value of 16 allows use of the D-channel for X.25 level 3 data transport uses. Other values are still being defined—such as particular values for SUCs for Frame Relay.

Terminal endpoint identifier. The TEI is a method of identification of the terminal (as opposed to the SAPI's use as determination of function). A TEI may be either assigned nonautomatically (or "fixed") or automatically by the switch. The range of values of 0 to 63 comprise the valid fixed TEI numbers. These are assigned by the terminal, with subscription agreement by the network. Automatic TEI values lie in the range of 64 through 126 and are assigned by the network in response to a request from the terminal. These automatic TEI negotiation messages are sent via Unnumbered Information (UI) frames over the management SAPI link (63) as a broadcast message (TEI value of 127). Automatic TEI negotiation is covered more fully later in this chapter.

Generally, automatic TEIs are used with multipoint BRI terminals. Fixed TEIs are used in point-to-point terminal configurations and in PRI situations. It may be useful to use automatic TEI negotiation in cases where cellular technologies are being used for physical layer connection to the network. This avoids the requirement to have prearranged the TEI between the terminal equipment and the network.

Control field

The Control field identifies the type of frame being transmitted (with the C/R field bit of the address field determining the mode of operation). There are three general types of frame formats: Information, Supervisory, and Unnumbered. As can be seen in Fig. 4.3, these can be distinguished by use of the lower 2 bits of the fourth byte of the frame. A 0 in the lowest bit indicates an information format. A 01 in the lower 2 bits indicates a supervisory format, and an 11 indicates

Control field bits (modulo 128)	8	7	6	5	4	3	2	1	Octet (byte)
I format	N(S)							0	4
	N(R)							P	5
S format	X	X	X	X	S	S	0	1	4
	N(R)							P/F	5
U format	M	M	M	P/F	M	M	1	1	4

N(S)   Transmitter send sequence number   M   Modifier function bit
N(R)   Transmitter receive sequence number P/F   Poll bit when issued as a
                                                  command. Final bit when
                                                  issued as a response.
S   Supervisory function bit   X   Reserved and set to 0

Figure 4.3   LAPD control field format. (*From ITU-T Recommendation Q.921.*)

an unnumbered format. In conjunction with the function encoding bits and the Poll/Final (P/F) bit, enough information is conveyed to allow full decoding of the control field.

P/F field bit. The P/F field bit is used as a polling indication in command frames. If this bit is set to a 1, it indicates that a response is expected from the peer. When used as part of a response frame, the bit is called the final bit and will be set to 1 when the frame is sent in response to a command with P bit set to 1.

Modulo. The HDLC family allows various modulo numbering schemes to identify frames. LAPD uses modulo 128. This scheme is used within the N(R) and N(S) fields of I-frames and within the N(R) of S-frames. In addition to the Received frame number [N(R)] and Sent frame number [N(S)], a number of internal variables must be kept to verify sequencing validity. These are referred to as V(S), V(A), and V(R).

Sequence numbers and windows. The V(S) and V(A) variables are used in frame transmission. V(S) contains the modulo 128 number of the next frame to be sent. From an implementor's point of view, this variable is used when an I-frame is transmitted to fill in the N(S) field. After transmission, the V(S) variable is incremented. The V(A) variable contains the sequence number of the next I-frame to be transmitted. When a valid N(R) is received, the V(A) variable is updated.

In LAPD, a number of I-frames may remain unacknowledged at any period. This number is referred to as k and indicates the "window" size. At initialization or when no I-frames are unacknowledged, V(S) will equal V(A). When one I-frame is still unacknowledged, V(S) will differ from V(A) by 1 (modulo 128). Because of the limitation of the value of k, V(S) may not differ from V(A) by more than k. Thus,

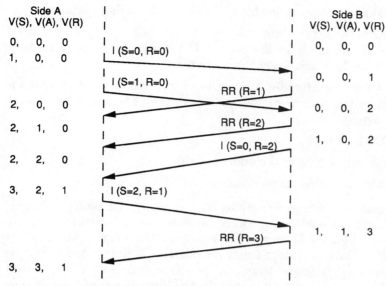

Figure 4.4 An example of LAPD sequence variable use.

upon transmission, an I-frame may be sent only if V(S) < V(A) + k (modulo 128).

It is possible for a N(R) field to acknowledge more than one I-frame at a time. A received N(R) is valid only if its value is such that V(A) ≤ N(R) ≤ V(S). An N(R) that is equal to V(A) indicates that no additional I-frames are acknowledged. An N(R) that is equal to V(S) indicates that *all* I-frames have been acknowledged, and V(A) < N(R) < V(S) indicates that one or more I-frames have been received (but not all). Figure 4.4 shows an example of sequencing and the updating of internal variables.

The third internal variable V(R) is the value of the *next* valid sequence number to be received in an N(S). If N(S) is equal to V(R) (assuming that the frame is otherwise valid), the received I-frame is accepted, and V(R) is incremented (modulo 128) to prepare for the next received I-frame. Thus, under idle conditions, V(S) is the sequence number of the *next* I-frame to be transmitted, V(R) is the sequence number of the *next* I-frame to be received, and V(A) is a boundary variable whose value minus 1 (modulo 128) indicates the *previous* transmitted I-frame (if any) that was acknowledged.

Invalid frames

Section 2.9 of ITU-T Recommendation Q.921 lists the specific conditions under which a frame is to be considered invalid and discarded without notification to the sender or other action. These conditions are as follows:

1. The frame is not preceded by, and followed by, a flag sequence.

2. The frame has less than 6 bytes between flags in the case of I-frames or S-frames. In the case of U-frames, the limit is a situation where there are fewer than 5 bytes.

3. The bit stream (after 0-bit removal on reception, prior to 0-bit insertion on transmission) does not contain an even number of 8-bit octets/bytes.

4. The FCS is in error.

5. There is only 1 byte in the address field.

6 The frame contains an unsupported SAPI.

Unfortunately, testing situations do not always follow these guidelines directly. In particular, many layer 2 test suites may insert a LAPB frame into the D-channel link as a method of generating an *invalid control field*. Although the frame would fail the second condition above (possessing fewer than the required number of bytes) and should be discarded without action, test suites exist that require consideration of the frame as an invalid control field (with a consequent reinitialization of the logical link).

The previous example is given as an indication of a situation that an implementor may find when testing for a specific certification of ISDN use. In any such cases, it is necessary to follow the requirements of the certification rather than the standards or specifications.

Specific Q.921 commands and responses

As a member of the HDLC family, Q.921 has a number of commands and responses. These are indicated in its control field. A full listing of these are in Table 4.2.

Information transfer.   The I-frame command is used to convey data from a higher-level entity to its peer in an acknowledged fashion.

Set asynchronous balanced mode extended.   The SABME U-frame command is used to set the peer into modulo 128 multiple frame acknowledged mode. Upon receipt, internal variables $V(S)$, $V(A)$, and $V(R)$ are reset to 0. Any outstanding I-frames (if any) are considered to be invalid and must be resubmitted by higher layers—which are notified by a Data Link Establish Indication (DL_EST_IN) primitive of the situation. The initiator of this command is acknowledged by receipt of an UA response frame.

Disconnect command.   The DISC U-frame command is used to discontinue multiple frame acknowledged mode. It may be acknowledged with a UA frame or a Disconnect Mode (DM) response frame. Any out-

TABLE 4.2 LAPD Commands and Responses

Application	Format	Commands	Responses	Encoding							
				8	7	6	5	4	3	2	1
Unacknowledged and multiple frame acknowledged information transfer	Information transfer	I (information)					N(S)				0
							N(R)				P
	Supervisory	RR (receive ready)	RR (receive ready)	0	0	0	0	0	0	0	1
							N(R)				P/F
		RNR (receive not ready)	RNR (receive not ready)	0	0	0	0	0	1	0	1
							N(R)				P/F
		REJ (reject)	REJ (reject)	0	0	0	0	1	0	0	1
							N(R)				P/F
	Unnumbered	SABME (set asynchronous balanced mode extended)		0	1	1	P	1	1	1	1
			DM (disconnected mode)	0	0	0	F	1	1	1	1
		UI (unnumbered information)		0	0	0	P	0	0	1	1
		DISC (disconnect)		0	1	0	P	0	0	1	1
			UA (unnumbered acknowledgment)	0	1	1	F	0	0	1	1

(Continued)

TABLE 4.2 LAPD Commands and Responses (Continued)

Application	Format	Commands	Responses	Encoding							
				8	7	6	5	4	3	2	1
Unacknowledged and multiple frame acknowledged information transfer	Unnumbered		FRMR (frame reject)	1	0	0	F	0	1	1	1
Connection management		XID (exchange identification)	XID (exchange identification)	1	0	1	P/F	1	1	1	1

SOURCE: ITU-T Recommendation Q.921.

standing I-frames (if any) are considered to be invalid and must be re-submitted by the higher layers. The higher layers are notified of this situation by a Data Link Release Indication (DL_REL_IN) primitive.

Unnumbered information command.   The UI U-frame command is used to send data submitted by higher layers without sequencing or acknowledgment required. Thus, information can be lost without no-tification. This type of command may be used whether or not multiple frame acknowledged mode has been established. Thus, it is particu-larly useful for management (SAPI 63) information which may be at-tempting to obtain a TEI (necessary for establishment of multiple frame acknowledged mode).

Receive ready command or response.   This RR S-frame may be used to indicate readiness to receive an I-frame, acknowledgment of re-ceived I-frames or the change of status in readiness to receive I-frames. An RR S-frame command with the P bit set to 1 is useful for polling the current status of the peer.

Reject command or response.   The REJ S-frame is used to explicitly ask for retransmission of I-frames by the peer. (Optional procedures allow for implicit retransmission upon timer expiration with no ac-knowledgment.) Since the REJ S-frame contains an N(R) field, it also acts as an acknowledgment of any frames up to the rejected number [indicated by N(R)]. It also implicitly clears any local busy condition (RNR, as follows).

Receive not ready command or response.   The RNR S-frame is used to indicate a current inability to accept additional I-frames. The N(R) field may contain acknowledgment of previous I-frames, and the P bit (in a command) can be used to solicit peer information.

Unnumbered acknowledgment response.   The UA U-frame response is used to acknowledge a mode setting command (SABME or DISC). Resetting of mode also clears local busy conditions.

Disconnected mode response.   The DM U-frame response is used to indicate that local multiple frame acknowledgment mode has been terminated.

Frame reject response.   The FRMR U-frame consists of a special set of fields that help to indicate the precise reason for rejection of a frame. These conditions include undefined control fields, invalid length for a frame, or an invalid N(R). The above tests for an invalid frame take precedence over conditions for which an FRMR is sent. Normally, within LAPD, only the network will send an FRMR re-sponse. (For X.25 layer 2, LAPB, the terminal may also send such.)

Exchange identification command or response. The XID U-frame is rarely implemented in current switches. It allows for exchange of parameter values or identification information. Transmission, or receipt, of an XID U-frame does not affect the state of the protocol.

Q.921 Interlayer primitives

Table 4.3 lists all the primitives associated with ITU-T Recommendation Q.921, with the omission of primitives that pass between the physical and data link layers (which are covered in Chap. 3). As is typical for ITU-T primitives, there are four general types of primitives. These are requests, indications, responses, and confirmations.

The DL_EST_RQ and DL_REL_RQ primitives from layer 3 are used to request establishment, or release, of multiple frame acknowledgment mode. The appropriate confirmations are sent in response to the request after the peer entity has acknowledged the situation. DL_EST_IN and DL_REL_IN primitives are used by layer 2 to indicate changes in mode initiated by the peer.

The DL_DA_RQ and DL_U_DA_RQ primitives are used by layer 3 to request transmission of acknowledged or unacknowledged data. The indications are used by layer 2 to pass such data to layer 3. Note that MDL_U_DA_RQ primitives are used for management requests (SAPI 63, broadcast TEI 127).

The MDL_AS_RQ and MDL_RM_RQ primitives are used by the management layer to assign or remove TEIs. Although it is possible for a TEI to be assigned spontaneously by the peer (management layer notified by a MDL_AS_IN), no corresponding MDL_RM_IN is possible. This allows simplification of the coordinating software in the case of use of fixed TEIs.

State transitions

As mentioned in earlier chapters, a protocol is composed of states, events, and actions. Events include primitives, as mentioned above, and timers. A state is the general environment which gives special meanings to events. Actions are done in response to events, possibly changing the state of the protocol. LAPD has eight major states with substates possible (in states 5, 7, and 8) concerning the current ability of the local, or remote, peer to transfer data. A graphical representation of the states and their transitions may be found in Fig. 4.5. The major LAPD states follow.

TEI unassigned state. This state (1) is the initial state for LAPD links that are not within a fixed TEI system. (It may also be used as a transitory state for fixed TEI systems.) This state is left when a

TABLE 4.3 LAPD Primitives.

Generic name	Type				Parameters		Parameter data contents
	Request	Indication	Response	Confirm	Priority indicator	Parameter data	
L3 ↔ L2							
DL_ESTABLISH	X	X	–	X	–	–	CES also exists.
DL_RELEASE	X	X	–	X	–	–	–
DL_DATA	X	X	–	–	–	X	Layer 3 PDU (peer-to-peer message).
DL_UNIT_DATA	X	X	–	–	–	X	Layer 3 PDU (peer-to-peer message).
M ↔ L2							
MDL_ASSIGN	X	X	–	–	–	X	TEI value, CES (TEI only in request).
MDL_REMOVE	X	–	–	–	–	X	TEI value, CES.
MDL_ERROR	–	X	X	–	–	X	Reason for error message.
MDL_UNIT_DATA	X	X	–	–	–	X	Layer management PDU (peer-to-peer message).
MDL_XID	X	X	X	X	–	X	Connection management PDU (peer-to-peer XID frame).

SOURCE: ITU-T Recommendation Q.921.

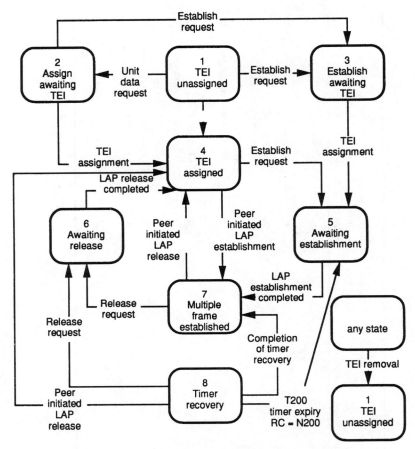

Figure 4.5   LAPD protocol states. (*From ITU-T Recommendation Q.921.*)

DL_U_DA_RQ, DL_EST_RQ, or MDL_AS_RQ is received. A
DL_U_DA_RQ causes a queuing of the UI frame, an MDL_AS_IN
primitive to be generated, and a transition to the assign awaiting TEI
state (2). A DL_EST_RQ causes an MDL_AS_IN primitive to be generated and a transition to the establish awaiting TEI state (3). An
MDL_AS_RQ causes a transition to the TEI assigned state (4). In
both cases, the generation of the MDL_AS_IN primitive should invoke either an MDL_AS_RQ primitive if fixed TEIs are in use or the
commencement of TEI negotiation if automatic TEIs are in use.

Assign awaiting TEI state.   This state (2) is a transitory state. An
MDL_AS_RQ will cause a transition to TEI-assigned state (4). A
DL_EST_RQ causes a transition to establish awaiting TEI state (3). A
failure of the physical layer in this state will cause a transition back
to TEI unassigned state (1).

Establish awaiting TEI state. This state (3) is left upon receipt of an MDL_AS_RQ primitive. A transition is made to awaiting establishment state (5), a SABME command is generated, and a timer (T200) is started to await acknowledgment of the SABME. A failure of the physical layer in this state will cause a transition back to TEI unassigned state (1).

TEI assigned state. The first three states allow queuing of unnumbered information (DL_U_DA_RQ) data but do not allow transmission (MDL_U_DA_RQ primitives are allowed in all states). This state (4) allows transmission of UI frames. The LAPD states can thus be combined into three groupings. States 1 through 3 are passive states to data transfer. States 4 through 6 allow UI data transmission. States 7 and 8 are considered to be in multiple frame acknowledgment mode, and all primitives are allowed. States 4 through 6 will go back to the TEI unassigned state (1) upon receipt of an MDL_RM_RQ primitive.

The TEI assigned state is left by receipt of a DL_EST_RQ primitive. A transition is made to awaiting establishment state (5), a SABME command is issued, and a timer (T200) is started to wait for acknowledgment of the SABME command. This state may also be left by establishment of multiple frame mode by the peer (i.e., upon receipt of a SABME command), upon which a DL_EST_IN primitive is sent, transition to multiple frame established state (7) is made, and the idle timer (T203) is started. The T203 timer is considered optional by some systems. A failure of the physical layer in any of the TEI assigned states (4 through 8) will cause a transition back to TEI assigned state (4). (An MDL_RM_RQ may be issued for some systems under these conditions.)

Awaiting establishment state. This transition state (5) is attempting to enter multiple frame established state (7). The exact possible transitions become too complicated in states 5 through 7 to allow full discussion here. The primary ones are that a UA is received in response to a SABME command. A transition to multiple frame established state (7) is made and the idle timer (T203) is started. If the T200 timer expires and the maximum number of retries (N200) has not been exceeded, a new SABME command is issued, and the timer (T200) is restarted. If T200 timer expires and N200 has been exceeded, a transition back to TEI assigned state is made, and a DL_REL_IN primitive is issued. For some systems, a DL_REL_IN may provoke a new DL_EST_RQ primitive from the higher layers.

Awaiting release state. Another transitory state, this state (6) is awaiting acknowledgment of a DISC command to leave the multiple frame established mode. Receipt of a UA or DM response frame will cause a direct transition back to the TEI assigned state (4). If the T200 timer expires and the maximum number of retries (N200) has not been exceeded, a new DISC command is issued, and the timer (T200) is restarted. If the T200 timer expires and N200 has been exceeded, a transition back to the TEI assigned state is forced without explicit acknowledgment from the peer, and a DL_REL_CF primitive is issued.

Multiple frame established state and timer recovery state. These states (7 and 8) are the primary states for acknowledged data transfer. Temporary transitions may be made to the timer recover state (8) if a polled command is not acknowledged before expiration of the T200 timer or if the idle timer (T203) is implemented. In a functioning system with active data transfer, the protocol will spend most of its time in state 7.

TEI management procedures

Properly speaking, TEI management procedures are not part of the data link layer. These procedures act on the UI frames conveyed via MDL_U_DA_RQ and MDL_U_DA_IN primitives (SAPI 63, TEI 127). Since proper OSI layering does not act upon the contents of data frames, such use is not really part of the data link layer. However, since Section 5.3 of the ITU-T recommendation defines these procedures, a discussion of them seems appropriate at this time.

There are four basic procedures. These are TEI assignment, TEI checking, TEI removal, and optional user equipment initiated TEI identity verification procedures. As mentioned above in the discussion of LAPD states, TEI assignment begins when layer 2 issues an MDL_AS_IN primitive to the management layer. If the TEI is fixed, the TEI will return immediately from the management layer by use of the MDL_AS_RQ primitive. Otherwise, TEI assignment procedures are initiated.

TEI assignment procedure. A ID_REQUEST primitive is sent by the user equipment. This primitive consists of the management entity identifier (value 0x0F) in the first byte, followed by 2 bytes of a random value used as a reference number (Ri). The last 2 bytes are a message-type identifier (value 0x01), and an action indicator (Ai) of value 127 (left shifted by 1 bit to fit the message format) to indicate that any TEI available can be assigned. The primitive is sent via an MDL_U_DA_RQ primitive, and a timer (T202) is started. Since the message is sent via the broadcast link (TEI 127), the Ri is used as an identifier to verify that a reply applies to the request outstanding.

The network will respond with either a TEI selection or a denial of the request. The ID_ASSIGNED (value 0x02) message-type identifier is used to assign a TEI, with the Ai field loaded with the assigned TEI value. When this MDL_U_DA_IN primitive arrives, a check is done on the Ri. If the Ri matches, further checks are done to ensure nonduplication of the assignment. Under simple error-free scenarios, the TEI is placed within an MDL_AS_RQ primitive, and the T202 timer is canceled. If a legal ID_ASSIGNED message is not received before expiration of T202, a new message is formatted up to a limit (called N202).

TEI check procedure.   The network initiates a TEI check procedure to check on TEIs in use or for potential duplicate assignments. Normal procedures will have the network layer send an ID_CHK_RQ message-type identifier (value 0x04) with an appropriate Ri and Ai (as defined above in the ID_ASSIGNED situation). The user equipment responds with a corresponding ID_CHK_RS message type (value 0x05).

TEI removal procedure.   The network may initiate TEI removal via an ID_REMOVAL message-type (value 0x06) which results in MDL_RM_RQ primitive(s). The user equipment may also internally remove (via MDL_RM_RQ primitives) TEI assignments based on local conditions. Various conditions for TEI removal are given in Section 5.3.4.2 of ITU-T Recommendation Q.921.

System parameters

Most systems use default specifications for system parameters as listed in Table 4.4. An optional parameter negotiation method is listed in Q.921 Appendix IV which makes use of XID frames to negotiate data link parameters. This table lists default window (k) sizes for signalling and packet links, as well as timer and counter values. Each of these timers has been discussed earlier in this chapter.

Specification description language charts and state tables

ITU-T Recommendation Q.921 gives three different forms of state table descriptions for an implementor to use. These are procedures described within the text of Q.921, a set of Specification Description Language (SDL) diagrams and a set of state transition tables. In general, the state transition tables are easiest to implement. However, the SDLs are useful for verification of optional procedures that may be implemented for a given piece of user equipment.

X.25 Layer 2 (LAPB)

ITU-T Recommendation X.25 does not really have a direct relevance to ISDN. However, one of the goals of ISDN is to be able to interconnect to existing networks. X.25 is a protocol that has existed and been implemented for a long time. There are many implementations in existence, and the beginnings of ISDN reflect the desire to incorporate those net-

TABLE 4.4 LAPD System Parameters

		k	T200	T201	T202	T203	N200	N201	N202
Point-to-point data link procedure on a D-channel at 16 kbps	Signalling (SAPI = 0)	1	1 s	N/A	N/A	10 s	3	260	N/A
	Packet communication (SAPI = 16)	3	1 s	N/A	N/A	10 s	3	260	N/A
Point-to-point data link procedure on a D-channel at 64 kbps	Signalling (SAPI = 0)	7	1 s	N/A	N/A	10 s	3	260	N/A
	Packet communication (SAPI = 16)	7	1 s	N/A	N/A	10 s	3	260	N/A
TEI assignment procedure (SAPI = 63)	User side	N/A	N/A	N/A	2 s	N/A	N/A	N/A	3
	ASP	N/A	N/A	1 s	N/A	N/A	N/A	N/A	N/A

SOURCE: ITU-T Recommendation Q.921.

works into that of ISDN. Q.931 allows for direct *out-of-band* (D-channel) connection of both circuit-switched and packet-switched X.25 routers.

Therefore, there are two stages to X.25 ISDN use. The first is connection to an X.25 router. Some public switches allow direct subscription to an X.25 router. We have already discussed the allowance within Q.921 of X.25 data to be delivered over the SAPI-16 logical link of the D-channel. This mechanism allows the transport of X.25 layer 3 (PLP) as a "nailed-up" (semipermanent) subscription offer. Such a subscription service may also be offered on B-channels. The advantage to user equipment is that the X.25 router is always available.

In most cases, the public ISDN does not offer direct subscription services for X.25 routing. In these cases, a connection is made to an X.25 router via the same mechanisms as for all other calls. The B-channel will then be used with X.25 layers 2 and 3. Two connections must be made. The first uses ISDN to establish the circuit to the X.25 router. The second uses the X.25 router to establish an end-to-end data connection for data transfer.

X.25 link layers

The X.25 protocol originally had two forms of the data link layer defined. These were known as LAPB and LAP. The LAP data link protocol was of the *primary-secondary* form of the HDLC family and set one end to the Asynchronous Response Mode. LAP also does not support the MultiLink Procedure (MLP) as a possible option. This data link protocol was not popular, since it mainly decreased options with few inherent advantages, and has been removed from the latest ITU-T Recommendation X.25.

LAPB is thus the data link layer protocol in possible use over an ISDN channel. LAPB also offers the MLP as a way of software "bonding" of multiple physical channels into one logical data stream. LAPB is a member of the HDLC family and shares many characteristics with LAPD.

Differences between LAPD and LAPB

Only 1 address byte is used within LAPB (as opposed to 2 for LAPD). Additionally, the C/R bit for LAPB is not split out of the address field—there are fixed address contents for DTE commands/responses and data communications equipment (DCE) commands/responses. These addresses vary between LAPB in a Single Link Procedure (SLP) mode and in MLP mode. The form of the LAPB addresses may be seen in Table 4.5.

LAPB defines only two timers (T1 and T3) with a timer parameter T2. For mapping purposes, in multiprotocol link layer modules, T1 of

TABLE 4.5   LAPB Address Fields

Operation type	Type	Direction	8 7 6 5 4 3 2 1
Single link	Commands	DCE → DTE	0 0 0 0 0 0 1 1
	Responses	DTE → DCE	
	Responses	DCE → DTE	0 0 0 0 0 0 0 1
	Commands	DTE → DCE	
Multilink	Commands	DCE → DTE	0 0 0 0 1 1 1 1
	Responses	DTE → DCE	
	Responses	DCE → DTE	0 0 0 0 0 1 1 1
	Commands	DTE → DCE	

SOURCE: ITU-T Recommendation X.25.

LAPB may be compared in purpose to that of LAPD T200. T3 may be compared in purpose to that of LAPD T203. LAPD, which supports additional protocol requirements (such as TEI management), has four timers. Additionally, LAPB has three system parameters (N1, N2, and k). LAPD has four system parameters (N200, N201, N202, and k). Generally, LAPB N1 may be mapped to LAPD N201, LAPB N2 may be mapped to LAPD N200, and the k values correspond equally as a measure of window size. N202 is not applicable to LAPB since TEI management is not part of LAPB.

LAPB does not directly support point-to-multipoint access to the switch. It also multiplexes logical links at layer 3, rather than layer 2 (as is used with the TEI/SAPI fields in LAPD). LAPD is limited to modulo 128 only. LAPB has a choice of modulo 8 or modulo 128 sequencing.

Finally, LAPB has a different set of criteria for framing signals at the physical layer. An abort signal is a sequence of seven to fourteen contiguous 1 bits, while LAPD considers it to be an abort sequence after seven 1 bits. A value of 7 bits can be used for both and still be valid. The number of contiguous 1 bits to indicate an idle channel, however, cannot be easily reconciled. LAPB requires 15 or more contiguous 1 bits while LAPD uses 8 bits. For most software implementations, this difference can be ignored, since knowledge of an idle channel is normally needed only at the physical layer.

Command/response specification

As mentioned above, there is a difference in the form of the C/R field between LAPD and LAPB. A more serious difference exists. In the case of LAPD, a piece of equipment always knows whether it is a DTE (piece of user equipment) or DCE (network switch). The roles of DTE and DCE are less clearly defined in many uses of X.25.

Generally, a piece of user equipment will behave as a DTE.

However, what should happen if two pieces of equipment are directly connected and can function as either DTE or DCE? An I-frame response would be considered illegal. It is very important that each side knows its particular protocol role.

An example of an algorithm used in this situation can be found in ISO 8208. Upon initialization of the link, a RESTART REQUEST packet is transmitted with the address field set as a command from a DTE. If a RESTART INDICATION is received with a cause field that is *not DTE originated,* the equipment remains in DTE mode. If a RESTART INDICATION is received with a cause field of DTE originated and no RESTART REQUEST is pending, the equipment goes to DCE mode. If a RESTART REQUEST is pending in this case, the INDICATION is ignored, and a RESTART REQUEST is sent after a pseudo-random delay (allowing for either of the first two cases to take place). If the RESTART REQUEST is responded to by a RESTART CONFIRMATION, the equipment remains in DTE mode. Other algorithms exist, but some method is needed to determine equipment roles. Such a procedure is known as *handshaking.*

Use within X.31

ITU-T Recommendation X.31 is known as *support of packet mode terminal equipment by an ISDN.* This standard provides recommended linkage between use of Q.931 to set up an X.25 link on the D-channel or a B-channel and that of in-band X.25 *call request* primitives. For layer 2, this means the establishment of an X.25 link via LAPD (SAPI 16) or the setup of the B-channel by use of LAPD (SAPI 0) for signalling. Different switches provide various services for the D-channel, in particular. Many offer only semipermanent (nailed-up) subscription access to the D-channel X.25 packet handler. Some require that the D-channel X.25 link be set up by separate signalling procedures. The various options/requirements of specific switches can make portability of equipment a serious problem.

Multilink procedure

A typical B-channel is defined as able to carry 64 kbps of information. This is an improvement over the bandwidth of analog lines but, especially with continuous algorithm improvements on analog coding standards, is not sufficient for many needs. A PRI line has the capability of handling multiple B-channels in combination. The exact method of integration of the channels may be implementation dependent; it may rely on the physical layer to keep the transmission of B-channels in sequence. There may also be higher-level methods of ordering of data in such combinations.

However, what about BRI? There are two B-channels potentially available (plus a 16-kbps D-channel logical link). The aggregate bandwidth of the B-channels is 128 kbps. This is sufficient for some video applications (with compression). Thus, methods of bonding the B-channel data streams together are of great interest to implementors and designers of ISDNs. Some approaches involve use of special hardware features that can automatically keep the separate B-channels synchronized. Software methods are also possible. One such method is the MLP, originally defined in ITU-T Recommendation X.25.

If X.25 provides a mechanism for doing the software bonding of B-channels, why is there still a search for another method? The main reason for this is efficiency. Hardware bonding is always more efficient. It is also less flexible and potentially more expensive. The X.25 MLP is oriented toward use of X.25 at layers 2 and 3. X.25 is an old protocol. It was devised when transmission capabilities were slow and unreliable. Due to the expected unreliability of the networks, X.25 has built-in redundancy that is reflected in the protocol. Layer 2 has a window mechanism and so does layer 3. Layer 2 has a sequencing header for I-frames. Layer 3 has a corresponding set of sequence numbers. Assuming that layer 2 has done its job correctly, the layer 3 checks should be unnecessary.

This redundancy is one reason why additional software methods of bonding are being investigated. Other reasons include the capability to transparently send various protocols using the same software methods. Still, many of the software methods are based upon X.25 MLP and, thus, provide a good place for implementors to study methods.

An MLP group is perceived as a single data link by layer 3. During transmission, data is parceled out to the various SLP data links available in equivalent amounts. This provides the mechanism for taking advantage of multiple SLPs to give access to a greater aggregate bandwidth for the data stream. Upon reception, the data from the various SLPs are resequenced and passed up to the network layer as if all the data arrived on a single data link.

MLP subscription. The first thing that is necessary for the multilink procedure to be used is to subscribe to this feature. This may be done in two primary ways. It may be requested during Q.931 out-of-band call setup or as part of in-band negotiation during call setups. It may be assumed as part of the connection service and invoked based on the address field of the link. (The MLP use of LAPB has a separate address code, as noted in Table 4.5.)

MLP coding. The address field of LAPB is different, depending on whether MLP is in use. Secondly, there are 2 bytes inserted as an *upper sublayer* to the data link layer. This means that 2 bytes of MLP

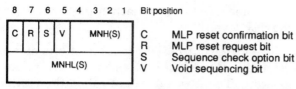

| 8 | 7 | 6 | 5 | 4 | 3 | 2 | 1 | Bit position |

| C | R | S | V | MNH(S) |
| MNHL(S) | | | | |

C MLP reset confirmation bit
R MLP reset request bit
S Sequence check option bit
V Void sequencing bit

MNH(S)   Bits 9 - 12 of 12-bit MLP send sequence number MNH(S)
MNL(S)   Bits 1 - 8 of 12-bit MLP send sequence number MNH(S)

Figure 4.6 X.25 Multilink Control field format. (*From ITU-T Recommendation X.25.*)

headers are inserted after the data link layer header and before the layer 3 packet within an I-frame. Figure 4.6 shows the Multilink Frame formats and the contents of the MultiLink Control (MLC) field.

Note that Fig. 4.6 differs in form from that of Table 2-10/X.25 as contained in the ITU-T recommendation. ITU-T documents sometimes list frames in bit transmission order. In other places, the bytes will be presented in software logical order. Normally, this will be explained in annotations on the figure. This can be confusing, however, and all such formats in this book will be presented in the software logical order.

This MLC is located between the data layer control field and the data layer information field (if any). It is possible to send what would be an I-frame without data that contains only an MLC. There are five sections within the 2-byte control field. These include a multilink sequence number, MN(S) and 4 control bits.

The void sequencing bit (V) and sequence check option bit (S) are always set to 0 in the X.25 MLP standard. They are present for potential unsequenced transmission and reception. If the V-bit is set to 1, sequencing is not required. This might be potentially useful if each data frame is completely self-contained. The S-bit may be used, when the V-bit is set to 1, to indicate whether the sequence number is being used as an identifier (as opposed to true sequencing where order of received or transmitted frames is important). A value of 0 for the S-bit indicates that sequence numbers have been assigned. A value of 1 indicates that sequence numbers are not valid. As noted at the beginning of this paragraph, these bits are not actively used in X.25 MLP.

The other 2 control bits, R and C, are used to request a reset of an MLP group of SLPs and to confirm resetting of the group. The use of these bits will be discussed shortly.

The multilink sequence number, MN(S), is used as an ordered identifier of each transmitted frame. The number is in modulo 4096 and therefore requires more than 1 byte to contain all possible values. This is done by sending the high-order 4 bits of the sequence number in the first byte of the MLP control field and the low-order 8 bits in the second byte. Unlike the sequence numbers of LAPB SLP, there is only a transmit sequence number. Delivery of each frame is verified by acknowledgment at the SLP level.

Internal MLP variables. The MLP control field contains 4 bits and a sequence number. Additionally, three internal variables are maintained by the protocol software. These variables are called the multilink send state variable, MV(S), the transmitted multilink frame acknowledged state variable, MV(T), and the multilink receive state variable, MV(R). Two constants, MW and MX, are also used in conjunction with the internal variables. MW is the window size for the multilink group—the number of frames that can be transmitted before the first is acknowledged. MX is used as a supplementary window in cases where unacknowledged data is being sent over the SLPs.

Windows and reception. MW defines the size of the potential window. The internal variable MV(R) corresponds to the lowest (modulo 4096) value of a sequence number for a packet to be received. Thus, if MV(R) is set to 0 and MW is defined to be 8, frames 0, 1, 2, 3, 4, 5, 6, and 7 may be received within the window. If frames 1 and 2 are received, followed by 0, MV(R) is updated to 3 *after* frame 0 is received. MV(R) is updated only after a frame with an MN(S) equal to the current value of MV(R) is received. MV(R) is then incremented once for each in-sequence frame that has been received, and all packets are forwarded (without the MLC field) to the network layer in sequence.

MX is a receive MLP window guard region. This value may be used to determine an allowable "lost frame" size. It is particularly useful if the SLPs are being used in unacknowledged mode. In such a case, it is possible for frames on the SLPs to be lost without notification. The guard region allows reception to be continued without deadlock after a frame has been lost.

For example, given an MW size of 8, it is possible for the transmitting side to send MLP frames with sequence numbers 0, 1, 2, 3, 4, 5, 6, and 7. The receiving side gets frame numbers 2 through 7. It then receives frame numbers 8 and 9. The frame numbers 0, 1, 8, and 9 occupy a spread of 10 frames. Since MW is only 8, the transmitting end must have already cycled the window—frames 0 and 1 have been lost and will not be retransmitted. With MX set to 2, or greater, the guard region will have been entered. A received frame that is not in the receive window but is still within the guard region indicates that frames have been lost.

If a method is not available to allow for such lost frames, reception is stuck. Only frames 0 through 7 are in the legal reception window. MX allows a maximum number of frames to be accepted as lost, with notification to the network layer of any frames lost in this manner. The internal variable MV(R) is then updated as if the first two frames were received. This allows continued reception. MX should never be defined as greater than MW. If unacknowledged data transfer is not supported at the SLPs, the guard region should never be entered.

An additional protection mechanism is provided by the use of a lost frame timer called MT1. MT1 is started, or restarted, whenever a frame in the valid window is received if there are other frames waiting for delivery to the network layer. If the MX guard region is entered, or no additional frames need to be delivered, the MT1 timer is stopped. If the MT1 timer expires, MV(R) is updated to the value of the MN(S) of the next frame awaiting delivery to the network layer, and the network layer is informed of lost frames.

Windows and transmission. The MW constant is also used for transmission. The internal variable MV(T) marks the first, unacknowledged, frame that was transmitted. MV(S) is the sequence number of the next MLP frame to be transmitted. Thus, frames numbered from MV(T) to MV(T) + MW − 1 (modulo 4096) can be transmitted before acknowledgment is required.

MLP does not have an explicit acknowledgment protocol of its own. It relies upon acknowledgment at the SLP level. Each frame, as it is transmitted, is allocated to one of the SLPs. This must be kept track of so that, when acknowledgment is obtained at the SLP level, the MV(T) variable can be updated. The MV(T) variable can only be updated after the frame which contained MN(S) equal to MV(T) has been acknowledged.

As an example, start MV(T) at 0. The MW constant is set at 8. The software can transmit frames with MN(S) values of 0, 1, 2, 3, 4, 5, 6, and 7. At this point, MV(S) is equal to 8. No more frames can be transmitted until frame 0 is acknowledged. Frames 0, 2, 4, and 6 were transmitted via SLP 1. Frames 1, 3, 5, and 7 were transmitted via SLP 2. For some reason, SLP 1 is not able to immediately transmit the data. So, SLP 2 acknowledges frames 1, 3, 5, and 7. The MV(T) variable cannot be updated yet because frame 0 is still unacknowledged. When SLP 1 acknowledges frame 0, the MV(T) can be updated to 2 because frame 1 was previously acknowledged. Two more frames can now be transmitted within the window region MW.

What if SLP 1 was never able to deliver the frames? If, after N2 (LAPD N200) retries, the frame is not acknowledged upon transmission, it is the responsibility of the MLP to retransmit the outstanding unacknowledged frames on other SLPs.

MLP initialization and resetting. After the connections for the various SLPs to be used as part of the MLP group are made, it is necessary to initialize the MLP. This includes setting all of the internal variables to values of 0 and initialization of the SLPs. Once at least one SLP has been initialized, a MLC is sent as part of an I-frame with the R bit set to 1. The timer MT3 is started and reception of an MLC with the C bit set to 1 is expected. If the MT3 timer expires before an MLC with C equal to 1 is received, another MLC with R bit set is transmitted.

It is possible to invoke the reset procedure at times other than at initialization. ITU-T Recommendation X.25 states that this shall be done when "deemed necessary," but no further guidelines are given. It is probable that an MLP reset would be appropriate at any time that the active network layer sends an X.25 layer 3 RESET packet.

The MLP entity requesting a reset will reset its MV(S) and MV(T) variables to 0, remove all pending frame transmission from the MLP and SLPs, and send an MLC with R bit set to 1. It starts the MT3 timer and does not send any additional frames, with the R or C bits set to 0, until confirmation is received. It will continue to process received frames until the MLC with the R bit set to 1 is received.

On the receiving side, receipt of an MLC with the R bit set to 1 will cause delivery of all received frames to the network layer, with notification of any lost frames. MV(R) is reset to 0 [and if it has not already been done, will reset MV(T) and MV(S) to 0]. If an MLC with R set to 1 has not already been sent, one is sent and the MT3 timer is started. At this point, both sides are in the midst of resetting.

Each side will expect two things: acknowledgment of the delivery of the MLC with R bit set to 1 by the SLP and reception of an MLC with R bit set to 1 by the MLP peer entity. Once this has been accomplished, each side will send an MLC with the C bit set to 1. It will wait for acknowledgment of delivery by the SLP and then reception of an MLC with C bit set from its peer. If this scenario completes before the MT3 timer expires, the reset is finished, the MT3 timer is stopped, and normal transmission and reception can be resumed.

If the MT3 timer expires before completion of the reset scenario, the process starts again. Any data received between reception of the MLC with the R bit set to 1 and the MLC with the C bit set to 1 is discarded. Similarly, an MLC with the R bit set to 1 is discarded in this interval. If an MLC with the R bit set to 1 is received after reception of the MLC with C bit set to 1, it will be considered a new reset procedure. An MLC should never have both the R and C bits set to 1.

Miscellaneous conditions. An SLP may be taken out of the MLP group at any time. This is usually done by disconnecting the SLP at the physical or data link layer. If voluntary, this process may be done without loss of data by not sending any additional frames on the SLP, putting the SLP into local busy, and then disconnecting the SLP. Any frames not yet received should be retransmitted by the peer on another SLP.

A group timer, MT2, may be optionally used for the condition where all SLPs are busy and received frames are still expected. If the timer expires before the "blocked" MLP frame with MN(S) equal to MV(R) is received, MV(R) is incremented to the next in-sequence frame to be received, the nonreceived frames are declared lost, and any intervening frames are delivered to the network layer.

Other Data Link Layers Used within ISDN

Q.921 must be used on the signalling D-channel within an ISDN. The bearer channels may use any protocols that they desire. One popular protocol, as discussed, is X.25. Another data link protocol used during rate adaption of terminals is V.120, which is a variant of Q.921 and is discussed in greater detail in Chap. 6. It would also be possible to use any other protocol over the bearer channels.

Software Design Issues

Some discussion has already taken place concerning HDLC family similarities. A software module requires object code and data space. Combining data link layers, if all are needed within a system, can save in maintenance, size, and general software complexity. With this in mind, an implementor needs to always consider commonality of function. Object-oriented programming attempts to provide common access and functions. This is one method of modularity. Using common blocks of code is another method. Consideration of full system needs also can improve the ability of the implementor in the design of particular sections of software.

Part 3 of this book discusses general implementation concerns. A discussion of concerns that are particularly important to the data link layer is also relevant, although these may apply to other layers in addition. One method of approaching this is through the analysis of data flow through the system. Where does the data come from? Where are they sent? How are they to be stored? What functions must be applied to the data? In the case of data protocols, these questions are vital for efficiency and throughput.

Module entry points.

Data comes into the data link layer through three entry points, as defined by ITU-T primitives. These are from the coordination/management *planes,* the network layer (or, in some cases, directly from higher layers), and from the physical layer. The primitives received and the form of the data are different, but they all feed information into the data link layer.

Data goes out of the data link layer through three general points also. These are, once again, to the coordination/management planes, the packet or other higher layer, and the physical layer. Each of these six entry points is a place where one can analyze data flow for the module.

Let us take the flow of primitives from the network layer to the data link layer as an example. There are three items of information within each primitive that may be used to identify the specific logical link. These are the TDM channel (D-channel, B1-channel, B2-channel for BRI), SAPI, and CES (an identifier defined in ITU-T documents as a Connection Endpoint Suffix). Use of this unique combination of identifiers allows the software to determine the logical link being addressed. (In the case of some primitives, the function may be applied to *all* links and identification of a particular logical link is not necessary.) Now that the exact logical link is known as well as the primitive and any other parameters associated with the primitive, analysis and action can take place.

A protocol that is defined via a state table (such as Q.921) acts in a way that is extremely conducive to having a central state handler. Each entry point takes the information necessary to uniquely identify the appropriate logical link, and the information is passed to the central state handler. The form of the state handler can vary, and a number of the possibilities are discussed in detail in Chap. 10.

The exit points may also be combined. The protocol, depending on current state and precise definition of event, may need to perform an action. This action will result in either internal state changes, or it will effect a primitive to the physical layer, upper layer, or management/coordination entity. A data link layer to physical layer function handler may, for example, examine the action desired, format a primitive, and then invoke the operating system function needed to send the primitive or call an appropriate handler for the primitive.

The general data flow, then, is as follows. A primitive arrives at an entry point. Identification of the appropriate logical link is made. The primitive is passed along to the central state handler. Based on state, environment, and exact events, actions may be performed. These actions invoke outgoing primitive formatters, and the state table protocol has completed its actions for the current event and state.

Operating system use

The primitives pass into, and out of, the data link layer. Each of them has its own requirements and takes a certain amount of time. The first requirement to be considered is whether synchronization of the primitive is required. That is, does a primitive require an immediate acknowledgment or response? All primitives perform a function (otherwise they would not be used). However, many primitives are asynchronous in action. Consider a DL_DA_RQ primitive passing from the network layer to the data link layer. First, the primitive must be parsed sufficiently to determine the correct logical link to which it applies. Next, the state handler must be invoked to determine the appropriate action at this time.

A number of different actions may happen. If the Q.921 protocol, for example, is in TEI_ASSIGNED state, a DL_DA_RQ primitive is illegal and will be discarded. If the Q.921 protocol is in MF_ESTABLISHED state, a DL_DA_RQ primitive is legal. However, because of window limitations or because of remote peer busy state, it may not be possible to transmit immediately. Even if it may be transmitted immediately, the action is not truly complete until it has been acknowledged by the peer. It is important that the rest of the system be allowed to continue running. The worst-case scenario must be applied to deciding whether the primitive should be handled in an asynchronous or synchronous fashion.

Function, or procedure (depending on naming conventions within the programming language), calls provide a synchronous interface. The function is invoked. It performs a number of actions. It returns. The return value (or possibly parameter values) can provide immediate feedback as to results from the function call. Management primitives often fall into this category. The question to be answered is, What is the longest time that the software will need to complete the operation?

An operating system message queue provides a break in the data flow. A primitive is sent from one entity to another. The entity is then free to continue processing data until it reaches a point where it can no longer do useful work. The task is swapped out on the processor. Another task waits for a primitive. It is inactive until the primitive arrives. Each task performs useful work as long as it can. This type of nonpreemptive task scheduling is very useful for many data protocols.

Let's examine the above example once again using operating system tasks and queues. The network layer entity is executing. It determines that a DL_DA_RQ is necessary. An operating system call is invoked to send the DL_DA_RQ primitive to layer 2. The network layer continues to process information until it finishes all actions associated with the event currently being processed. A data link layer task is idle and swapped out by the operating system. It is waiting for a primitive. The DL_DA_RQ primitive arrives via the operating system queue. The data link layer task starts to execute the protocol. The window is not full, so a PH_DA_RQ primitive is formatted and sent to the LLD associated with the TDM channel via a synchronous function call. This function acts on the request for the LLD to program the hardware to start transmitting the frame. It returns. The data link layer task finishes any internal work needed and calls a management function to start a T200 timer. Finally, it is finished with the primitive and is swapped out to wait for another primitive.

During this entire period of time, the software did not have to wait for any "outside" event. When it was necessary to wait for an event, the task was swapped out. Even though the PH_DA_RQ primitive was not completed in this period of time, all necessary work was done to have the hardware asynchronously continue to transmit the frame until completion. Similarly, the timer was started—the software did not wait for the timer to expire. Each action within the task was synchronous and was able to complete (although some actions may be asynchronous themselves).

In the upward direction, there is a slight difference. The LLD interacts with the hardware device directly. Events from the hardware device happen at irregular intervals, but the time spent in the LLD must always be kept as short as possible. Further discussion about LLD design occurs in Chap. 11. Briefly, events that occur at the hardware/physical level will either interrupt the LLD or be noticed during a high-priority polling task. When the event does occur, it must be handled quickly so that the LLD is able to process the next event. This need for speed requires a break in the data flow. Thus, an operating system queue is likely to be required for PH_DA_IN primitives.

A PH_DA_RQ does not require asynchronous primitive handling primarily because a synchronous function call from the data link layer does not interfere with the time-critical hardware event handler. It is possible that certain critical regions of this function must disable interrupts—but in a very brief manner. Note that if a high-priority polling task is used for hardware events, the software must be designed to be preemptible to allow the polling task to operate as needed.

We have seen why multitasking is very useful for an ISDN system. Operating system queues are mandatory—at the very least between hardware device handlers (LLDs) and other layers. This may be part of the operating system, or task handling system, or it may be implemented with semaphores.

Other functions that are useful within an operating system environment are timer and buffer, or data space, management. Whenever the data flow is interrupted, buffers are required to save space for specific data. These interruptions may occur as part of operating system queue delays, or they may occur as part of the protocol (for example, saving a frame for retransmission in case the transmission was unsuccessful). Operating systems often provide a consistent method of allocating and deallocating space for buffer use. Once again, however, anything that a general operating system provides can also be implemented in a manner unique to the needs of the software.

Timers are needed in most data protocols. We have noted that T200, T202, and T203 are needed by Q.921. Other equivalent timers are needed by LAPB of X.25. An I.430 LLD requires a T3 timer for activation requests. Other layers also have their own unique timers.

Operating systems often provide common timer mechanisms. Timers require hardware assistance to provide reasonable accuracy.

Data structures

Upon entry to the data link layer, the specific logical link is determined. We mentioned the specific state and environment as being important to the protocol for determination of appropriate action. What types of data must be kept? How can they be accessed? How are they initialized?

The data structures for a protocol state machine require five types of information. These may be called identification, state, current event, environment, and modifiable parameter data. The exact form of the data may vary from one programming language to another.

Identification data.   Each entry point must be able to identify the correct logical link. For layer 3 to layer 2 primitives, as mentioned in the above example, this data includes TDM channel, SAPI, and CES. For a physical layer to data link layer primitive, the information is TDM channel, SAPI, and TEI (for Q.921). In order to save data space, broadcast primitives (TEI equal to 127) may be allocated a separate data structure. Management layer primitives may be identified according to unique members associated with the software system. This suggests that the structure may have an identification of its own (for example, logical link number). These five items of data (TDM channel, SAPI, TEI, CES, and LLN) allow identification of the appropriate logical link for any primitives entering via the three entry points.

State data.   Each logical link must retain information concerning its current state. State includes the protocol state table states as well as possible substates for states 5, 7, and 8. These substates include, for state 5 (AWAITING ESTABLISHMENT), whether the establishment is for the first time, to reestablish a previously established link, or establishment during a pending release period. The state tables for Q.921 refer to these substates as 5.0, 5.1, and 5.2, respectively. Many of the actions for these three substates are similar in nature. Thus, it may make more sense to have one general state (AWAITING ESTAB-LISHMENT) and *flags* to indicate special preconditions that are indicative of a substate. This saves protocol space by allowing most of the state processing code to be common and substates checked only in required circumstances.

There is a similar situation for states 7 and 8. Behavior is modified depending on the current status of the local or remote end. If the peer has sent an RNR, indicating it is not currently able to accept data, a condition known as *remote busy* is true. If local conditions do not allow acceptance of data, a condition known as *local busy* is true and must be maintained until the condition changes. Finally, Q.921 during states where multiple frame acknowledged transfer is possible (which includes multiple frame established state as well as the timer recovery state) may be in the process of recovering from a REJ frame due to incomplete data transfer. These three possible conditions give eight (2^3) possible permutations. The Q.921 state tables describe these as states 7.0 through 7.7 and 8.0 through 8.7. For the same reasons stated above, the implementor will find it useful to maintain the status of these conditions as separate data items from the state.

Current event data.   In the discussion on entry points, it was pointed out that there are many benefits associated with a common central state handler. However, for this to be possible, all the information that was parsed from the event to be able to call the handler must be saved. An incoming frame has several items of information associated with it. What is the Control Field? What is the status of the C/R bit (or state of the Address field in the case of LAPB)? If appropriate, what is the status of the P/F-bit? If sequence numbers are part of the Control field, what are their values? During parsing of the frame (for PH_DA_IN primitives), was there an error? What was the form of the error?

In the case of primitives arriving from layer 3, specific parameters are associated with the primitive. This information, such as length of data, address of data, identification of buffer space, or other data allocation information, must be saved before the state handler is called. The primary reasons for the saving of current event information is to provide common structures that can, in turn, save on duplicated code and increase efficiency of the system.

Environment data.   Environment data is distinct from state data in that it keeps track of continuing events. For example, if a timer is running, note must be made of that fact. If a timer has a counter associated with it, that information must be retained in association with the logical link. The internal sequence variables associated with acknowledged transmission must be stored. Finally, data queues must be kept for data currently awaiting acknowledgment or transmission. Data queues are discussed in greater detail later in this chapter.

Modifiable parameter data.   This type of data includes system parameters such as window size, modulo of sequencing (for Q.921 this is restricted to modulo 128), and peer role (DTE or DCE) for equipment that can act in either role. A decision must be made early in the design of the system as to whether these data can be modified for each call. It is more efficient to have fixed values. In fact, such items can be ignored as data if the system has only one set of permissible values. However, the entire system design may need to change if requirements later change. This makes it a serious system consideration.

Data queues

The primary purpose of a data link layer is to provide efficient, reliable data transport. It does this in interaction with the physical layer and the higher layers. Each frame to be sent must be kept until it is transmitted and, in the case of acknowledged data transfer, acknowledged. Each frame that is received properly must be forwarded to the appropriate upper layer. Received frames do not need to be kept.

Data can be transmitted only under certain circumstances. The link must be in an appropriate state and other conditions (such as the remote peer ready to receive and the transmit window not saturated) must be true. Thus, transmission will not always be possible at the time a DL_DA_RQ (or DL_U_DA_RQ) primitive is received. The data must be saved. The order of transmission must also be preserved. This means that queues must be established. Two queues are needed. One is used for UI-frames and the other is used for I-frames.

The queue used for UI-frames is primarily used as a method of storing data until the state has changed to allow transmission. Additionally, storage of UI-frames in one location allows simple deletion of frames in cases where the link is disconnected. As is true for all queues, functions are needed to put data onto the queue and to remove single items from the queue. A separate function to discard all frames is also useful, although repeated uses of the *get* function can be substituted.

The queue for I-frames is more complicated than the one needed for UI-frames. Any queue, created as a linked list or as a circular array of structures, requires *head* and *tail* pointers. Data are obtained from the head and added onto the tail. In the case of a queue used for I-frames, an intermediate pointer is required. Normally, data which is retrieved from a queue is no longer kept in the queue. Thus, a get operation effectively removes the data from the data structure in addition to retrieving the item for use. The I-frame queue has different requirements.

All I-frames must be stored until they are acknowledged. It would be possible to maintain two data queues for I-frames—one for transmitted frames and one for frames not yet transmitted. A more effective method is to use three pointers in the queue. Two pointers reflect the traditional get and put functions. The third pointer is used to maintain a separation between I-frames in different states of transmission.

For example, start off the queue as empty. The get, put, and ack pointers are all void. Five DL_DA_RQ primitives arrive. The get pointer points to the first item on the queue, as does the ack pointer. The put pointer points to the next available space on the queue for further I-frames. One frame is transmitted. The get pointer is incremented to enable access of the next frame. The ack pointer is left in place. Another get operation takes place and the get pointer is incremented once again. Now, the first I-frame is acknowledged. Only at this point does the ack pointer get incremented, with the attendant removal of the data from the queue.

The use of three pointers with the I-frame queue has an even more important use than to save duplicated data space and copying. This is for the case of retransmission. If we continue the example given above, we may say that after the first frame is acknowledged, another frame is gotten from the queue for transmission. The get pointer is, once again, two entries past that of the ack pointer. If a REJ frame is received from the peer, preparation for retransmission of the I-frames requires only a decrementing of the get pointer. This is a very efficient method in terms of data space and time.

Note that all of the data queues contain only the data as sent by upper layers. Headers are added, with appropriate sequence numbers, at the time of transmission.

LLD interactions

The data link layer is responsible for interactions with the physical layer. They will often be done via an LLD. Each type of hardware will have its own LLD. Thus, a common method of interaction is useful in the design of an ISDN system. For PH_DA_RQ primitives, data requires an address, a length, a data space (or buffer) designation, what to do with the data after transmission, and a priority (for BRI, these priorities are known as Class 1 and Class 2 to distinguish the signalling use of a channel of higher class). Data buffer uses, and protocols, are discussed further in Chap. 11.

The physical layer must be active before data can be transmitted or received by the data link layer. The PH_AI and PH_DI primitives allow the LLD to inform the data link layer of the current state of the physical layer. A PH_AR primitive is needed to request activation of the physical layer if it has not been activated by the peer. However, the ITU-T documents do not explicitly state at just what point the PH_AR primitive should be sent to the physical layer.

Since the physical layer must be active before data can be transmitted, it is reasonable to expect the first attempt to send data to cause a PH_AR primitive as a side effect. However, if this is done, what should be done with the data that caused this effect? The Q.921 tables do not indicate a preferred solution. One method is to allow the LLD to queue up any data transmission requests until line activation is achieved. This works well if line activation is achieved in a normal situation. Test scenarios may expect different algorithms, and allowance must be made for this.

Multiple data link layers

Some mention has been made about the possibility of combining different data link layer protocols into one module. If this is done, the data structures need to hold information concerning the supported protocol. Additional data may also be needed. This is useful for Q.921, LAPB, and V.120 since they operate very similarly. It is also of potential use for Frame Relay, which is based on Q.921. A rule of thumb for combining data link layers (or other protocol layers) is for at least 75 percent of the logic to be common. If less than this is in common, the added complexity of checking for appropriate protocols is likely to exceed savings.

Summary of Chapter

The data link layer provides a reliable data transport mechanism for higher layers. Many of the data link layers directly used by, or referenced for support of, ISDN are members of the HDLC family and have features in common. The primary data link layer used by ISDN is Q.921, also known as LAPD. This protocol is needed for transport of signalling information for BRI and PRI. It may also be used for other purposes. The bearer channels may contain any protocol, but X.25 was incorporated into ISDN recommendations early because of the desire to interconnect into existing networks.

An ISDN system requires design of the entire system to provide effective service. General design issues are covered in Part 3 of this book. There are, however, a number of design issues that are of special relevance to the design of the data link layer. These issues include potential operating system use, data structures, entry points, and other aspects of interacting with other protocol layers.

5

Network Layer, Call Management

Layer 3 of the OSI model is called the network layer. It is primarily a means of providing routing information. It is the final layer of the *chained layers* which pass from node to node and are potentially interpreted by switches and routers in the network. The ITU-T Recommendation Q.931 provides a standard for specific implementations of call signalling and other *out-of-band* information for an ISDN. ITU-T Recommendation Q.932 gives a generic description of *supplementary services* for an ISDN. These services include such items as keypad protocols, feature key, and functional procedures. Each switch will provide services, within the framework of Q.931 and Q.932, specific to their own needs and tariffed services.

Layer 3 makes use of layer 2 to provide data transport capabilities. It also interacts with management and higher-layer entities to provide a general service to applications. As the highest layer of the chained layers, it is likely to be part of a basic ISDN software module, with application-specific materials and software outside of this block. It is possible that such may occupy a different processor system.

Architectural Position within ISO Model

The network layer is positioned, within the ISO model, between the data link and the transport layers. It is quite possible for the transport layer to be merged into other high-layer application layers. Thus, we will primarily talk about interactions between layers 3 and 4, with the understanding that the upper layer may actually be some form of combination of layers 4 through 7. Sometimes, the ITU-T documents will refer to the "upper layers" rather than a specific layer. This is an acknowledgment of possible merging of those layers.

ITU-T Recommendation X.25 merges the OSI layers together to a certain extent. This is probably largely due to the period of time in which it was developed. It can fill the *call control* functions that are associated with the network layer. It can also duplicate the data link layer responsibilities. This aspect of X.25 is called the PLP. In this chapter, we will discuss some aspects of X.25 call control functions.

Use of the Network Layer within ISDN

In general, the network layer is involved with getting the connection made—with directing the data flow to where it is desired. Q.931 (I.451) is a way to do this as an out-of-band signalling mechanism for ISDN. *Out-of-band* refers to the fact that the signalling mechanism is not performed on the same link where data transfer is to take place (on the D-channel, this may be logically separated rather than physically separated). Within in-band call control, for which X.25 Call Control is an example, signalling parameters are passed over the same layer 2 data stream as are the data transport services.

An ITU-T Recommendation X.31 call (covered more fully in Chap. 9) coordinates between the out-of-band setup of a link and the in-band control of a link. This aspect of out-of-band versus in-band control is an important concept within ISDN. Use of interswitch protocols such as Signalling System 7 (SS7) allows noninterruption of data channels. The out-of-band signalling, or control, of the bearer channels gives transparency to the data streams. There is a switching network. Its purpose is to provide connectivity. There is a set of protocols over the bearer channels. Its purpose is to provide data transport services and to support applications in their needs. By splitting the two into separate mechanisms, a step is made toward the ease of interoperability of networks.

Q.931

The set of ITU-T Recommendations Q.930 (I.450), Q.931 (I.451), and Q.932 (I.452) give the basic structure of ISDN call signalling. These documents give the structure, the states and actions, and the general purposes for each activity within the state. We will first cover the general message structure of a Q.931 message, or packet. Next, a short discussion of the various states will be made (with emphasis on those states relevant to the user side of the user-network interface). Third, a discussion of example messages will occur. These examples are chosen with particular use for the setting up of bearer channels and their capabilities.

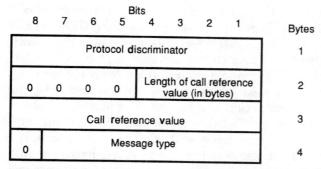

Figure 5.1   Q.931 message packet header. (*From ITU-T Recommendation Q.931.*)

Message structure

The message structure of a Q.931 packet is a simple one. It is composed of three parts. These parts are the header information, a message type (if appropriate), IEs associated with the message type. Allowance is made for various national, and switch-specific, needs by giving the concept of codesets. Codesets allow a change in interpretation of the message types and associated IEs.

Header information.   As can be seen in Fig. 5.1, the header of a Q.931 packet is composed of three basic fields. These fields are the protocol discriminator byte (or octet, as they are usually referred to in ITU-T documents), a CRV, and a message type. The protocol discriminator for Q.931 is a hexadecimal value of 8. Various ranges of values are defined in Q.931 as being part of the ISDN code range.

Values 0 through 7 are assigned to user-information types of messages. In such cases, the rest of the header may not follow the normal Q.931 pattern. Value 8 is reserved for the Q.931 (I.451) interpretation. Hexadecimal values 10 through 3F are to be reserved for other network layers (such as X.25). The hexadecimal values of 40 through 4F are defined for national use. As an example of such use, Germany 1TR6 protocol (a widely used pre-ITU-T ISDN protocol) uses values of hexadecimal 40 and 41 for its protocols. Finally, hexadecimal values 50 through FE are reserved for various network layer protocols. It should be noted that reservation of values does not necessarily imply that an active use is to be made of them. These are values reserved for potential use as part of an evolving family of protocols.

The CRV is used to identify a call that is in progress. It uniquely identifies (over a particular interface between the user equipment and network) a connection made through the network.

The CRV has two parts. These are the call reference length and the call reference value. Most switch versions use a call reference length of 1 for a BRI (2 for a PRI). The allowable range, however, is 1 through 3 bytes in length. There is a special value for a CRV: zero. A value of zero indicates a global call reference—a message not associated with a particular call. For certain messages, a global call reference can also be indicated by giving a call reference length of zero.

The CRV is given in the next series of bytes. In the first byte, the upper bit has a special designation as a *flag* bit. The flag bit is used to designate the side that originated this particular call. A 0 bit indicates that this is a message *from* the side that originated the call. A 1 bit indicates that this is a message *to* the side that originated the call. This allows verification of proper messages being sent and received for particular states. It is possible for the same CRV (with the flag as a discriminator) to be used for two active calls—one originated in each direction. The CRV is allocated by the originator of the call.

The last byte of the header information for Q.931 messages is the message type. The message type designates the general purpose of the packet within the particular codeset. A message will include a variety of IEs. Each IE conveys information about a certain aspect pertaining to the message.

Message types.   A message type is currently defined to occupy a single byte. As seen in Table 5.1, a number of values are defined within codeset 0 of Q.931. The value of 0 as the initial message type indicates that this is a direct escape to a nationally specific message type. It is also possible, as is discussed a little later in the section on codesets, to escape after having done codeset 0 specific actions. The upper bit of the message type is reserved to be 0, for the present, to allow possible future extension of the message type to a length of more than 1 byte.

Information elements.   IEs are associated with message types. The same IE can be used with different message types, but its presence, or absence, is defined with the message. Three conditions are relevant to decoding a Q.931 message. If an IE is mandatory, is it present? If it is present (and allowed for), does it convey accurate information? If it is allowed (or required) and correct, it will influence the actions represented by the message.

An IE may either be a single octet or a variable-length bundle. The upper bit of the IE describes its form. A 1 indicates that this is the only byte associated with the IE. If this is the case, the IE may convey information inherently by its presence, or it may be divided into an identifier and a contents field. If the upper bit is a 0, a variable-length IE is to be presented. This first byte will contain the identifier for the IE. The next byte will contain the length. The proceeding x bytes (x defined by the contents of the length byte) are the information fields associated with the IE. Each IE defines its own fields. IEs

TABLE 5.1   Q.931 Message Types

8	7	6	5	4	3	2	1	Description
					Bits			
0	0	0	0	0	0	0	0	Escape to nationally specific message type*
0	0	0	-	-	-	-	-	Call establishment messages
			0	0	0	0	1	- ALERTING
			0	0	0	1	0	- CALL PROCEEDING
			0	0	1	1	1	- CONNECT
			0	1	1	1	1	- CONNECT ACKNOWLEDGE
			0	0	0	1	1	- PROGRESS
			0	0	1	0	1	- SETUP
			0	1	1	0	1	- SETUP ACKNOWLEDGE
0	0	1	-	-	-	-	-	Call information phase messages
			0	0	1	1	0	- RESUME
			0	1	1	1	0	- RESUME ACKNOWLEDGE
			0	0	0	1	0	- RESUME REJECT
			0	0	1	0	1	- SUSPEND
			0	1	1	0	1	- SUSPEND ACKNOWLEDGE
			0	0	0	0	1	- SUSPEND REJECT
			0	0	0	0	0	- USER INFORMATION
0	1	0	-	-	-	-	-	Call clearing messages
			0	0	1	0	1	- DISCONNECT
			0	1	1	0	1	- RELEASE
			1	1	0	1	0	- RELEASE COMPLETE
			0	0	1	1	0	- RESTART
			0	1	1	1	0	- RESTART ACKNOWLEDGE
0	1	1	-	-	-	-	-	Miscellaneous messages
			0	0	0	0	0	- SEGMENT
			1	1	0	0	1	- CONGESTION CONTROL
			1	1	0	1	1	- INFORMATION
			0	1	1	1	0	- NOTIFY
			1	1	1	0	1	- STATUS
			1	0	1	0	1	- STATUS ENQUIRY

*When used, the message type is defined in the following octet(s), according to the national specification.

SOURCE: ITU-T Recommendation Q.931.

should always be presented in ascending order within a codeset as part of a message.

Codesets. Codesets allow particular IEs to be interpreted according to national or local network needs. The codeset *shift* operation is defined by use of a single-octet IE. This single-octet IE is hexadecimal 9*x*. The nine signifies a shift operation. The *x* is the codeset that is to be used. The lower nibble has 4 bits available. The upper bit of this nibble indicates whether the operation is *locking* or *nonlocking*. A nonlocking operation (signified by a 1 in this field) is valid only for the next IE. After this, the codeset reverts to codeset 0. A locking operation indicates that all following IEs are to be interpreted according to the new codeset. A locking shift operation can be used only to change to a higher-numbered codeset.

For example, a message starts its interpretation in codeset 0. A nonlocking shift operation occurs. The next IE is interpreted according to codeset 5 (indicating national use), for example. The following IEs (if any) are interpreted according to codeset 0 again. A locking shift operation occurs back to codeset 5. Finally, a locking shift operation occurs for codeset 6 (referring to local network needs). The message interpretation ends.

Network and user states

ITU-T Recommendation Q.931 starts its discussion of the call signalling protocol with a description of the possible states for the network (designated N-states) and user equipment (designated U-states) protocols states. In Q.931, the state defines the progress of the call. The states can be split into four categories. These categories are idle, active, call becoming active, and call becoming idle. State 0 is the idle state. (However, in the case of circuit-switched calls, this may mean a nonactive situation for a call that has been suspended.) State 10 is the active state. States 1 through 9 can be interpreted as states leading from the idle state to the active one. States 11 through 19 (with the exception of state 17) can be interpreted as states leading from the active state to the idle one.

Q.931 divides the states into network and user equipment states. The general situation is very similar for both. Some states are used only by one endpoint. Some actions and timers are different. Another division is made between circuit- and packet-switched calls. Most of these divisions are irrelevant to the designer of ISDN user equipment. The following state descriptions are of user states. A description of the network states would be very similar.

Null state (U0). No call exists. This is the idle state. No CRV has been allocated for the call and only call origination (or resumption) is relevant.

Call initiated (U1).   An outgoing call has been initiated. The local equipment has allocated a CRV and a SETUP has been transmitted to the switch. At this point, acknowledgment from the switch is expected. Depending upon whether all necessary call information has been conveyed (en-bloc signalling) or information is being sent in parts (overlap sending), the expected message from the switch will transfer the software to state U2 or U3.

Overlap sending (U2).   The switch has acknowledged that a call has been requested and is now expecting more information. More information will be expected from the upper layers and then sent to the switch. This is not possible for packet-mode calls.

Outgoing call proceeding (U3).   The switch has now received all necessary identification information to attempt to connect the call. If the call information is valid, the next transition will be to state U4.

Call delivered (U4).   The outgoing call request has been delivered to the other end. Further transitions depend upon the reactions of the peer entity.

Call present (U6).   A message has been received notifying the local end that an incoming call connection is desired. This state may be optional, depending upon specific implementation, along with the transition to the following state (U9). This depends on whether upper-level intervention is needed to acknowledge an incoming call.

Call received (U7).   The local upper layers have been notified of the incoming call and an acknowledgment of this has been sent to the network. Further transitions depend on the actions of the upper layers.

Connect request (U8).   The upper layers have decided to accept the call. A message is sent to the switch to indicate this, and the user is waiting for confirmation that the call is now active.

Incoming call proceeding (U9).   The user equipment can implement a process of negotiation of further information from the switch. This can be useful in certain special calls (such as callback situations). This state indicates that the user equipment has now received all information necessary to effect call establishment.

Active (U10).   The call is now complete and the bearer channel is capable of being used.

Disconnect request (U11).   The user has initiated a teardown of the call. This state allows for the possibility of the disconnection to be refused. (State U19 does not allow the possibility of refusal.)

Disconnect indication (U12).   A request from the peer has indicated a desire to disconnect the call. This state is unnecessary if the user equipment has not provided an option for the upper layers to refuse, or delay, the disconnection. This state may prove useful for applications that want to send user-to-user information as part of a responding message to a DISC message-type indication.

Suspend request (U15). The user has requested the suspension of an active call. If the request is acknowledged by the switch, the CRV will be released, and the call will revert to state U0. However, it may be resumed (via state U17) by the use of a special IE without transition through states 1 through 9. This is not relevant for packet-mode calls.

Resume request (U17). The user has requested resumption of a previously suspended call. This is not relevant for packet-mode calls.

Release request (U19). The user has requested release of the call and is waiting for acknowledgment. If, after timer expiration, the network does not acknowledge, the user equipment will unilaterally release the call.

Overlap receiving (U25). An incoming call has been acknowledged and any additional information (if any) can be forwarded from the network.

Detailed message-type analysis

There are a number of message types defined in Q.931. These are broken into four categories. These categories are call establishment, call information, call clearing, and miscellaneous messages. Each message type is defined as a particular code (shown in Table 5.1). It is also described in terms of its IEs and whether the individual IEs are mandatory or optional. Practically speaking, the actuality of the decision as to whether the IE is mandatory, or optional, is dependent on the specific network implementation.

As examples of this structure, we will discuss two different message types for circuit mode connections. (Packet-mode and other services vary slightly.) These message types, SETUP and DISCONNECT, represent call establishment and clearing messages. The general divisions to the messages are shown in Tables 5.2 and 5.3. All other message types are examined in the same manner within ITU-T Recommendation Q.931.

SETUP message type. The SETUP message type lists the three header bytes and 18 IEs. Each byte, or IE, contains a reference number to the explanation of the field within the ITU-T recommendation. Another piece of information is the direction—is this IE sent by the network to the user, the user to the network, or both? The type is defined as mandatory or optional. All mandatory elements should exist

TABLE 5.2 SETUP Message Contents

Information element	Reference	Direction	Type	Length
Protocol discriminator	4.2	Both	M	1
Call reference	4.3	Both	M	2–*
Message type	4.4	Both	M	1
Sending complete	4.5	Both	O	1
Repeat indicator	4.5	Both	O	1
Bearer capability	4.5	Both	M	4–12
Channel identification	4.5	Both	O	2–*
Progress indicator	4.5	Both	O	2–4
Network specific facilities	4.5	Both	O	2–*
Display	4.5	$n \rightarrow u$	O	2–34 or 2–82
Keypad facility	4.5	$u \rightarrow n$	O	2–34
Signal	4.5	$n \rightarrow u$	O	2–3
Calling party number	4.5	Both	O	2–*
Calling party subaddress	4.5	Both	O	2–23
Called party number	4.5	Both	O	2–*
Called party subaddress	4.5	Both	O	2–23
Transit network selection	4.5	$u \rightarrow n$	O	2–*
Repeat indicator	4.5	Both	O	1
Low-layer compatibility	4.5	Both	O	2–18
High-layer compatibility	4.5	Both	O	2 –4

Note: *All optional IEs have detailed notes on use in ITU-T Recommendation Q.931.
SOURCE: ITU-T Recommendation Q.931.

TABLE 5.3 DISCONNECT Message Contents

Information element	Reference	Direction	Type	Length
Protocol discriminator	4.2	Both	M	1
Call reference	4.3	Both	M	2–*
Message type	4.4	Both	M	1
Cause	4.5	Both	M	4–32
Progress indicator	4.5	Note 1	O	2–4
Display	4.5	$n \rightarrow u$	O	2–34 or 2–82
Signal	4.5	$n \rightarrow u$	O	2–3

Note: *All optional IEs have detailed notes on use in ITU-T Recommendation Q.931.
Note 1: Included in the network-to-user direction if the network provides in-band tones.
See Annex D for usage in the user-to-network direction.
SOURCE: ITU-T Recommendation Q.931.

in all true ISDN SETUP messages. The optional elements almost always have notes to describe the circumstances under which they should be used. These notes are of a generic description, and the implementor should reference the specification for the switch to be supported. The last item of information is the length of the IE.

The first two IEs listed for SETUP are each single-octet IEs. One is a Sending Complete IE to specifically indicate that all information associated with the setup of the call is included in this message. The other is a Repeat Indicator. The Repeat Indicator allows repetition of an IE that would normally only be included once. The position of this optional IE is before the Bearer Capability IE, indicating the possibility of presenting multiple bearer capabilities to the switch for acceptance of the call. If, for example, it would be acceptable to establish either a V.110 circuit call or a V.120 packet call, both bearer capabilities could be presented as alternatives.

The next IE, the Bearer Capability, is one of the most important mechanisms available to the user to determine, via Q.931, the compatibility of endpoints. Due to its importance, the Bearer Capability IE will be discussed in greater detail later in this chapter.

The channel ID IE is mandatory in the network to user direction. If present in the user to network direction, it will indicate a desired channel. This may be exclusive (if not available, release the call) or preferred. Most switches will not allow the user equipment to determine the channel. Some switches require this for access to the D-channel X.25 link.

The Progress Indicator IE may be added to the SETUP message if information concerning interworking is known to the switch but not to the originating equipment. For example, if the connection is not end-to-end ISDN, this information may be very useful to the other end. Such information that is added by the switch must also be relayed to the originating equipment (but via a PROGRESS message type rather than a SETUP).

The Network-Specific Facilities IE allows specification of particular equipment or routing. The Display IE provides a means for the network to add additional information that may be of interest to the end user.

The Keypad Facility IE is usually mutually exclusive with the called party number to indicate the address of the desired endpoint. Note that, in some instances, the CDN IE and the Keypad IE may both be present. In such a case, the CDN will give addressing information, and the Keypad IE may provide additional input useful (such as credit card billing information) to the other end.

The Signal IE may be used to convey information similar to that which we are used to with analog lines. It can give *tone* information. Ringback tones, busy tones, alerting messages, and so forth translate digital information into a form that can be converted to audible signals. Normally, this IE is used only for voice connections.

The next four IEs are actually a group. These are the Calling Party Number, the Calling Party Subaddress, the Called Party Number, and the Called Party Subaddress. The numbers are used, along with headers identifying the address scheme, to identify the endpoints that are to be in communication. The Calling Party fields are for the originating end. The Called Party Fields are for the terminating end. The subaddress IEs allow further identification in case the endpoint addresses a network or multiparty link.

The Transit Network Selection IE is used in a similar way to the network-specific facilities—allowing requests for specific routing.

The Repeat Indicator IE is given along with the Low Layer Compatibility (LLC) IE. This, once again, indicates that multiple possible indications can be given. Normally, if repeated, the IEs are given in descending order of preference. For many switches, the LLC IE (if present) must duplicate the information in the Bearer Capability IE. If this is true, why have both? The primary purpose of having both is to allow methods of negotiation.

The final IE mentioned in the SETUP message table is that of the High Layer compatibility IE. This IE is not acted upon by the network (nor should be the LLC). It is passed to the other end transparently for possible negotiation of compatibility.

DISCONNECT message type. The DISCONNECT message type is much simpler than that of the SETUP. It has four possible IEs past the header information. The one new IE (not explained within the SETUP message type) is the Cause IE. Why is the connection to be cleared? It may be due to facilities no longer available or a desire to normally hang up the call. The Cause IE allows this information to be conveyed.

The other three IEs associated with DISCONNECT are the Progress Indicator, Display, and Signal. These are used in very similar ways as that of the SETUP message.

Detailed information element analysis

A list of Q.931 IEs can be found in Table 5.4. This table shows the codes, descriptions, reference within the ITU-T Recommendation Q.931, and the maximum length. The IEs form a "pool" of information available for use within a message type—they are not normally used by themselves. This linkage between a message and IE allows more precise analysis of its use.

In our analysis of IEs, we will use the Bearer Capability and the Called Party Number IE as examples. These are items of information useful in almost every call. The Bearer Capability IE is also one of the most complex. A listing of the Bearer Capability IE (BC IE) is given in Fig. 5.2 and the Called Party Number IE is given in Fig. 5.3.

Bearer Capability IE. The first thing that can be noted in the Bearer Capability IE is that only the first 4 bytes are mandatory. These bytes are the IE identifier (value 4), the IE length, and 2 bytes associated with the type and speed of the connection. All other bytes are optional.

What does optional mean in this situation? Does it mean that the implementor can randomly choose which bytes to include? In practical terms, this is not possible. There must be a method of identification of the bytes. This is done by use of the extension bits and of the

TABLE 5.4 Q.931 Information Element Identifier Coding

8	7	6	5	4	3	2	1	Name	Section reference	Maximum length (bytes)*
1	:	:	:	-	-	-	-	Single octet information elements:		
	0	0	0	-	-	-	-	Reserved		
	0	0	1	-	-	-	-	Shift†	4.5.3/4.5.4	1
	0	1	0	0	0	0	0	More data	4.5.20	1
	0	1	0	0	0	0	1	Sending complete	4.5.27	1
	0	1	1	-	-	-	-	Congestion level	4.5.14	1
	1	0	1	-	-	-	-	Repeat indicator	4.5.24	1
0	:	:	:	:	:	:	:	Variable length information elements:		
	0	0	0	0	0	0	0	Segmented message	4.5.26	12
	0	0	0	0	1	0	0	Bearer capability†	4.5.5	32
	0	0	0	1	0	0	0	Cause†	4.5.12	10
	0	0	1	0	0	0	0	Call identity	4.5.6	10
	0	0	1	0	1	0	0	Call state	4.5.7	3
	0	0	1	1	0	0	0	Channel identification†	4.5.13	‡
	0	0	1	1	1	1	0	Progress indicator†	4.5.23	‡
	0	1	0	0	0	0	0	Network-specific facilities†	4.5.21	‡
	0	1	0	0	1	1	1	Notification indicator	4.5.22	3
	0	1	0	1	0	0	0	Display	4.5.16	34/82
	0	1	0	1	0	0	1	Date/time	4.6.15	8
	0	1	0	1	1	0	0	Keypad facility	4.5.18	34
	0	1	1	0	1	0	0	Signal†	4.5.28	3
	1	1	0	0	1	0	0	Information rate	4.6.3	6
	1	1	0	0	0	1	0	End-to-end transit delay	4.6.2	11
	1	1	0	0	1	0	1	Transit delay selection and indication	4.6.9	5

TABLE 5.4 Q.931 Information Element Identifier Coding (Continued)

1	0	0	0	1	0	0	0	Packet layer binary parameters	4.6.4	3
1	0	0	0	1	0	0	1	Packet layer window size	4.6.5	4
1	0	0	0	1	0	1	0	Packet size	4.6.6	4
1	0	0	0	1	0	1	1	Closed user group	4.6.1	7
1	0	0	0	1	1	0	0	Reverse charge indication	4.6.8	3
1	1	0	1	1	0	0	0	Calling party number	4.5.10	‡
1	1	0	1	1	0	0	1	Calling party subaddress	4.5.11	23
1	1	1	0	0	0	0	0	Called party number	4.5.8	‡
1	1	1	0	0	0	0	1	Called party subaddress	4.5.9	23
1	1	1	0	0	1	0	0	Redirecting number	4.6.7	‡
1	1	1	0	1	0	0	0	Transit network selection†	4.5.29	‡
1	1	1	0	1	0	0	1	Restart indicator	4.5.25	3
1	1	1	1	0	0	0	0	Low layer compatibility†	4.5.19	18
1	1	1	1	0	0	0	1	High layer compatibility†	4.5.17	5
1	1	1	1	1	1	1	0	User–user	4.5.30	35/121
1	1	1	1	1	1	1	1	Escape for extension§		

All other values are reserved. The reserved values with bits 5–8 coded 0000 are for future information elements for which comprehension by the receiver is required

*The length limits described for the variable-length IEs take into account only the present ITU-T standardized coding values. Future enhancements and expansions to this recommendation will not be restricted to these limits.
†This IE may be repeated.
‡The maximum length is network dependent.
§This escape mechanism is limited to codesets 5, 6, and 7. When the escape for extension is used, the IE identifier is contained in octet-group 3, and the content of the information element follows in the subsequent octets.
SOURCE: ITU-T Recommendation Q.931.

111

Bits								Bytes
8	7	6	5	4	3	2	1	
0	0	0	0	0	1	0	0	1
	Bearer capability information element identifier							
Length of the bearer capability contents								2
1 ext	Coding standard		Information transfer capability					3
1 ext	Transfer mode		Information transfer rate					4
1 ext	Rate multiplier							4.1 (multirate)
0/1 ext	0 1 Layer 1 ident.		User information layer 1 protocol					5*
0/1 ext	Sync/ asynch	Negot.	User rate					5a*
0/1 ext	Intermediate rate		NIC on Tx	NIC on Rx	Flow control on Tx	Flow control on Rx	0 Spare	5b* (V.110)
0/1 ext	Hdr/ no hdr	Multi frame	Mode	LLI negot.	Assignor /ee	In-band neg.	0 Spare	5b* (V.120)
0/1 ext	Number of stop bits		Number of data bits		Parity			5c*
1 ext	Duplex mode	Modem type						5d*
1 ext	1 0 Layer 2 ident.		User information layer 2 protocol					6*
1 ext	1 1 Layer 3 ident.		User information layer 3 protocol					7*

ITU-T Recommendation Q.931 has many additional notes

Figure 5.2 Q.931 Bearer Capability IE. (*From ITU-T Recommendation Q.931.*)

identification bits located in bits 6 and 7 of the first byte of each sub-field. Thus, it is possible to include only bytes 5, 5a, and 7. Byte 5 is identified by the bits 01 in bits 6 and 7 of the byte. Byte 7 can similarly be distinguished from byte 6 by use of these identifying bytes. Bytes 5b through 5d can be omitted by use of the extension bit—marking 5a as the last byte of the subfield.

Byte 3 is a description of the coding standard and the purpose of the connection. The coding standard may be ITU-T, ISO, a national standard, or a public network standard. The ITU-T recommendation makes note of the warning to not use other standards if the ITU-T coding is sufficient. In this way, the ITU-T allows possible divergences but tries to restrain the designers of networks.

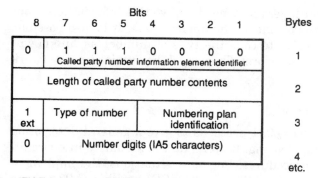

Figure 5.3 Q.931 Called Party Number IE. (*From ITU-T Recommendation Q.931.*)

Byte 4 is used to indicate the type of connection. Is it circuit mode or packet mode? What is the speed of the connection? If it is multirate, byte 4.1 can be used as a multiplier of the base rate (64 kbps). These 2 bytes must be present in each Bearer Capability IE and, since this is a required IE for SETUP, must be part of each connection.

The subfield designated by bytes 5 through 5d is associated with layer 1 information. This physical layer information is necessary if special hardware must be used to ensure compatibility. A voice circuit will need byte 5 to determine the appropriate digital-to-analog translator hardware algorithm. The voice circuit will not need bytes 5a through 7.

A data connection may make use of bytes 5 through 5d to specify the physical needs of the circuit. The exact interpretations of the following bytes will depend upon the content of byte 5. If byte 5 indicates V.110 rate adaption, the interpretation of the following bytes will differ from that of X.31, for example.

Layers 2 and 3 may be specified by use of bytes 6 and 7. These are useful only for interworking protocols (such as V.120) that may use multiple protocols upon the link.

Called Party Number IE.   As can be seen in Fig. 5.3, the Called Party Number has fewer fields than the Bearer Capability IE, but the total number is variable. The length field determines the size of the IE completely. The third byte (first byte after the length) determines the type of number and the numbering plan. The type of number discusses the network within which the number identifies a subscriber. It may be an international number, a national number, an abbreviated number, and so forth. The numbering plan, itself, identifies how the number will be decoded.

It is important for the implementor to determine use of the equipment. If international or national numbers are to be allowed, information from the upper layers may need to give this information. That is, byte 3 may need to be sent by the upper layers. It is also possible, for some switch networks, to code the type of number as "unknown" and allow the switch to interpret. This capability varies from network to network.

Circuit-switched call control

ITU-T Recommendation Q.931 proceeds to give a general description of call connections and clearing. It splits up the scenarios into circuit- and packet-switched procedures. For most situations, the implementation is very similar between the two. The SDLs within Q.931 give a more thorough explanation of the various options and error responses. A brief discussion of the possible message interchanges follow.

Outgoing call requests. The outgoing call starts with the formatting of a SETUP message. Within this message may be included a CHAN ID IE. If so, the IE may declare either an exclusive choice of B-channel, a preferred choice, or that any channel is acceptable. Lack of a CHAN ID may be considered equivalent to any channel being acceptable (some switches may require the IE).

At this point, the states diverge depending on whether en-bloc or overlap sending is involved with the setup. If overlap sending is being used, the network will respond with a SETUP_ACK to indicate further information may be sent. Eventually, a CALL_PROC message will be sent back to the user to notify the equipment that no further information is needed, or expected, to place the call. Note that the network may refuse the call at any point, giving the cause IE within a REL_COM message.

The network, if possible, will deliver the SETUP request to the terminating side. An ALERT will be sent to the originator to show that the call has been delivered. Finally, the terminating end will either accept the call (with a CONN message) or decline the call (with a REL or DISC message). If accepted, both sides are now in the active state.

Incoming call requests. The network will forward requests from the other endpoint. The first SETUP message for an incoming call will be sent in a UI-frame. This is to allow passage of a message without a valid TEI. If the equipment does not have a signalling TEI, this UI-frame will give notice that the line is to be used and that the user equipment must start TEI negotiation.

Within the SETUP message forwarded by the network, a CHAN ID IE should be present. Under most conditions, a specific B-channel will be listed. Some networks allow the user equipment to pick the appropriate channel, and this must be relayed to the network in a CHAN ID IE in the next message returned to the network.

If the user equipment allows for overlap receiving, a SETUP_ACK will be sent to the switch to indicate that more information may be relayed. At some point, however, the user equipment will notify the upper layers and send back an ALERT, CALL_PROC, or CONN message. The active state is entered when the upper layers decide to accept the call, a CONN message is sent to the switch, and the network responds with a CONN_ACK.

Call clearing. Call clearing takes place when the negotiation is interrupted or when an active call is torn down. A REL_COM message will be sent by the switch for point-to-point terminals if an error in procedure is noticed. Various collision scenarios are provided for within Q.931, but eventually the idle state is regained when a REL_COM is obtained from the network.

Timers on the user side

There are a total of 16 timers defined for use on the user equipment side of the ISDN. These timers vary in purpose and duration, and many are optional. Timers are set up to put a limit on the time waited for an event.

T301. This timer is started after an ALERT message is received from the switch. If a CONN message is received before the timer expires (suggested time 3 min at minimum), the timer is stopped. Otherwise, the call is cleared. This timer is optional unless Annex D procedures are implemented—which are concerned with symmetric call connections for specified B-channels.

T302. The overlap receiving timer is used after a SETUP_ACK is returned to the network. An INFO message is expected with additional connection information. If the INFO message is not received within the timer period (suggested time of 15 s) and more information is needed, the call is cleared. Otherwise a CALL_PROC message is returned to the network. This timer is optional and used only if overlap receiving is implemented.

T303. An outgoing SETUP message starts this timer. When one of the expected replies is received, the timer is stopped. Expected replies include CALL_PROC, ALERT, SETUP_ACK, and REL_COM (if refused). The suggested duration of this timer is 4 s. If the timer expires before a suitable message arrives, a second SETUP is sent, and the timer restarted. If the timer expires a second time, the call is cleared with a REL_COM message. This timer is optional unless Annex D procedures are implemented—which are concerned with symmetric call connections for specified B-channels.

T304. Timer T304 (suggested duration 30 s) is an optional timer to be used when in overlap sending state. It is started upon transmission of an INFO message and restarted upon each subsequent INFO message transmission. The timer is stopped when a CALL_PROC, ALERT, CONN, or DISC message is received. If the timer expires, a DISC message is sent.

T305. This is a mandatory timer. It is started when a DISC message is sent. If the expected REL or DISC message is not received before the timer expires (suggested duration 30 s), a REL message is sent (and state U19 is entered).

T308. This timer is also mandatory. It is started when a REL message is sent. A REL or REL_COM is expected within the suggested 4-s duration of the timer. If the timer expires, a second REL message is sent (with cause value 102), and the timer is restarted. If the timer expires a second time, the link is placed in a maintenance state, and the CRV is released.

T309. This optional timer allows the data link to be dropped temporarily without losing calls. It is started when a DL_REL_IN primitive is received and stopped when a DL_EST_IN arrives. If the DL_EST_IN does not occur within the suggested 90 s, all calls are cleared locally and the CRVs are released.

T310. This timer is used between the time a CALL_PROC message is received and an ALERT, CONN, DISC, or PROG message is received. If the 30- to 120-s timer expires, a DISC message is transmitted. This timer is optional unless Annex D procedures are implemented—which are concerned with symmetrical connections for specified B-channels.

T313. This mandatory timer (suggested duration of 4 s) is started when a CONN message is sent. It is stopped when a CONN_ACK is received. If the CONN_ACK is not received before the timer expires, a DISC message is sent.

T314. This timer is not currently required but may be in the future for user equipment that allows segmented messages. It is started when a message segment is received and is stopped when the last message segment is received. If all the message segments are not received before the timer expires (suggested time of 4 s), the message is discarded.

T316. This timer is started when a RESTART message is sent. The suggested period for the timer is 2 min. If it expires, another RESTART message is sent and the timer is started again. This may be repeated several times—depending on the user equipment behavior desired.

T317. This timer is associated with internal clearing of CRVs after a RESTART message is received. It should be at least as long as T316 in duration. If the internal clearing cannot be done within this period, a maintenance warning should be given in the user equipment.

T318 and T319. These timers are used for RESUME and SUSPEND requests on active calls. Each timer is 4 s long, and they are canceled when an appropriate acknowledge or reject message is received. If T318 expires before a RESUME_ACK message is received, a REL message is sent (cause value 102). If T319 expires before a SUSPEND_ACK is received, the call returns to active state, and the user is notified.

T321. This timer is mandatory if backup D-channels are used on multiple interfaces. If the D-channel fails, the timer is started (suggested time of 30 s). If the timer expires before a layer 3 message is received (indicating that the D-channel is active again), a DL_EST_RQ primitive is issued on both D-channels.

T322. This 4-s timer is used for cases where a STATUS_ENQUIRY may be sent by the user equipment. It may be used repeatedly, with expiration of the timer indicating a need to reissue the STATUS_EN-QUIRY message. It is canceled when a STAT, DISC, REL, or REL_COM message is received.

Q.931 annexes

Q.931 contains a number of annexes for further clarification of some protocol issues and to give optional procedures for certain types of equipment. Annex A contains SDLs for the U- and N-state transitions. Annex B refers to the exact procedures necessary at the network and user equipment to determine compatibility of the equipment and service provided. This will be discussed in greater detail shortly.

Annex C discusses transit network selection, as per the IE discussed as part of the Bearer Capability IE. Annex D gives procedures for symmetric call operation, and Annex E details the procedures for network-specific facility use. Annex F, as mentioned briefly in the discussion of timer T321, describes the use of a backup D-channel in cases where multiple interfaces are in operation.

Annex G describes the use of progress indicators. Annex H notes the methods used for message segmentation if desired. Annexes I and J deal with the low-layer information coding and negotiation. This will also be discussed later in greater detail. Annex K concerns "procedures for establishment of bearer connection prior to call acceptance," and Annex L covers optional procedures for bearer channel change. The annexes end with a description of cause values, example message sequences, and a listing of all currently defined ITU-T IEs and message types (whether they are in Q.931, Q.932, Q.933, or Q.952).

Compatibility checking

Annex B of Q.931 presents the details for procedures of compatibility and address checking. These checks fall into three categories: user-to-network, network-to-user, and user-to-user checking. A certain minimum amount of user-to-network checking will always take place within Q.931. Other checking may be done at various points of the protocol.

User-to-network checking. The network primarily checks against services provided. The Bearer Capability IE will be checked. It may not provide full-service checking because certain details are not handled by the network. These would include such items as actual protocols used over the bearer channels.

Network-to-user checking. The first check is against the Called Party Number and (if present) the Called Party Subaddress. If the addressing matches, the call is forwarded to the upper layers. If it does not match, it may be ignored on a point-to-multipoint system. For a point-to-point configuration, a mismatch indicates an erroneous incoming call and should be rejected.

The user equipment also checks the Bearer Capability IE to see if it is able to handle the service indicated. It may either ignore, or explicitly reject, the call if the service is not supported.

User-to-user checking. If a LLC IE is present, the user equipment will use this to determine compatibility of services. If not, the Bearer Capability IE will be used. The user equipment may also check the High Layer Compatibility IE (if present).

In the case of interworking, a PROG IE may also be included by the switch. This information is used in deliberations of compatibility. For example, if the connection is not end-to-end ISDN, the equipment may not be able to support 56K adaption. In this case, the call must be rejected. If the incoming SETUP message is on a broadcast link for a point-to-multipoint passive bus situation, it may be ignored if other terminals on the bus have varying capabilities. If on a point-to-point configuration, or if all terminals on the passive bus have similar capabilities, the call may be rejected.

Low-layer information coding and negotiation

Annexes I and J cover procedures for when a user specifies low-layer capabilities during call setup and how negotiation can be done. Annex I covers the first area and splits the procedures into three types. Type I information is information that is only concerned with terminal capabilities. This information is conveyed in bytes 5 through 7 of the Bearer Capability or LLC IE. Type II information is used by the network to determine whether the requested transport mechanism is available. The information is coded in bytes 3 and 4 if it is a circuit-mode call and in bytes 3, 4, 6, and 7 if it is a packet-mode connection. Type III information is used by the network to determine terminal capability and interworking requirements. Byte 5 is particularly important in this use.

Type I information may be included in the LLC IE or the Bearer Capability IE. If it is included in the Bearer Capability IE, it is examined by the network and the remote user. Type II and III information must be included in the Bearer Capability IE if interworking may be needed. Thus, for most instances, the Bearer Capability and LLC IE must duplicate information within the SETUP message.

Note, however, that the network may change some elements of the Bearer Capability IE due to interworking requirements. For example, if the network provides an A-law-to-μ-law conversion, byte 5 may have been altered by the network before being delivered to the user equipment. If the LLC IE is also used, the Bearer Capability IE contents will now conflict since the LLC IE is passed transparently. It is suggested, therefore, that only elements not used for interworking be used within the LLC IE and that the Bearer Capability IE always take precedence.

As Annex J points out, not all networks will allow the LLC IE to be transmitted. If a network does not, and the LLC IE is used, the network will send back a STATUS message indicating that the IE was unrecognized. If the LLC IE is transmitted by the network, the out-of-band signalling bit may be set to indicate that negotiation is possible.

Negotiation occurs by either modification of the LLC (or Bearer Capability contents) or as a choice of prioritized repeated IEs. The endpoint equipment will return its choice in an LLC IE included with the CONN message. Negotiation may take place out-of-band (Q.931), in-band (according to the particular protocols chosen), or a combination of the two.

Q.932 Generic Supplementary Services

Supplementary services are those services that are not necessary in the basic setting up and tearing down of bearer channel connections. They may apply globally to the user equipment, or they may apply only to a specific call. They are broken into two general categories. These are considered to be stimulus and functional protocols. Stimulus protocols are independent of the state in which a connection is at the time of the stimulus. A stimulus may be considered as a digital analog of pressing a key on an analog phone. Functional protocols have their own internal states, and events will have different responses depending on the sequence of events.

Stimulus protocols are broken into two types called keypad and feature key management. The keypad protocol is based on the Keypad and DISP IEs. These may be included in the SETUP and INFO message types. These apply only in the user-to-network direction. It is applicable to both BRI and PRI.

Feature key management makes use of the Feature Activation IE (FA IE) and the Feature Indication IE (FI IE). This may be included in the SETUP and INFO message types from the user to the network. It may be present in various message types from the network. An FA IE may be sent by the user at any time, and an FI IE may be sent by the network at any time. Certain feature key aspects are related to noncall appearances. These feature keys are associated specifically according to subscription information associated with the terminal and acted upon by the network. In order for this to be done on a terminal-by-terminal basis, the user equipment is assigned a Service Profile Identifier (SPID) that identifies the equipment and the various services that are to be provided by the network.

For both keypad and feature key management protocols, the network can request additional information by sending a message with an Information Request (IRQ) IE. The network will set timer T312 at this point and not cancel it until the desired information has been relayed by the user equipment. Networks may support multiple forms of supplementary services. If a feature is available via functional protocols, such will take priority in use. This subscription time (services applied to and tariffed by the network) service is identified by the SPID, and the user is notified of specific stimuli and their effects at the time of subscription. That is, the precise stimuli are network specific and dependent. For example, for one user, FA IE of value 1 may mean one service—for another user, the same service may be invoked with an FA IE of value 5.

Keypad protocols

The keypad protocol described in Q.932 (I.452) makes use of the Keypad IE and DISP IE. It is very similar to what currently happens on an analog network. A key is pressed. The network interprets the key and performs an action. In ISDN the Keypad IE is employed by the user equipment to request an action from the switch. The network performs an action and provides feedback by use of a DISP IE or, if it is an audible voice call, may provide in-band tone signalling. The Keypad IE may be sent during the SETUP message or a subsequent INFO message. The DISP IE is used by the network to provide feedback and to request additional information from the user equipment.

Feature key management

The feature key management protocol is very similar, since it is a stimulus protocol, to that of the keypad protocol and can be used in conjunction with the keypad protocol. The user equipment uses an FA IE to ask for a service. The network uses an FI to indicate status of service. The user may send the FA IE as part of a SETUP or INFO message type. The network may send the FI IE as part of SETUP, SETUP_ACK, CONN, CALL_PROC, ALERT, INFO, DISC, REL, or REL_COM. The FA IE will be associated with a CRV, if available and applicable. Otherwise a dummy CRV is used. (For some feature key management messages, the dummy CRV is created as a zero-length CRV.) Multiple FA IEs may be responded to by the network with multiple FI IEs contained in a single message or by a series of messages, each with an FI IE. If the latter case happens, the user equipment should treat each FI IE as a match for FA IEs in the order that they were presented to the network.

A service may be deactivated by the user in one of two ways. A second message with a FA IE (same identifier) may be used to "toggle" the state of the service, or a separate identifier may be used as an explicit disabling of the feature. If a call is cleared while a service is in use, the network will either send a FI IE within a call clearing message or within a non-call-related INFO message after the call has cleared.

Functional protocol

Functional protocols have their own states and events. The generic functional protocols described in Q.932 are broken into two categories. The first category is a "separate message approach" and has separate message types for specific functions such as HOLD or RETRIEVE. The second category makes use of the FI IE. Unlike the stimulus protocols, functional protocols can be used in a symmetric fashion. Functional protocol messages may be valid at all points of a call state and may be associated with a particular call, groups of calls, or multiple invocations within a single message.

Separate messages category. Q.932 specifies only a single matched set of messages in this category. These are the HOLD, HOLD_ACK, and HOLD_REJ messages and the matching set of RETRIEVE, RETRIEVE_ACK, and RETRIEVE_REJ. Other specifications may extend this message set to other types of functions such as TRANSFER, CONFERENCE, and so forth. Many of these services are used primarily for voice calls.

A user can request that a call be placed on hold by issuing a HOLD message. The network either accepts with a HOLD_ACK or rejects the request with a HOLD_REJ. The network can also initiate the holding of a call with an incoming HOLD message. The RE-TRIEVE message is sent by the user or network to reconnect the call and to be accepted with a RETRIEVE_ACK or a RETRIEVE_REJ message.

Functional protocols have their own set of states. The states for the HOLD/RETRIEVE set are as follows: IDLE, HOLD_REQ, CALL_ON_HOLD, RETRIEVE_REQ, HOLD_IND, and RETRIEVE_IND. The functional substates are independent of the Q.931 state, but certain messages are valid only in specified Q.931 states. The HOLD function during call origination is valid only in states U2, U4, and U10. That is, a HOLD is valid while the call is proceeding, after it has been delivered, and after the call is active. It may also be used in states U7, U8, and U9 for the terminating end if it is a point-to-point configuration.

Since the functional protocol is symmetric, it is possible for the network and the user to specify requests at the same time. This situation is called a collision. The network requests always have precedence over the user requests. For example, if the network sends a HOLD and the user sends a HOLD, the user equipment enters the HOLD_IND substate, and the network ignores the HOLD received from the user equipment. Timers may be used to check for acknowledgment (positive or negative) of responses.

Common information element category. The functional protocol presented in ITU-T Recommendation Q.932 is designed to allow an evolution in future standards, with some specific features given in the ITU-T Q.95x series of recommendations. It makes use of an FI IE, which can be included within a REGISTER, FACILITY, or other Q.931 message. For example, an FI IE can be included as part of a SETUP message, and the ability to support this service can be a part of the compatibility checking done by the network and other equipment. An FI IE can be associated with a call reference and is, in this way, connected with a particular call and bearer service. It can also be used independently from any active calls.

Bearer connection independent supplementary services procedures. These procedures are not directly associated with a call reference. Three categories exist which pertain to this. These are for point-to-point connection-oriented calls, connectionless point-to-point, and broadcast services. Such a service used on a point-to-point connection-oriented call begins by REGISTER messages being exchanged by the equipment and network. FACILITY messages are then used to exchange information. Finally, one side releases the pseudo-connection (no SETUP/CONN exchange has ever taken place) by issuing a REL_COM message (cause value 16).

Connectionless transport. Connectionless situations use a similar procedure, but the CRV is always a dummy (or null) one. In the case of broadcast use, a DL_U_DA_RQ primitive is used on the broadcast link (SAPI equals 127). It is possible to include called party number and called party subaddress information to the contents of the FACILITY message. If done, it is used by the receiver in the same manner as a SETUP message for compatibility issues.

Connection endpoint identifier. For procedures which temporarily deactivate a call or prepare to transfer a call, it is necessary to "reserve" the bearer channel that is in use—or will be in use. The network uses a connection endpoint identifier (CEI) for this. A user can implicitly reserve a channel by using a HOLD or RETRIEVE function (which returns a CEI for later use). User equipment can also explicitly obtain CEIs by use of the Facility IE. CEIs are canceled when the call is disconnected or released or when a RESTART message includes the associated channel.

Message types. ITU-T Recommendation Q.932 defines eight messages specific to Q.932. These messages are FACILITY, HOLD, HOLD_ACK, HOLD_REJ, REGISTER, RETRIEVE, RETRIEVE_ACK, and RETRIEVE_REJ. Each message is fully defined in form and content, as are Q.931 messages. Messages (such as FACILITY and REGISTER) may be also defined in Q.931. IEs include those from Q.931 as well as some new ones specifically for Q.932.

Information elements. These are coded in the same fashion as IEs defined in ITU-T Recommendation Q.931. There are nine variable length IEs defined in Q.932. These are the EXTENDED FACILITY, FACILITY, CALL_STATE, INFO_REQ, NOTIFICATION_IND, FEATURE_ACT, FEATURE_IND, SPID, and EID. The SPID and EID will be covered in more detail shortly.

One interesting detail of the Facility IE is that the IE can "embed" Q.931 IEs within itself. This is done by use of a Q.931 IE's tag (value of hexadecimal 40). This tag is followed by a length and enables the following specified bytes to be parsed as regular Q.931 IEs. Other Q.932 IEs are specified as in Q.931. The Facility IE, in addition to Q.931 embedding, has a complex internal structure that is given by an abstract syntax within ITU-T Recommendation Q.932.

User service profiles and terminal identification

User service profiles are useful for identification of specific terminals on a multipoint user-network interface. A service profile refers to the various information that a network keeps concerning a piece of equipment. This includes the address number, features supported, and other such user-specific information. The SPID is a number that uniquely identifies a type of terminal equipment and is given to the user at subscription time. A User Service IDentifier (USID) identifies a particular active access on an interface. The Terminal IDentifier (TID) is unique to a USID. Two or more pieces of user equipment may subscribe to the same service profile and, if used on the same access interface, may be given the same USID. However, each piece of equipment will be given its own TID. An Endpoint IDentifier (EID) incorporates the USID and TID as part of its parameters to uniquely identify the terminal and access point.

USID access protocols. The terminal equipment may have its USID/TID assigned at subscription time. It may be requested by the network. It may also be requested by the terminal equipment. The network needs this information so that it can address all of the terminals using a particular USID or a specific terminal within a USID or can exclude a terminal within a USID.

If the terminal equipment needs to obtain a USID/TID, it follows a set procedure. The equipment first obtains a TEI (if not fixed at subscription time). After the signalling link is established, a message is sent to the network with a SPID (specific messages vary between ISDNs). Often the message is an INFO message. The network uses this TEI/SPID combination as a unique key for assigning a USID (corresponding to the SPID) and TID (corresponding to the TEI and able to be used in broadcast procedures).

The network can also elicit initialization by sending an INFO message with the INFO_RQ IE. In the case of collision with a request from the terminal equipment, the request from the network is ignored. After the EID is obtained, all broadcast SETUPs received from the network are checked against the EID. If the EID does not match, the SETUP is ignored. An invalid EID within a nonbroadcast SETUP is treated as an illegal message.

X.25 Call Control

We have already discussed the data link layer of ITU-T Recommendation X.25. This included LAPB and MLP. X.25 covers both the data link layer and the network layer. This network layer is divided into three categories of messages. These categories are call control, PLP for data transfer, and maintenance messages. At the data link layer, groups of data are called frames. At the network layer, the groups are referred to as packets. Data and maintenance aspects of X.25 are discussed in Chap. 6 with other bearer services. The call control portion of X.25, however, is relevant here.

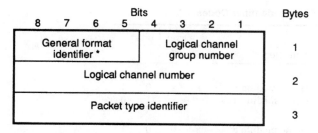

* - coded XX01 (modulo 8) or coded XX10 (modulo 128)

Figure 5.4 X.25 packet header bytes. (*From ITU-T Recommendation X.25.*)

X.25 offers two forms of data connections. These are called virtual calls (VCs) and PVCs. A PVC is a dedicated logical channel available for data transfer at any time which does not use call setup or teardown procedures. A PVC may be reset to make sure that both ends are in initial states. A VC is transient. It is established, and it is released.

Format of X.25 layer 3

X.25 layer 3 has a minimum of 3 bytes in each packet. These bytes (as seen in Fig. 5.4) include the General Format Identifier, the Logical Channel Identifier, and the packet type identifier. The General Format Identifier (GFI) allows for identification of modulo sequencing and the purpose of the packet. Generally, the lower 2 bits of the GFI are used to identify the modulo sequencing, and the upper 2 bits are an identifier of type of message. Table 5.5 lists the GFI codes. The packet type identifier is used in a manner similar to that of a message type in Q.931. The Logical Channel Identifier (LCI) provides a method for logical multiplexing of channels.

Q.921 allows for logical multiplexing of links at the data link layer. X.25 also allows for logical multiplexing—but at the network layer. These logical channels each have a logical channel group number and a logical channel number. Together, these items of data are called the LCI. Rather than using a separate channel dedicated for signalling, each LCI is established during call setup.

Restart of the physical channel

Before any channels are used, it is often useful to make sure that the protocols are in compatible states. This is called a restart procedure. Within X.25, this procedure may be used by a DCE or a DTE to clear all virtual calls and reset any permanent VCs. A RESTART_RQ packet is sent by the DTE to request a restart, and timer T20 (default time limit of 180 s) is started. A packet received from the DCE is

TABLE 5.5   X.25 General Format Identifier Codes

	Octet 1 bits*			
General format identifier	8	7	6	5
Call setup packets				
Sequence numbering scheme modulo 8	x	x	0	1
Sequence numbering scheme modulo 128	x	x	1	0
Clearing packets				
Sequence numbering scheme modulo 8	x	0	0	1
Sequence numbering scheme modulo 128	x	0	1	0
Flow control, interrupt, reset, restart, registration, and diagnostic packets				
Sequence numbering scheme modulo 8	0	0	0	1
Sequence numbering scheme modulo 128	0	0	1	0
Data packets				
Sequence numbering scheme modulo 8	x	x	0	1
Sequence numbering scheme modulo 128	x	x	1	0
General format identifier extension	0	0	1	1
Reserved for other applications	†	†	0	0

*A bit which is indicated as x may be set to either 0 or 1, as indicated in the text.
†Undefined.
SOURCE: ITU-T Recommendation X.25

called a RESTART_IN—but the packet-type identifiers are identical. (The LCI is not used in this type of packet.)

The terminology used within ITU-T Recommendation X.25 can be confusing when read within the context of more recent documents. Distinct names are used for primitives depending on direction of transmission. The packet-type identifiers, however, are the same. Thus, CALL_RQ and CALL_IN primitives are identical in coding at the packet level but indicate opposite directions.

The originating side can receive either a RESTART_CF (restart confirmation) or a RESTART_RQ/RESTART_IN packet. The latter case is considered to be a form of collision, but no distinction is made between the two situations. A timer may be used to resend the packet. Calls are neither set up nor accepted while confirmation is pending.

Virtual call service

The idle state of X.25 layer 3 is called the ready state and is indicated by p1 in diagrams. Although there is no explicit mention of it, this state requires that the data link layer be established. Therefore, in an implementation which separates the data link layer from the network layer, an additional state (or flag structure member) is required. A more precise initialization scenario is that the data link layer is established (DL_EST_CF or DL_EST_IN), a restart exchange may be sent, and the

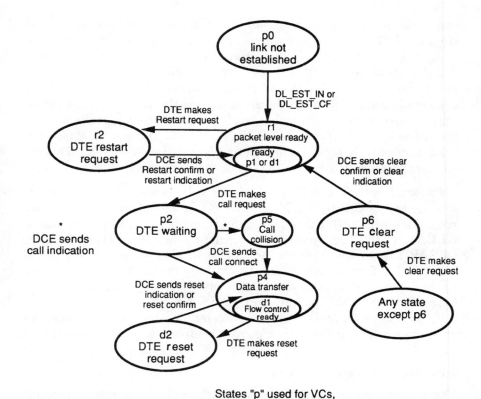

States "p" used for VCs,
states "d" used for PVCs or VCs within p4 state

Figure 5.5 X.25 DTE state transitions. (*From ITU-T Recommendation X.25.*)

physical channel is now ready for virtual services. A DTE will make a
transition from nonlink active state to waiting for the link-to-link established state to beginning restart procedure (r1) to waiting for restart
confirmation (r2) to the ready state (p1 for virtual calls). Figure 5.5
gives a summary of DTE-side state transitions for X.25 layer 3.

Call setup. A calling DTE will issue a CALL_RQ packet and enter
the DTE waiting state (p2). If the outgoing CALL_RQ is accepted
(CALL_CONN with the same LCI), the logical channel enters the
data transfer state (p4). An incoming call (from the DCE) will create
an internal state transition at the DTE—causing notification of upper
layers. If the upper layers accept the call, a CALL_ACCEPT packet is
generated, and the DTE enters the data transfer state (p4). If the call
is refused, a CLEAR_RQ (CLEAR_IN from the DCE for outgoing
calls) primitive is issued. Timer T21 (default time period of 200 s) is
used as a safeguard for call requests from the DTE.

Collisions occur when equivalent call setup messages are received
from peer entities with the same LCI. In such cases, the DCE will

cancel the connection coming into the DTE and accept the CALL_RQ from the DTE. In order to minimize collisions, an LCI allocation table is agreed upon with the X.25 router. This table (discussed in greater detail in Annex A of X.25) allows for ranges of LCI values. One group is allocated to PVCs. A second group is allocated only for incoming calls (calls coming through the DCE toward the DTE). A third group may be used for either direction but is allocated from opposite ends of the range by the DTE and the DCE. The fourth group is used only for allocation of LCIs for outgoing calls (originated by the DTE). This allocation scheme greatly reduces the chance for collisions.

Call teardown.   The DTE may initiate call clearing at any time (including before CALL_CONN is received). It will enter the DTE clear request state (p6), start the T23 timer (180 s), and wait for a CLEAR_CONFIRM packet to reenter the ready state (p1). A CLEAR_RQ packet from the network will also act as confirmation. When a DCE sends a CLEAR_IN packet, the DTE will notify the upper layers, (optionally after approval of the upper layers) send a CLEAR_CONFIRM packet, and go back to the ready state (p1). Collisions of clearing messages are acceptable—but the DTE must respond to a colliding clear message as if confirmation had occurred.

X.25 Layer 3 call control packets

The packet type identifiers for the various X.25 packets, including call control packets, are shown in Table 5.6. There are four packet types used within X.25 call control plus the restart and restart confirmation packets. The packet type identifiers are in the third byte of the X.25 packet. The lower-order bit is set to 0 for data packets and 1 for all other packet types. The data, RR, RNR, and REJ packets have a field (two in the case of data) for modulo 8 sequence numbers. Modulo 8 is the default modulo numbering system for X.25. Modulo 128 can also be used as an option. In this case, the sequence numbers of byte 3 are filled in by 0s, and additional bytes are used for the sequence numbers. Data transfer packets are discussed in Chap. 6.

Address block formats.   The A-bit field of the GFI indicates the form of the call addressing block. All DTEs should implement the non-TOA/NPI format, which is indicated by the A-bit set to 0. This format allows up to 15 binary coded decimal (BCD) address nibbles for both the calling and called address fields. (If the total number of address nibbles is odd, the low-order nibble of the last byte has of value 0.) The TOA/NPI address format (A bit set to 1) allows longer addresses with each address identified as to type and numbering plan in the first two BCD nibbles of each address field.

TABLE 5.6 X.25 Packet Type Identifiers

Packet type		Octet 3 bits*							
From DCE to DTE	From DTE to DCE	8	7	6	5	4	3	2	1
Call setup and clearing									
Incoming call	Call request	0	0	0	0	1	0	1	1
Call connected	Call accepted	0	0	0	0	1	1	1	1
Clear indication	Clear request	0	0	0	1	0	0	1	1
DCE clear confirmation	DTE clear confirmation	0	0	0	1	0	1	1	1
Data and interrupt									
DCE data	DTE data	x	x	x	x	x	x	x	0
DCE interrupt	DTE interrupt	0	0	1	0	0	0	1	1
DCE interrupt confirmation	DTE interrupt confirmation	0	0	1	0	0	1	1	1
Flow control and reset									
DCE RR (modulo 8)	DTE RR (modulo 8)	x	x	x	0	0	0	0	1
DCE RR (modulo 128)†	DTE RR (modulo 128)†	0	0	0	0	0	0	0	1
DCE RNR (modulo 8)	DTE RNR (modulo 8)	x	x	x	0	0	1	0	1
DCE RNR (modulo 128)†	DTE RNR (modulo 128)†	0	0	0	0	0	1	0	1
	DTE REJ (modulo 8)†	x	x	x	1	0	0	0	1
	DTE REJ (modulo 128)†	0	0	0	1	0	0	0	1
Reset indication	Reset request	0	0	0	1	1	0	1	1
DCE reset confirmation	DTE reset confirmation	0	0	0	1	1	1	1	1
Restart									
Restart indication	Restart request	1	1	1	1	1	0	1	1
DCE restart confirmation	DTE restart confirmation	1	1	1	1	1	1	1	1

(Continued)

TABLE 5.6 X.25 Packet Type Identifiers (*Continued*)

Packet type		Octet 3 bits*							
From DCE to DTE	From DTE to DCE	8	7	6	5	4	3	2	1
Diagnostic†									
Diagnostic†		1	1	1	1	0	0	0	1
	Registration								
	Registration request	1	1	1	1	0	0	1	1
Registration confirmation		1	1	1	1	0	1	1	1

*A bit which is indicated as x may be set to either 0 or 1 as indicated in the text of the recommendation.
†Not necessarily available on every network.
SOURCE: ITU-T Recommendation X.25.

Software Design Issues

There are a number of software design issues specific to the network layer of ISDN. Many of the types of issues are similar to those of the data link layer. One issue that is specific to the network layer is due to the fact that it is the highest layer of the chained layers—the network layer must interact with the less-known application layers. The exact primitives will depend upon the needs of the applications and the specific options supported by the ISDN. Other system aspects, such as module entry points and data structures, are equivalent to those needed at the data link layer.

Interlayer primitives

The primitives between the data link layer and the network layer are defined in Chap. 4. Table 5.7 gives example primitives that may exist between the higher layers and the network layer. Parameters depend on the configuration of the system and the options being used. Additional primitives may also be useful. (For example, if Q.931 states U6 and U9 are implemented, N_CALL_PROC_RQ and N_ALERT_RQ primitives are needed for transitions between U6 to U9 and U9 to U7.)

Primitives presented follow the ITU-T and ISO architectural guidelines. Just as data link layer parameters begin with DL (or MDL for management primitives), network layer primitives begin with an N. The four types of primitives are represented: REQUEST, INDICATION, RESPONSE, and CONFIRMATION. Management primitives depend on architectural requirements. Timer primitives will be needed (MN_TM_IN and MN_TM_RS) and potentially so will other management and coordination functions.

The primitives, and associated parameters, of a piece of equipment define its possible applications. Each system will have different requirements and, therefore, different primitives and parameters. The design of the interface defines the system's use. The following primitives are examples of base primitives that may be useful in many ISDN systems. The first few primitives explain variations that are possible and the added capabilities that are available with different configurations. A minimal system is a single-purpose system. It is used for one logical connection and supports one bearer service, and most information is hardcoded into the application. This may be useful for certain systems. Other systems need to be more flexible.

N_CONN_RQ and N_CONN_IN.   The connect request and connect indication primitives are symmetric in use. Each needs some type of "handle" to identify the logical call. Additional required information

TABLE 5.7 Examples of Network Layer/Upper-Layer Primitives

Primitives		Mimimum parameters	Additional useful parameters
L4+ → L3	L3 → L4+		
N_CONN_RQ		Called address	Call handle ID, calling address, subaddress information, service type, user-to-user information
	N_CONN_IN	None	Call handle ID, called address, calling address, subaddress information, service type, user-to-user information
N_CONN_RS	N_CONN_CF	None	Call handle ID, user-to-user information
N_DISC_RQ	N_DISC_IN	None	Cause, user-to-user information
N_SUS_RQ		Call handle ID	
	N_SUS_CF	Call handle ID, CEI	
	N_SUS_NAK	Call handle ID	Cause
N_RESUME_RQ		CEI	
	N_RESUME_CF	Call handle ID, CEI	
	N_RESUME_NAK	CEI	Cause
N_REGISTER_RQ		None	
	N_REGISTER_CF	None	
	N_REGISTER_END	None	Result code
N_DATA_RQ	N_DATA_IN	Data reference, length	Call handle ID, data attributes
N_EXP_DATA_RQ	N_EXP_DATA_IN	Data reference, length	Call handle ID, data attributes
N_KEYPAD_RQ	N_KEYPAD_IN	Data reference, length	Call handle ID
N_RESET_RQ	N_RESET_IN	None	Call handle ID, cause
N_RESET_RS	N_RESET_CF	None	Call handle ID, cause
N_SS_RQ	N_SS_IN	Variable	Variable
N_MISC_RQ	N_MISC_IN	Variable	Variable

includes the called address. If an implementation supports multiple bearer services, a method is needed to convey that information. If both out-of-band and in-band call setup techniques are used within the same system, this information must be identified. The system may require subaddress information. User-to-user information may need to be passed from one layer to the next.

A minimum system requires a method of passing the called address. This would limit the system to one active call and one bearer service type. There would be no support of additional optional parameters (such as user-to-user information). In this way, an N_CONN_RQ(address) primitive initiates a call setup. All other information in the system would be fixed. For many systems, additional information would be needed.

The N_CONN_IN primitive has the same requirements as the N_CONN_RQ primitive. The minimum requirements are even smaller, however, since there is no requirement to pass the address information. Thus, a minimum N_CONN_IN would indicate that the only call possible on the system has arrived (with a fixed bearer service, local address, and no subaddress). Minimum services allow equivalent events as for a Plain Old Telephone System (POTS) line. Additional parameters allow the ISDN to be used in a versatile manner, with passed through information (such as calling party number).

N_CONN_RS and N_CONN_CF. Minimally, no parameters are required—if only one fixed bearer service call is possible on the equipment. The addition of a handle to identify the call allows multiple call instances to be invoked. The passage of the bearer channel ID allows application resource allotment. If compatibility negotiation is supported, the final bearer service type may be needed.

N_DISC_RQ and N_DISC_IN. A minimal system does not require any parameters. Useful parameters include a call handle ID and cause reasons. A system may also need N_REL_RQ and N_REL_IN primitives if it is desired to distinguish the manner in which a call is terminated. User-to-user information is also useful in cases where "fast select" data transfer is desired.

N_SUS_RQ, N_SUS_CF, and N_SUS_NAK. These (and the next two groups) primitives are examples of methods to add services past basic services. A SUSPEND message type may be acknowledged or rejected. A method needs to exist to pass this information back to the higher layers. This primitive is not useful unless more than one call instance can exist in a system—therefore, a call handle ID is required for this type of primitive. The N_SUS_CF must have a CEI passed back to allow later use of the N_RESUME_RQ primitive. If an N_SUS_NAK is passed back, there must be some cause reason associated with the primitive.

N_RESUME_RQ, N_RESUME_CF, and N_RESUME_NAK. The N_RE-SUME_RQ makes use of the CEI passed as part of the N_SUS_CF primitive. In return, after successful resumption of the call, a new call handle ID must be returned. The N_RESUME_NAK should be able to return a cause reason.

N_REGISTER_RQ, N_REGISTER_CF, and N_REGISTER_END. This type of supplementary service primitive allows registration of a terminal with the network. The N_REGISTER_RQ primitive is responded to with an N_REGISTER_CF to indicate that registration has begun. The N_REGISTER_END primitive notes that registration is complete. The N_REGISTER_END primitive can include a result code to indicate successful or unsuccessful conclusion of the primitive. A fourth primitive may be desired if an explicit rejection is needed. Note that registration brings back a segmented series of messages from the network. Some method of communicating this information to the upper layers may be needed.

N_DATA_RQ and N_DATA_IN. These are examples of primitives used with the network layer for data transfer. What types of information are needed for such a primitive? First, if more than one connection is possible, a call handle ID is needed. Next, the data must be included. The format of the data primitive depends on the overall architecture of the system. One method is to pass a data pointer and a length. Another is to pass a specific buffer reference number and length. The length may be any value including zero. (A zero-length data block may be of use in file transfer protocols.) Some data transfer network layers may allow special attributes to be attached to the data. X.25, for example, allows segmentation of data, end-to-end data acknowledgment, and special data attributes to be associated with the data. If such is supported, provision must be made to pass the information between layers.

N_EXP_DATA_RQ and N_EXP_DATA_IN. A separate primitive for expedited data is useful in several ways. In the first place, different parameters may be associated with the primitive. Next, the coordinating entity may route such a primitive differently than for a N_DATA_RQ primitive. It is also useful for special applications and nonstandard usage of the network layer. An N_EXP_DATA_RQ may be used to skip the network layer and be translated directly into a DL_U_DA_RQ primitive to pass data to the other side without OSI layer intervention.

N_KEYPAD_RQ and N_KEYPAD_IN. These primitives are examples of methods of passing additional information from Q.931. In the cases of overlap sending or receiving, these primitives map directly to the use of KEYPAD message types. In other instances, they can be used to create INFO messages to carry keyboard data during a voice connection, for example. The exact use of network layer primitives always depends on the architecture of the system.

N_RESET_RQ and N_RESET_IN. These are examples of primitives that are useful for a particular type of network layer. The X.25 PLP allows the logical channel to be reset at any time. As primitives, they give the upper layers access to the full capabilities of a lower layer. An N_RESET_RQ (or N_RESET_IN) needs a cause value (plus the call handle ID, if more than one simultaneous call is supported).

N_RESET_CF and N_RESET_RS. Request and indication primitives require acknowledgment. The N_RESET_CF and N_RESET_RS primitives are examples of acknowledgment across the higher layers for a specific need of a lower layer—in this particular case, that of X.25. Only a call handle ID is needed for such a primitive.

N_SS_RQ and N_SS_IN. Primitives can always be grouped according to function. For example, the N_DATA_RQ and N_DATA_IN primitives could handle *all* types of data transfer in a system with special parameters to identify the particular form and requirements of the primitive. This type of "unioned" primitive allows growth of the system by use of special parameters. An N_SS_RQ primitive, for example, can be designed to make use of HOLD and RETRIEVE functions on a system. If the primitive structure is designed sufficiently well in advance, the interface can be extended to support other services (such as CONFERENCING) without changing the interface structure.

N_MISC_RQ and N_MISC_IN. In a manner similar to that of the N_SS_RQ and N_SS_IN primitives above, a system should always have a way to add miscellaneous primitives without changing the global architecture. It is a good idea to explicitly design such a capability into the system before it is needed. Protocols and standards change slowly but steadily. Products exist in the market for a period of time. An architecture that is sufficiently open to allow later expansion will be of service for maintenance and upgrading purposes.

Module entry points

Data comes into the network layer through three entry points, just as in the data link layer. The primitives coming from and going to the data link layer are defined in Chap. 4. The other two are from the coordination/management planes and the upper layers. The primitives associated with the coordination/management planes and the upper layers are, as discussed above, dependent on the exact network layer and the primitives and operations supported.

Data goes out of the network layer through three general points also. These are, once again, to the coordination/management planes, the data link layer, and the upper layers. Each of these six entry points is a place where one can analyze data flow for the module.

Let us take the flow of primitives from the upper layers to the data link layer as an example. There are two categories of primitives going between the upper layers and the network layer. One category of primitive applies the entire system. For example, an N_START_RQ primitive might be used to reinitialize the user equipment. Most primitives are associated with a particular call. If multiple call appearances are possible, each call must be identified with a unique identifier. In this book, we have called such an identifier a call handle ID. There are other possibilities for a unique identifier, but they are only possible for groups of data from various layers. It is better, from an OSI layering methodology, to use an identifier that does not take direct advantage of data from the other layers.

An implementor might be tempted to use the CRV as a unique identifier across the network layer to upper layer boundary. This can work for all Q.931 calls. It would not work for use of a data transport mechanism such as X.25 or even V.120, which uses its own bank of CRVs. Two types of identifiers could be used—one for Q.931 and one for the data transport mechanism. Each connection-based primitive would then have to explicitly identify the use of the connection. For these reasons, mapping of an identifier per connection is easier.

As mentioned before, a protocol that is defined via a state table (such as Q.931) acts in a way that is extremely conducive to having a central state handler. This information is repeated for the network layer to emphasize the commonality of structure of the protocol layers. Each entry point takes the information necessary to uniquely identify the appropriate logical channel, and the information is passed to the central state handler. The form of the state handler can vary, and a number of the possibilities are discussed in detail in Chap. 10.

The exit points may also be combined. The protocol, depending on current state and precise definition of event, may need to perform an action. This action will result in internal state changes, or it will effect a primitive to the data link layer, upper layer, or management/coordination entity. An upper layer to network layer handler may, for example, examine the action desired, format a primitive, and then invoke the operating system function needed to send the primitive or call an appropriate handler for the primitive.

The general data flow, then, is as follows. A primitive arrives at an entry point. Identification of the appropriate logical link is made. The primitive is passed along to the central state handler. Based on state, environment, and exact events, actions may be performed. These actions invoke outgoing primitive formatters, and the state table protocol has completed its actions for the current event and state.

Operating system use

The general principles of message-handling architecture are the same in the network layer as they are in the data link layer. The primitives pass into, and out of, the network layer. Each of them has its own requirements and takes a certain amount of time. The first requirement to be considered is whether synchronization of the primitive is required. The needs of the data link layer to network layer are discussed in detail in Chap. 4. The needs of the coordination/management entity to network layer are also similar.

The main point of difference in the network layer rests in its interaction with the upper layers. This interface between the chained layers and the upper layers depends greatly on the precise implementation and functionality needed. There are two main categories of interfaces. One is a coprocessor interface. The other is a standalone interface.

Coprocessor use.   The coprocessor interface means that the chained layers are operating independently on a processor while the application and all attendant OSI layers are present on a separate processor. A mechanism is needed to transport primitives and associated parameters (including data, if necessary) across the coprocessor boundary. It is also necessary to determine whether primitives need to be asynchronously handled. A synchronous interface can make use of special areas of memory which can be called *mailboxes*. An asynchronous interface requires additional queuing methods similar to those of an operating system message queue.

Whichever method is used, there must be communication between the two processors. A dual-port RAM, or shared memory, interface is very efficient for data and primitive transfer. A serial data link can also be used (via a UART or system data bus). In this case, data must be accumulated until a complete transaction is ready for processing. This implies a protocol between the processors.

The shared memory interface approach can be very simple. Two mailboxes are used. One is for primitives in the network layer to upper layer interface. The other is for primitives in the downward direction. A bit, or field, within the mailbox area is used as a semaphore flag to indicate that the mailbox is ready for delivery. Polling, or interrupts, can be used to check on the status of the semaphore flag. For synchronous use, the semaphore flag can be changed to a different value to indicate that the information from the mailbox has been used. For asynchronous use, a separate step is necessary before setting the semaphore flag—a copy of the primitive and attendant data must be placed into a message queue.

In the previous chapter, criteria for synchronous versus asynchronous use were presented. One of the primary criteria is the speed of the transaction. Let us suppose that we put a software break between the network layer and the upper layer coprocessor interface. Furthermore, call this software break the coordinating/management entity. There can be operating system queues between this coordinating/management entity and the network layer. This minimizes the actual processing that needs to be done and allows use of a synchronous interface. A second advantage to this is that the coordination entity can handle bookkeeping details such as mapping call handle IDs to appropriate network layer identifiers and to route more generic primitives based on the condition of the link.

Standalone interface. At some point, communication with the external world is necessary. This may be via a display and/or keyboard for a piece of user equipment. It may be a UART with an asynchronous protocol (such as the AT command set) or synchronous protocol (such as X.21 or V.25 bis). There is no explicit synchronization with another processor, but there is a protocol to be studied. It may be a carriage return character received over a serial data link. It may be a hardware register bit that has been set by a button hit on the front panel display.

Some event will indicate that a transaction is ready to begin. At this point, we are in a situation that is similar to the coprocessor situation. Data have arrived. They must be processed. The use of a coordinating/management entity protocol funnel becomes even more useful. As we can see, it not only provides a method of being able to use a synchronous interface but also provides an entry point independence.

Timers. Timers are needed in most data protocols. We have noted that T305, T308, and T313 are minimally needed by Q.931. Other equivalent timers are needed by X.25 call control (and, as we shall see in the next chapter, PLP). It is most convenient if the network layer can use a common structure to access timers as issued in the data link layer.

Data structures

Upon entry to the network layer, the specific logical channel is determined. The types of data are very similar between the network and data link layers, as discussed in the previous chapter. One additional type of data is needed. This is a subset of what we called modifiable parameter data. This is configuration data. Configuration data, for Q.931, would include any subscription data obtained from the network when the access interface was obtained.

The data structures for a protocol state machine require five types of information. These may be called identification, state, current event, environment, and modifiable parameter data. The exact form of the data may vary from one programming language to another.

Identification data. Each entry point must be able to identify the correct logical link. For layer 2 to layer 3 primitives, as mentioned in the previous chapter, this data includes TDM channel, SAPI, and CES. For an upper layer to network layer, a unique identifier must be used which, in this book, is being called the call handle ID. This same unique ID may be used by management layer primitives. These four items of data (TDM channel, SAPI, CES, and LCI, or other channel identifier) allow identification of the appropriate logical link for any primitives entering via the three entry points.

State data. Each logical link must retain information concerning its current state. State includes the protocol state table states as well as possible functional protocol supplementary service states for Q.931. Each logical channel also has a CRV (and associate length) and bearer service information. Signalling state information also includes negotiated information such as TEI and USID/TID.

Each network layer entity will have its own data. So, for example, there might be a set of state data (and current event data, and so forth) for a Q.931 network layer and a separate body of data for an X.25 network layer entity. X.25 is also a state-based protocol and requires noting of the d, p, and r states.

Current event data. Just as in the discussion on the data link layer, the network layer can be designed to incorporate a single central protocol state handler. However, for this to be possible, all the information that was parsed from the event to be able to call the handler must be saved. The task is similar to that of the data link layer, but the precise data and form of the data are different. For an X.25 network layer entity, data must be saved in certain cases for resegmenting or retransmission. For a Q.931 network layer entity, data are composed largely of the primitives and their parameters.

Therefore, for Q.931, saving of current event data is not necessary. Data is either acted upon or must be capable of regeneration (such as a new SETUP message after the timer has expired). For this reason, once the interface has been identified (in the case of multiple access interface units) data can be passed from function to function without long-term storage in internal data structures. Although information is needed in addition to the primitives and their parameters, it does not need to be internally stored. It may be useful, however, to parse out message types and IEs into a data structure more appropriate to event handling. This, once again, may be passed as arguments to subroutines or functions.

Environment data. Environment data is distinct from state data in that it keeps track of continuing events. For example, if a timer is running, note must be made of that fact. If a timer has a counter associated with it, that information must be retained in association with the logical link. The internal sequence variables associated with acknowledged transmission must be stored.

Finally, some limited copies of saved data must be available. For example, an incoming broadcast SETUP cannot be acted upon until after the signalling link has obtained a TEI and is established. The data contents must be saved temporarily until then. An N_CONN_RQ or N_KEYPAD_RQ primitive may be sent by the upper layers before the link is ready. They must be saved. These differ from the data queues needed by the data link layer in that we are talking about single primitive images.

Modifiable parameter data. This type of data includes system parameters such as window size, modulo of sequencing (for X.25), SPID, local equipment number, LCI range, en-bloc operation, and peer role (DTE or DCE) for equipment that can act in either role. A decision must be made early in the design of the system as to whether these data can be modified for each call. It is more efficient to have fixed values. In fact, such items can be ignored as data if the system has only one set of permissible values. However, the entire system design may need to change if requirements later change. This makes it a serious system consideration.

Integration with supplementary services

In the discussion on supplementary services, it was noted that there are two general categories of messages. One is call related and, for functional protocols, has additional state information not directly related to Q.931. The other is not associated with a CRV and affects the general environment of the system.

A call-related supplementary service message must be handled via the same type of central state handler as are other Q.931 messages. This is due to restrictions on the appropriate states for supplementary services messages. One method is to allow routing of the message to the appropriate specific state handler and then, before parsing of Q.931 messages, checking for supplementary service messages. Another method is to allow full routing of the message and treating the message type as an addition to the basic call management message types. If the latter case is implemented, it is important to make sure that mandatory IE checking takes the supplementary service message type into account.

A non-CRV-related supplementary service message should *not* go through the central state handler. This is because no logical channel is associated with the message, and it should not directly affect the state of any active call. (The message may, as a side effect, cause changes to some or all of the active calls.) Thus, the data link layer to network layer entry point must route such messages to a separate handler. It is possible to design such a supplementary service state handler in a manner that can be used for both call- and non-call-related messages. This is done by using a dummy data structure for the non-call-related messages. The state handler can be the same but with different calling locations and environments.

Switch and Country Variants

ITU-T Recommendations Q.931 and Q.932 are not specifications. They are standards that can be used to create ISDNs that can interoperate. Each country, region, or switch manufacturer is able to implement the parts that are applicable to their own needs. Current movement of such is to bring together two specifications. One is for Europe. This is addressed by the ETSI, an entity that was created in 1988 to help unify telecommunications standards within the European communities. The other is for North America. This specification is created as a consensus of primary switch and equipment manufacturers for use within North America. Both specifications are in use in other areas based upon the networks that are best suited to the local needs.

Although some consolidation of specifications is starting to happen, the results are still unclear. It may take a number of years before all ISDNs fall into one of the two specifications. It is also possible that the two specifications will eventually unite into a single global specification. In the meantime, there are close to a dozen different specifications being used throughout the world for ISDNs. An implementor needs to know the specific requirements for the equipment that is being designed. User equipment can be designed to work with a single specification or with multiple specifications. If multiple specifications are needed, their differences must be taken into account by the implementor.

Variances are provided for by the ITU-T documents. There are optional states, optional responses to given events, and mandatory versus optional protocol syntactic and semantic differences that come into consideration. In this section, these major categories of differences will be discussed.

Artificial testing differences

It would be good to be able to say that an implementor could look at one set of specifications, compare them to another, and immediately be able to know the differences between them. Unfortunately, this is not possible. Some specifications are created as a delta, or set of changes, from that of the ITU-T recommendations. Some specifications are based on that of another set of specifications. Others are written completely from scratch, in a form similar to that of the ITU-T recommendations. Others are written in a form distinctive to the needs of the particular ISDN. In other words, they do not have a common base from which to be compared—except in source. All (except for some ISDNs specified before the ITU-T recommendations were ready) are based on the ITU-T documents, but the form varies tremendously.

ISDN specification hierarchy. ISDN standards form a pyramid with three steps. The bottom level is composed of ITU-T and ISO documents. These are the international documents that provide a framework to allow for interworking of networks. The next step is that of the national, regional, and manufacturer specifications. These specifications are based on the ITU-T and ISO documents and may be derived directly from some other set of documents, but they cannot be directly compared. The third step is that of the testing and certification laboratories.

This top step is what user equipment must pass in order to be qualified to connect to the network. Thus, this provides judgment as to whether the equipment is ISDN compatible. It is completely possible to implement ISDN equipment that can connect to the network and behave properly in every way—and yet not pass the conformance testing.

Some of these differences arise from treatment of error conditions. In effect, testing requires the equipment to not only handle errors from other equipment but also errors from the network. Such errors may never occur in a working network, but they are tested anyway. What does the equipment do with an illegally formatted message? How does the equipment recover from a duplicated IE created by the network? What happens if the network starts to send illegal framing signals? Does the equipment properly reject a duplicated TEI assignment? Such are a subset of the tests to which equipment may be subjected.

Another difficulty in the testing is evolution of test procedures. A specification is created. After creation, the network is implemented, and the testing criteria for user equipment is documented. Equipment manufacturers want to be able to use their equipment on the network. A test suite is developed. Neither the switching equipment nor the test suite will be a full mirror of the specification. Some protocol areas will not have been implemented yet. Some physical considerations may affect timer values. The network may allow certain optional IEs and messages, but if the test suite is not implemented to handle them, the use of them may cause the equipment to fail the tests.

Artificial specification diversions. The above problem causes artificial diversions in specifications. For example, one network accepts an optional IE as part of a SETUP message. Another network test suite does not handle the testing of this optional IE. Is it because it is forbidden or not handled? The implementor does not know—but is required to remove the optional IE to pass the test. In this way, a difference between two networks has been created.

Artificial differences may also arise in the other direction. TE is designed to work with one network. An optional timer is not implemented and is not needed for passage of certification. The equipment is made ready for use on another network. On this equipment, the optional timer is required. However, the implementor does not know whether having the optional timer for the first network will cause tests to fail. What does the implementor do? The implementor makes the timer mandatory for the second network and forbidden for the first. In this manner, it is known that the software will continue to pass the tests for both networks—but it is quite possible that the timer could be present for both.

Finally, it also true that mistakes happen in all types of implementations. The network implementation may vary from the specification. The test suite may vary from the network implementation or from the specification. New releases of the network (or the specification) may eliminate the discrepancy, but the equipment implementor has to work with the current tests. Sometimes, especially if other equipment has already gone through testing, test suite mistakes are not politically possible to change. In any event, the testing is the judge of the equipment, and the software must perform as required. And this may also create de facto diversions of network specifications.

National or network requirements

Many variations between specifications are created by the specific needs of the network. For example, some networks require a special coding of the TERM_CAPAB IE as part of the SETUP message. The presence, or absence, of such an IE is a difference. The precise coding may also vary.

Use of address fields vary between ISDNs. One network may require addresses to be contained within a Called Party Number IE. Another may require use of the Keypad IE. If subaddresses are supported, the header information may change as may the format of the address itself. The support of subaddresses, in itself, is a variation.

Finally, certain message types and IEs may be required only in a particular ISDN. The NATIONAL_FACILITY IE is one such item. Others may be optional—and vary in form from other ISDN implementations.

Network capabilities

This category of differences is really related to the physical require-ments of the network. For example, Q.931 gives a suggested default packet size (N201 at the data link layer). It also gives suggested win-dow sizes. Both the window size and packet size make a difference in the resource needs of the network. A larger packet size requires more buffer space in the network. A larger window size requires additional data queues to be apportioned.

A network may or may not also implement point-to-multipoint sup-port. If such is not supported, there is no need to keep tables of SPIDs, USIDs, and TIDs. Thus, there are additional resources needed to support the service. Some networks will not support certain IEs (such as the LLC) because they cause messages to increase signifi-cantly in size. Many network requirement differences fall into the cat-egory of resource requirements.

Another area of differences is in the allocation of bearer channels. All switches support the provision for "any channel" requests. However, some require specification of a D-channel link if data link support of X.25 PLP (SAPI 16) is needed. Others will allow specifica-tion of particular exclusive B-channels. This area of differences is a matter of resource control rather than resource allocation. The algo-rithms needed by the network are significantly more complex if either the TE or the network can allocate bearer services (though, in the end, the network has final control).

Finally, there is the area of physical interfaces. Some networks re-quire layer 1 specification in the SETUP message (Bearer Capability IE) to allocate the appropriate physical connections. Others leave this as a transparent attribute of the channel.

Pre-ITU-T ISDN implementations

Some countries and manufacturers were eagerly waiting for the ITU-T to release sufficient documentation to implement ISDNs. In order to make use of the digital networks being built, some preliminary guesses were made as to the final composition of the ITU-T recom-mendations. Since the recommendations are created from a consensus of requirements, these preliminary implementations were of great in-terest and general use. However, since they did predate the ITU-T recommendations, they do not always follow the adopted standards. Two such implementations are Germany's 1TR6 and AT&T's DMI.

Germany's 1TR6, in particular, has been widely implemented with-in Germany. Currently, both 1TR6 and the ETSI-specified Euro-ISDN are supported with a likely migration of equipment eventually to the ETSI specification. The 1TR6 specification follows the same type of structure as Q.931 but varies significantly in details. For example, the protocol header byte differs from that of Q.931. The SETUP message does not contain a Bearer Capability IE. Instead, it makes use of a Service INdicator (SIN) IE. Many other protocol elements diverge.

AT&T's DMI protocol was an effort to combine various bearer services into specific "modes" of operations. Two signalling techniques were provided. One technique provided a functional protocol similar to that of Q.931. Another technique provided stimulus signalling protocols similar to that found on existing analog networks. Four different transport modes were provided—providing capabilities similar to that of the bearer capability within Q.931. These transport modes supported 64-kbps *clear channel,* 56-kbps (still needed in many cases of interworking) rate adaption for synchronous/asynchronous terminals, and multiplexed virtual channels.

Germany's 1TR6 was closer to the final Q.931 standard. AT&T's DMI provided methods of interworking and maintained usefulness of past standards. Both are still in existence and in use, and both provided valuable input to the standards process.

Signalling requirements

The ITU-T documents specify how a signalling channel is to be used in call manipulations. They do not specifically state how many such are to be used for a UNI. So, some ISDN implementations have interpreted the standards to mean that a separate signalling link should be used for each bearer channel to be supported. This means that two logical links must be established on the D-channel, and two TEIs must be obtained. The ITU-T recommendations do not explicitly deny this option, nor do they require it.

In a related fashion, the ITU-T recommendations state that 1 to 3 bytes may be used for the CRV (2 for PRI). Therefore, a network may require 1-byte CRVs, or it may require 2-byte CRVs—or it may allow either.

Line activation and TEI negotiation

The ITU-T Recommendation I.430 describes the procedures for line activation for a BRI. However, it does *not* define just when activation and deactivation should occur. Thus, it is left to the designers of the network to determine this requirement.

In Europe, most networks require that the line be activated only when a call is in progress. This allows some savings in power and other network resources. A piece of user equipment will activate the physical layer when an outgoing call is placed. The network will activate the physical layer when an incoming call is ready for delivery. Normally, deactivation will be handled by the network.

In the United States, and many other locations, the state of the line is not relevant except for the fact that it must be activated before a call can be sent. Thus, user equipment will usually activate the line after it has been powered up. This simplifies the call setup procedure. The network will normally keep the line active even when there are no calls present.

This difference affects other aspects of the software. Line activation (physical layer establishment) is at the lowest level of the OSI protocol stack. Line activation must be achieved before frames can be transmitted. Frames must be able to be transmitted before a TEI can be obtained (in the case of automatic TEI negotiation). Therefore, line activation and TEI management are fully interlinked.

Generally, any time the line is deactivated (or any other "persistent deactivation" situation), the user equipment will release its TEI (or both TEIs, in the cases of those networks where there are two signalling links in use). Thus, each time the line is reactivated, a (potentially different) TEI must be obtained. In the point-to-multipoint access situation, a new exchange of SPID and USID/TID information is required.

The designer of an ISDN decides how to allocate resources. These requirements are based on general network needs, tariffing situations, and general consumer laws. Reestablishment of a link carries a significant amount of overhead. Maintenance of a link also has overhead (possible power requirements, maintenance RR messages at the data link layer, and so forth). Currently, the requirements fall into the above two categories. However, ongoing power-sensing requirements may split the situation into three categories.

Bearer services supported

The first area of differences to be mentioned is that of support of D-channel X.25 PLP. Some switches offer this service as part of the network. Other networks offer a switched version to an X.25 router. Some require the service to be nailed-up and always available. Others require a SETUP message (indicating the D-channel exclusive). All these variations are an aspect of varying bearer services.

Another area of difference concerns support of voice services. Chapter 6 discusses some aspects of voice support. However, different physical layer decoding schemes are used for voice digitization, and a network will use the system adopted for the rest of the network.

Other services are optionally supported. Europe tends to use ITU-T Recommendation V.110 for rate adaption. North America uses ITU-T Recommendation V.120 in general (some interworking situations use V.110 as a method of convenience). Some networks support "on-demand" X.25 services. Others don't. Much of the variation of support depends on how closely the network requires the full description of the bearer service within the Bearer Capability IE. If a full description is required, the network has closer control over the services available.

Global CRV support

In our discussion about supplementary services, mention was made about the use of dummy, or global, CRVs. These types of messages are used for messages not directly oriented toward a particular call. Supplementary services are not the only area which may make use of global CRVs. For example, INFO messages may be used to transmit general information about the state of the network (such as a display message indicating general congestion).

Some networks support global CRVs, and others do not. Support of global CRVs depends on the data structure architecture of the network. Use of global CRVs implies a structure associated with the UNI in addition to the call accounting structures that are associated with each active call or CEI (if on hold). Some networks will support only a particular set of message with a global CRV (such as INFO, RESTART, RESTART_ACK, and STATUS).

Q.931 states supported

If overlap sending is not supported by the network, state U2 (overlap sending) is not required of the user equipment. Many networks do not support overlap sending or receiving. En-bloc requests are more efficient in switch resource usage—requiring the user equipment to accumulate sufficient information before sending the SETUP message. A lack of overlap receiving indicates that there is no need for state U25 (overlap receiving).

A more complicated case involves the states U6 (call present) and U9 (incoming call proceeding). These states allow the user equipment to wait for upper-layer acknowledgment before making internal state transitions. This is an aspect of user equipment design as well as network design. Therefore, most networks will support user equipment supporting U6 and U9, but much equipment will not allow upper-layer intervention. This effectively short circuits the U6 and U9 states and progresses directly to state U7 (call received).

A similar situation exists with state U12 (disconnect indication). This may not be implemented in some networks; however, the situation is largely transparent to the network because it only affects internal transitions. The implementation of U12 allows user equipment, in theory, to ignore or delay a disconnect of the line. This may be useful for last minute data transfer across the bearer channel but, generally, not useful in a standalone device.

The last two states of uncertain implementation are U15 (suspend request) and U17 (resume request). Some networks do not allow suspension of calls. Some consider it to be a supplementary service. Others allow the process for voice circuits only. A SUSPEND request is very similar to a supplementary service HOLD request. Both take up network resources while the call is suspended. However, the HOLD request does differ from the SUSPEND request in that another call can take place while the original one is on hold. That is, with HOLD services, there might be two calls on hold and two calls active, taking up four connection slots on the network. User equipment that has suspended a call cannot normally initiate another call on the same channel because the channel is still marked as not available by the CEI.

Q.931 coding differences

There are two categories of differences in Q.931 coding between various network specifications. These can be called internal protocol and peer interaction differences. Internal protocol is associated with error conditions, how to treat anomalies, and expected syntactic variations. Peer interaction differences are in relation to the messages sent from one part of the system to another.

What happens if a Q.931 message has repeated instances of an IE (without a valid REPEAT INDICATOR)? Well, the recommendation is not fully clear on the matter. It is possible to ignore repeated instances. It is also possible to retain only the last received duplicated IE. Another possibility is that such a violation causes the entire message to be discarded. All such options have been implemented within various networks.

What about within the message or IE? IEs should be in ascending order within a codeset. Once again, what should be done with the IE, or message, if one appears out of order? The same three possibilities stated above may take place. What if an IE is longer than the user equipment supports (for example, user-to-user information)? Should it be truncated or discarded? Truncated information may be misleading or dangerous. Throwing away the information may cause a situation from which it is difficult to recover (because it is difficult to notify the peer that this is the precise reason without tearing down the call).

There is also the category of variations of supported headers within IEs. Is an "unknown" address type sufficient? Does the network (or user equipment) require specific types of information within the IE? If an IE is optional, should an error be treated as a "who cares"?

The peer interaction differences are directly related to internal protocol use—except that they are within the creation of peer messages.

We have already noted some differences. Is a TERM_CAPAB IE needed or desired? Should the CDN IE be used or the Keypad IE? Many IEs have a default value—does a null length IE constitute a valid default, or should the default be coded explicitly?

There are also differences that are associated with the system, rather than just the network layer. Should an ALERT be automatically sent in response to a SETUP (associated with whether U6 and U9 are implemented)? If overlap receiving is not implemented, should a CALL_PROC ever be sent by the user equipment?

Timers supported and timer values

Certain timers, as we have seen, are considered to be mandatory by the ITU-T recommendations. However, in the specifications, such timers may still be considered optional—or optional ones may be mandated. Timeout values are suggestions and may be altered to fit the needs of the network. Timers T305 and T318 seem to be particularly likely to be changed.

Cause values

Cause values give information about the reason why something has happened. They may be associated with DISCONNECT, RELEASE, PROGRESS, RELEASE_COMPLETE, RESUME_REJECT, SUSPEND_REJECT, STATUS, and INFORMATION message types. Certain cause values are expected from the user equipment, and certain cause values are expected to be sent from the network. Some test suites enforce these values and verify that only proper values are sent or received.

In one respect, what is the importance of the validity of the cause value? An inappropriate cause value negates the value of the IE. Thus, it is important that user equipment send a cause value that will be of use to the network, or peer equipment, to adjust to the circumstances. What if the network sends a bad cause value? This is something over which the user equipment has no control. However, some networks require specific parsing of any cause values received from, or through, the network.

Reactions required by the various specifications once again fall into the three categories of rejection, ignoring of the IE, and dismissal of the message as a whole. An implementor must allow for any, or all, of these possible needed responses.

Error condition responses

Cause value discrepancies are one category of error conditions. Others fall into the area of duplicated IEs, erroneous IE contents, unexpected IEs, out-of-order IEs, mandatory IEs that are not present, message types not valid for the current state, and so forth.

In many instances, the ITU-T Recommendation Q.931 suggests a particular action. In other areas, it is much less explicit. These open areas may be, and are, interpreted in varying ways. As discussed above in several places, general response categories include ignoring of the invalid segment, dismissal of the entire event, and a specific reaction to the event. If a specific action is taken, it is often a RELEASE_COMPLETE message with a particular cause value.

Optional message support

Optional messages fall into two major categories. These categories are optional service support and optional facilities support. Optional service support includes whether SUSPEND and RESUME are considered supplementary services or basic. It also includes whether LLCs are passed through the network or USER_TO_USER information is allowed.

Optional facilities support is concerned with categories such as point-to-multipoint support, overlap sending and receiving, and optional network and user states. Both categories may depend strongly on the effect on network resources. They are different in that one category affects the abilities of the application, and the other affects only internal protocol procedures.

Supplementary services

Supplementary services, by their very nature, will differ between specifications. ITU-T Recommendation Q.932 (and associated Q.95x Recommendations) specifies a number of common services that *may* be implemented as part of a supplementary services offering. None are mandatory, and each can be implemented in a variety of ways— stimulus versus functional, specific functional IEs, and so forth.

Many test centers do not certify supplementary services. It is difficult to create a test suite that adequately tests the exact mixture of services that may be provided. It also depends more heavily on the local network access point. Such services may not be sufficiently widespread, or consistent, to make a standard test suite possible.

Ongoing ISDN Evolution and Ease of Use

ISDN standards continue to evolve. Areas of evolution include increasing bandwidth, interworking, bearer service identification, and ease of use. Bandwidth issues are being addressed by the ITU-T in its expanding series on B-ISDN and Frame Relay. Interworking is being

addressed in an evolving series of I.5xx documents. Bearer service issues (covered more fully in the next chapter) are being addressed primarily from the user equipment direction. Ease of use is an area that
is directly involved with implementation and design of ISDN user
equipment—in interaction with the network.

A user with ISDN equipment currently needs to know a lot about
the equipment and network before actually making use of the equipment. First, the user needs to know what network specification is
being used with the equipment. In some areas of the world, this is not
a problem because there is only one possibility. In North America,
however, there are four possible specifications that may be involved—
and the specification used depends directly on the decisions of the
local network access provider. Users may not be given a choice—and,
in contact with service personnel, they may not find it easy to even
obtain the information about the network.

Next, users need to know physical configuration information. If
they are configured for a point-to-multipoint access point, they need a
SPID (for some systems, they need a SPID on a point-to-point access),
and the equipment must be configured to do the initial SPID/USID-
TID negotiation.

The next step (even harder) is to find out if the equipment is fully
compatible with the network specification. What options are turned
on by the network? Will LLCs be passed through the system? What
bearer capabilities are permitted? How does the equipment interact
with the network?

Finally, there is subscription information to be programmed into
the equipment. What is the local number? Is there a subaddress? If
so, what is it and how is to be coded? Are TEIs fixed or to be negotiated at call setup time? What is the assigned SPID (if needed)? What
services are supported with the SPID? Are supplementary services
available for this SPID? If so, which ones?

Let's examine all of these issues and questions. As an ISDN implementor, you may know the importance of each of these questions and
how the answers interact with the network. However, these are all
questions that have to be answered *by the user* in order to have the
equipment work properly with the network. This is completely unac-
ceptable if ISDN is to gain wide acceptance and use. Plus, it is totally
redundant information.

The network knows every answer that the user must program into
the equipment. It knows the called party number, or numbers, for
that particular UNI. It knows whether point to multipoint is supported for the UNI or not. It certainly knows the network specification
(and even release number of the network software) and what services
it will provide.

With the possible exception of the SPID (in the case of multiple SPIDs supported on an access point), the network knows exactly what information is to be associated with the equipment. In this way, it is the same as current analog equipment. You plug the equipment into the access interface and use the equipment. Everyone *else* needs to know your number to call you—but you don't tell the phone its calling number (except possibly for outgoing data services). If your phone is meant to work with analog special supplementary services, it will work (if the features are subscribed to) or not.

So, if the network already knows all of this information, why does the user need to program the equipment? Primarily because there hasn't been a standard way to tell the network to give the user equipment the information. The network knows the information, and the user equipment needs to know the information. So, the user has been used as a method of conveying information from the network to the equipment. Seems a bit inefficient doesn't it?

Standardization efforts are now underway to eliminate this inefficient and confusing user requirement. One method is by use of the REGISTER message type (defined in Q.932). The REGISTER message type can potentially retrieve all the information from the network automatically. The one piece of information that it needs is the SPID (remember that we discussed the fact that network does not know this fact about the terminal). The REGISTER message is responded to with a FACILITY message (or multiple segmented FACILITY messages). The network information is taken from the FACILITY messages. Recent North American specifications are supporting this message. Global standardization of this kind of feature would help tremendously in ISDN deployment.

Summary of Chapter

The network layer, or layer 3, of the OSI model provides a method of routing information to various endpoints. ITU-T Recommendation Q.931 is the primary reference for out-of-band signalling protocols for ISDN. It is possible to also have in-band signalling protocols within an ISDN. The call control portion of ITU-T Recommendation X.25 was discussed as an example of this type of call management software.

Layer 3 is the top layer of the chained layers, and as such, it must communicate with the application layers. Variations in application and network needs produce a generic type of interface but with varying parameters and requirements. Supplementary services may add to the capabilities of the network and user equipment. Such services can be designed in a stimulus (non-state related events) or functional manner.

The network layer also is the area, within the chained layers, which has the most variation between specifications. Many of these differences are due to resource requirements. Others are due to the needs of the local tariffing situation or regulations of the industry.

The signalling protocol is useless without some type of bearer service. The next chapter discusses how bearer services are set up, used, and implemented.

6

Bearer Services

What do you have when you have a network without bearer services? Not very much. You may get some lights blinking on and off. There are even possibilities (with ISDN or some of the central office analog services) of getting the caller ID of the equipment that is calling you or some other display information or user-to-user information. The signalling protocol of ISDN does allow transfer of data over the signalling channel while in the process of making or ending a call and during it. For most purposes, however, this is not enough. The main purpose of the signalling protocol is the establishment of a bearer channel over which bearer services may be provided.

Bearer services are the base of three types of services defined by the ITU-T. They are called bearer, tele-, and supplementary services. The bearer service is the underlying mechanism used to provide data transport. Data can be defined as any information. Thus, speech, unrestricted data, video, music, and other data types yet to be defined are all valid data types to be supported by bearer services. One requirement of the bearer service is that the data be transported in real time and without alteration of the data.

Teleservices make use of bearer services in combination with upper-layer protocols to allow application access to user features. It is possible to look at bearer services as providing the equivalent of layers 1 through 3 of the OSI protocol stack and the teleservices as combining this with the higher layers 4 through 7. Speech is a bearer service. Telephony is a teleservice. The bearer service provides a mechanism for transporting data that contains a type of data. The teleservice translates that information into a form that a user can use. A teleservice situation may be in the form of a keypad and a transceiver—or it may be an integrated software package on a computer with remote device access (remote energy monitoring, for example).

Supplementary services have already been discussed in the previous chapter. Supplementary services *add* to the capabilities provided by the bearer service or teleservice. They may be considered as a supplement to the lower layers or the higher layers or both. They cannot exist without the supporting bearer services or teleservices.

ITU-T Recommendations I.112 and I.140 define the terminology and attributes associated with services. Attributes are broken into two main classifications. These are service attributes and network attributes. A service attribute is associated with how a connection is to be used for purposes of communication. Communication implies a protocol, or format, that both ends use in agreement. The network attribute is associated with the form of the connection. What is the physical format, speed, symmetry, method of switching (if any) that is needed for the network to provide the communication capabilities desired?

Bearer Service Types

ITU-T Recommendation I.230 gives the general overview of bearer services supported by an ISDN. These bearer services are broken into two categories: circuit mode and packet mode. ITU-T Recommendation I.231 goes into greater detail on the circuit-mode bearer services and ITU-T Recommendation X.31 (I.462) covers the VC and PVC bearer service category for packet mode. Two other packet-mode bearer services, connectionless and user signalling, are marked as being "for further study."

Of the eight circuit-mode bearer services defined, three are considered "essential." These three are 64-kbps unrestricted data, speech, and 3.1-kHz audio. The other five services are considered to be "additional" and may be provided by ISDN on a local basis.

ITU-T Recommendation I.231 goes into much greater detail concerning the attributes and services associated with each of the circuit-mode bearer services. The attributes are split into three categories. These are information transfer, access, and general attributes.

There are seven information transfer attributes. The first three are called the information transfer mode, information transfer rate, and information transfer capability. These correspond directly to the same fields of bytes 3 and 4 of the Bearer Capability IE. Byte 5 may also be specified, directly or indirectly, through the various attributes. The remaining four information transfer attributes are structure, establishment of communication, symmetry, and communication configuration. These attributes are concerned with what the user of a bearer service can expect from the network but are identified by use of the contents of bytes 3 and 4 (and, in some cases, 5) of the Bearer Capability IE.

Access attributes include the access channel and access protocol. These refer to which channel (D or B, normally) is to be used for what purpose and how the protocol is to be structured. The final category of general attributes includes the supplementary services provided, quality of service (QOS), interworking possibilities, and operational and commercial aspects. ITU-T Recommendation I.250 is sometimes referred to by each listing for the supplementary services, but the other attributes are left as "for further study."

Essential circuit-mode bearer services

The following will cover the three essential bearer services. Essential implies that this should be provided by all ISDN providers on an international basis.

64 kbps unrestricted, 8 kHz structured. This bearer service is circuit mode, 64 kbps, and unrestricted. This corresponds to byte 3 of the Bearer Capability IE as 10001000 (or hexadecimal 88) and byte 4 as 10010000 (or hex 90). The remaining bytes of the Bearer Capability IE are left for precise protocol definition, if desired. The term *8 kHz structured* indicates that the transmission is accompanied by an 8-kHz timing that allows each endpoint to feel assurance that data will always arrive in 8-bit octets (bytes). This is the bearer service most often associated with circuit-mode terminal adaption, facsimile, circuit-switched X.25, and other data transfer modes.

Sometimes, the user equipment will only parse the Bearer Capability IE as far as byte 4 for this service. The other information associated with the bearer service is then assumed (if only one data service is supported) or negotiated by use of some in-band protocol scheme.

64 kbps unrestricted, 8 kHz structured usable for speech. This bearer service is circuit mode, 64 kbps, and speech (G.711 A-law or μ-law). This corresponds to byte 3 of the Bearer Capability IE as 10000000 (or hexadecimal 80) and byte 4 as 10010000 (or hex 90). Byte 5 is defined as 10100010 (or hexadecimal A2) for G.711 μ-law (used in North America, Japan, and some other parts of the world). It may also be defined as 10100011 (or hexadecimal A3) for G.711 A-law (used in Europe, Australia, and some other parts of the world). The remaining possible bytes of the Bearer Capability IE are left unused—so, for this service, the Bearer Capability IE will have a length of 5.

Byte 5 is important for cases where the connection is made between service areas. For example, a voice connection from the United States to France would require a conversion of the digitization of the voice signal. Knowing that the connection is to be used for speech allows use of internetwork mechanisms such as intermediate analog transmission networks, echo cancellation, and so forth. This conversion means that the data are *not* guaranteed to be exactly the same as originally produced but *are* guaranteed to be within the quality range expected of voice service. This allows the network more options.

64 kbps unrestricted, 8 kHz structured usable for 3.1-kHz audio.   This bearer service is circuit mode, 64 kbps, and 3.1-kHz audio. This corresponds to byte 3 of the Bearer Capability IE as 10010000 (or hexadecimal 90) and byte 4 as 10010000 (or hex 90). Byte 5 is required when the connection may pass national boundaries. Analog intermediary circuits may be used if necessary, but other speech-appropriate methods which may cause degradation of the signal are not permitted.

Additional circuit-mode bearer services

Five other circuit-mode bearer services are defined within ITU-T Recommendation I.231. Four of these are combinations of B-channels, and one is used for alternating speech and data capacity over the same bearer channel. These are all considered to be services that may be appropriate for local ISDNs but are not considered essential for international service.

64 kbps alternate speech/unrestricted data, 8 kHz structured.   This bearer service is circuit mode, 64 kbps, and unrestricted. This corresponds to byte 3 of the Bearer Capability IE as 10001000 (or hexadecimal 88) and byte 4 as 10010000 (or hex 90). Byte 5 should also express the appropriate G.711 voice encoding. This combination of "data" (byte 3) with byte 5 (G.711) is sufficient to allow notification to the network of the use of this bearer service.

Practically speaking, some method must be used to delimit unrestricted data from speech. This is not yet defined. This type of bearer service might be particularly useful for group situations where short comments may be desired but a dedicated B-channel for speech is more than needed.

2×64 kbps unrestricted, 8 kHz structured.   This bearer service is circuit mode, 2×64 kbps, and unrestricted. This corresponds to byte 3 of the Bearer Capability IE as 10001000 (or hexadecimal 88) and byte 4 as 10010001 (or hex 91). The remaining bytes of the Bearer Capability IE are left for precise protocol definition, if desired. It may be noted, at this point, that the ITU-T Recommendation Q.931 gives four specific multiple bearer channel rates. All other combinations must be done with byte 4 as 10011000 and the use of byte 4.1 as a multiplier.

Thus, the 2×64 kbps is distinct from a multirate information transfer rate with a multiplier of 2. The former may be two associated B-channels controlled by the same signalling CRV. The exact nature of the association is unknown. The latter would indicate that the two bearer channels are bonded in some manner for the same service.

384 kbps unrestricted, 8 kHz structured.   This bearer service is circuit mode, 384 kbps, and unrestricted. This corresponds to byte 3 of the

Bearer Capability IE as 10001000 (or hexadecimal 88) and byte 4 as 10010011 (or hex 93). The remaining bytes of the Bearer Capability IE are left for precise protocol definition, if desired. This bearer service, and the following two, vary only in the amount of bandwidth allocated to be associated with the call.

1536 kbps unrestricted, 8 kHz structured. This bearer service is circuit mode, 1536 kbps, and unrestricted. This corresponds to byte 3 of the Bearer Capability IE as 10001000 (or hexadecimal 88) and byte 4 as 10010101 (or hex 95). The remaining bytes of the Bearer Capability IE are left for precise protocol definition, if desired.

1920 kbps unrestricted, 8 kHz structured. This bearer service is circuit mode, 1920 kbps, and unrestricted. This corresponds to byte 3 of the Bearer Capability IE as 10001000 (or hexadecimal 88) and byte 4 as 10010111 (or hex 97). The remaining bytes of the Bearer Capability IE are left for precise protocol definition, if desired.

Bearer Service Negotiation

Negotiation is mainly used with unrestricted data bearer services. Negotiation implies that both sides can offer, or accept, alternatives. The bearer service can be negotiated within the signalling channel (out-of-band) in two primary ways. The first is via the repeat indicator associated with the Bearer Capability IE (or the LLC IE) within the SETUP message. This requires some mechanism for the endpoint to return the accepted bearer service. This would normally be within a LLC IE in a final CONN message. The other possibility is for the other end to revise the bearer service and return it within the LLC IE in the CONN message. If the altered service is not acceptable to the originator, a DISC message would be sent to tear down the connection. This second method assumes that the network is satisfied with the original Bearer Capability IE and is able to pass the LLC IE (though this is true for the first also).

It is also possible to use the Bearer Capability IE in its minimal fashion. This means that only bytes 3 and 4 are used. All other information is determined by some (agreed upon one hopes) in-band protocol after the connection has been made.

Types I, II, and III information revisited

As mentioned in the previous chapter, Annex I of ITU-T Recommendation Q.931 discusses the coding of low-layer information. This information may be contained in the Bearer Capability IE and the LLC IE. Type I information is coded in bytes 5 through 7 and is intended to be of use only by the terminals at each end of the connection. However, because of interworking needs, the network may sometimes alter, or add, physical layer information (bytes 5 and 5a).

Type II information is coded in bytes 3 and 4 (plus 4a if appropriate) for circuit-mode connections. It may be coded in bytes 3, 4 (and 4a if appropriate), 6, and 7 for packet-mode connections. Type II information is used by the network to ensure appropriate connections and available network facilities. The information given in the above discussion on the circuit-mode bearer services is Type II information.

Type III information is information that may be used by the network to provide additional services. This information is included in byte 5 of the low-layer information. It is also used for interworking information (such as connections that are not end-to-end ISDN).

Examples of bearer service information

The previous main section detailed bearer service information associated with the main types of bearer service. For speech and audio types, little variation is possible. For unrestricted data applications, several major variations are possible with many minor variations within each type. Annex I of ITU-T Recommendation Q.931 contains examples of the Type II and Type III information needed for various bearer service uses.

Two examples contained in this annex are of particular note. The first is concerned with packet-mode use of the low-layer information bytes. The second is for interworking with a Public Switched Telephone Network (PSTN) using 3.1-kHz audio.

In the packet-mode example, it is noted that a series of information is needed (if the packet switch is part of the ISDN, as opposed to being an X.25 router at the end of the main connection). The transfer mode is packet, capable of unrestricted information and an undefined transfer rate. This translates to a value for byte 3 of 10001000 (hex 88) and for byte 4, a value of 11000000 (hex C0). X.31 rate adaption is required and (if rate adaption is being actively used) so is the user rate. This translates into bytes 5 and 5a. Assume (for simplicity in the example) that no modem rate adaption is needed. This would give byte 5 a value of 10101001 (hex A9). Bytes 6 and 7 are used to convey the information that layer 2 is using LAPB and layer 3 is X.25 PLP. This gives us values of 11000110 (hex C6) and 11100110 (hex E6) for bytes 6 and 7. Note that if nonstandard modulo operations or nonstandard packet sizes are used, bytes 7a and 7b would be used. Use of additional bytes means that the extension bit of byte 7 would be set to 0 to indicate additional bytes in the field.

The interworking of 3.1-kHz audio data is very similar to the previous section's description. The additional information necessary is that of bytes 5 and 5a—indicating data rate and user rate. Imagine a modem on a traditional analog network. It is operating at a particular baud rate. If it connects into an ISDN, the connection must know the general data rate of the connection (probably 56 or 64 kbps). If the ISDN gives data back to the analog network, it must know the appropriate user rate (and signalling code).

Voice

Voice (or speech) is still the primary use of most connections across the world. Thus, it is extremely important to support voice in any integrated network. The early chapters discussed the differences between analog data and digital. As part of an ISDN, voice needs to be conveyed as a type of digital signal also.

Another interesting side effect of the need to carry voice over digital transmission connections is that of the 64-kbps bearer channel size. Why is the standard bearer channel at this particular rate? It turns out that this is a form of historical accident based on the need to transmit voice.

Digital voice encoding

The analog voice signal is normally limited to 3.4 kHz in bandwidth. In order to convert this into a reconvertible digital form, it is necessary to sample a signal at twice the bandwidth (Nyquest's sampling theorem). Thus, rounding upward to make sure of clarity, a voice signal is sampled 8000 times per second. Each of these samplings is then given a quantifier to express the signal value. It was found that adequate speech encoding could be done with 12 bits—giving $\pm$ 2048 levels.

No matter how many bits are used (although a limit would be reached that the ear could not discern), some distortion is found in the analog-to-digital conversion. Twelve bits are sufficient for most purposes. However, it turns out that distortion is much more important at low volumes than at high volumes. (Consider the effect of shouting in a noisy room as opposed to whispering.) So, low volumes need more information than high volumes. This implies a logarithmic scale for encoding.

ITU-T Recommendation G.711 details two methods of Pulse Code Modulation (PCM) of voice signals. These are known as the μ-law and the A-law and are named after parts of the mathematical formulas used in calculating appropriate coding values. (The mathematical details will not be gone into in this book.) Each coding method makes use of 8 bits per sample. The exact decoding is different for each law but carries about the same information. (In practice, table lookups of values are normally used rather than the mathematical formulas.)

So, we have 8 bits of information from samples taken 8000 times per second. This gives a bandwidth of 64 kbps. Voice signals actually do not vary that much from one sampling to another. This allows other methods of encoding—based primarily on deltas, or changes, in the signal value rather than the absolute value. This would allow use of fewer bits needed per sampling, and this is basically what ITU-T Recommendation G.721 does for Adaptive Differential Pulse Code Modulation (ADPCM). This recommendation allows voice signals to be encoded in 32 kbps.

Taken from the opposite direction, given a bandwidth of 64 kbps, an ADPCM encoding method can be used to encode a higher bandwidth of analog signal. ITU-T Recommendation G.722 does this to allow encoding of 7-kHz audio over a 64-kbps digital transmission circuit. This allows digital encoding of a better "quality" of sound. This is not really good enough for CD-quality music but still is an improvement for applications that desire high-quality speech.

Mixing voice and data

One of the bearer services defined for circuit-mode ISDN is that of mixing voice and unrestricted data on the same circuit. Use of G.721 can allow the "stealing" of 1 or 2 bits per sample—allowing an 8- or 16-kbps unrestricted data stream along with the voice. This is one method of mixing voice and data. Other methods involve embedded control bytes in the data stream to allow mode switching between voice and data. Most of these methods must be incorporated into the hardware support for the circuit and are, thus, only of application interest to an ISDN software systems designer.

Facsimile

Facsimile, or fax or telefax, services have become an expected data service of most offices and many homes. This service allows the possibility of transmitting copies of a written form of data. Just as voice continues to be an ongoing part of any evolution of networks—so do the written and visual media. One could, perhaps, say that evolution of data networks is a continuing progression of being able to extend the human senses. At any rate, networks must support the standard extensions that are currently in place.

Facsimile services are one form of data that can be called formatted data. Formatted data have a particular method of encoding so that the data can be reexpressed into the original form. Facsimile services are one such service; audio, video and teletex services are others. As such, we are able to break the service into two categories once again. The chained layers provide connectivity and data transport. The higher layers provide interpretation and application access.

When an image is examined (or produced) line by line to store the image, it is considered to be rasterized. This is used widely in many different services such as television, general video, copying, and facsimile services. The equipment scans over a point—usually in a horizontal pattern (from one side to the next, then progressing back to the beginning of the next line). Similar to the way that voice signals are sampled during digitization, each position is sampled as the pattern progresses. This leaves a bit-mapped representation of the image that has been scanned. Different amounts of data can be obtained from each sample. If color is needed, the various hue and value components of the color may be saved. In the case of a graduated image, it may save multiple values for each scanned point. This is called a *gray scale*. It is also possible to just save two possible values for each point—a binary system of yes or no, in this case called black or white. If the resolution (number of samples per unit dimension scanned) is high enough, a binary sampling system is sufficient to provide equivalent effects to gray-scale coding systems.

Presently, using digital technologies, it is possible for such an image to be created, and stored, in different ways. If a screen image from one computer is transmitted to another computer without an intermediary paper copy, is this a facsimile service or a video service? To a certain extent, this is a philosophical concern. To an implementor, however, the service and mechanism are based upon the protocols and not upon the actual form of use. If a facsimile standard is used in the creation, transmission, and recreation of the image, it is a facsimile service.

Facsimile standards

The first patented version of facsimile services was created in 1843 by Alexander Bain. His method used a master-slave synchronized version using pendulums. By keeping the transmitting and receiving pendulums in synchronization, each sample could be read and transcribed in the same place. Think of a situation where two people are talking on a phone. You tell each person to start at the (for example) upper left corner of the sheet. Both -start to move their arms with a stylus across a piece of paper in fairly good synchronization such that each person reaches the edge of the paper at the same time. The "transmitter" yells "now" each time that the stylus passes over a piece of dark point. The receiver makes a mark at each corresponding point. The image is thus transmitted by use of common synchronization. The earliest patented methods used a mechanical version of this. Later developments incorporated an *internal synchronization* method such that, in effect, "edge of page" became part of the data transmitted.

Facsimile services, until the 1960s, were used primarily as a specialized private data service offered by various companies throughout

TABLE 6.1   ITSS Facsimile Standards

Characteristic	Group 1	Group 2	Group 3	Group 4
Recommendation	1968 T.2	1976 T.3	1980 T.4	1984 T.5
Signal form	Analog	Analog	Digital	Digital
Sample type	gray scale	gray scale	pel	pel
Horizontal resolution	Analog	Analog	7.7 lines/mm	200 pels/in (low resolution, class 1) 300 pels/in (high resolution, classes 2 & 3)
Vertical resolution	3.85 lines/mm	3.85 lines/mm	7.7 lines/mm	200 pels/in (low resolution, class 1) 300 pels/in (high resolution, classes 2 & 3)
Transmission time for A4	6 min	3 min	<1 min	<10 s
Supporting network	PSTN	PSTN	PSTN	ISDN
Compression algorithms	N/A	2/3-level spectrum compression	MH default, MR optional	MR
Service use	Facsimile	Facsimile	Facsimile	Class 1—fax Class 2—fax with receive from telefax Class 3—fax and telefax

the world. The use increased, particularly in international situations, such that an international standard was needed. The CCITT (now known as the ITU-T) developed such a standard beginning in 1968 for a standard called *Group 1*. Table 6.1 gives a listing of the various standards that have been developed over the years.

The first standard, Group 1, was developed as a low-speed analog transmission technique. Using a Frequency Modulation (FM) technique, it allowed gray-scale transmission at a resolution of four lines per millimeter on the vertical direction. (The horizontal direction does not have a precise number of points per distance since it uses analog techniques.) A page defined as the ISO A4 size (about 8.25 by 11.7 in, or 210 by 297 mm) takes about 6 min to transmit.

The second standard, Group 2, was published as a recommendation in 1976. This is still an analog standard but uses bandwidth compression techniques to reduce transmission time to about 3 min for an ISO A4 page. It continues to support gray-scale values (without the digital method of delineating horizontal resolution, gray scale is almost mandatory) and is still in use in many areas.

The first digital standard, Group 3, was released as a recommendation in 1980. Digital methods allow a horizontal resolution to be enforced, and thus, black and white values are supported rather than gray scale. Horizontal resolution is about 200 samples per inch. A run-length compression algorithm is used to reduce the number of bits transmitted. For example, if 15 black points in a row are encountered at the scanning end, it would be possible to send 15 bits indicating black. It is also possible to send a code indicating black followed by a multiplier of 15. The first method requires 15 bits. The second method may be used with as few as 5 bits. It should be noted that the digital standards are defined in terms of *pels* (picture elements). A pel has a binary value as opposed to *pixels,* which may have multiple values associated with gray-scale or color information. This first digital standard was defined in expectation that most transmission would occur over analog lines using modems for digital-to-analog conversion. At 4800 baud, a page can be transmitted (on average) in less than 1 min.

Group 4 facsimile use of ISDN

When the CCITT (now ITU-T) started to create the series of recommendations that would form the foundation of ISDN, facsimile services were considered as only one of a family of services needed. In reality, facsimile services are now highly used and at an increasing pace. This may give way to other services in the future as ISDNs evolve and interconnect with additional possible services. However, in 1984 when Group 4 services were produced, facsimile was considered as a part of a group of services. These services included what is known as teletex. Thus, the 1984 recommendation broke Group 4 into three classes. The first class was to be used for facsimile only. The second class allowed reception of teletex with facsimile service, and the third allowed full integration of teletex and facsimile services. This concept did not turn out to be immediately useful but provides an inherent possibility for interworking of the two services.

A Group 4 facsimile system can be broken into three main components. These are the input system, the communication interface, and the output system. The first part of the input system is to gather the raw data—to collect the pels that are part of the input data stream. The next part is to compress the raw data so that the information can be sent efficiently and, with Group 4 standards, in a manner that allows for error correction. The last requirement for the input system works with the encoding mechanisms used for error control. This is a system of buffers used for retransmission, flow control, and allowance for temporary discrepancies in user equipment speed.

The communication interface can be an ISDN. Indeed, at the speeds desired for Group 4 fax, a digital network is needed. However, the bearer service, itself, really amounts to unrestricted digital data transmission—although it is possible to convey additional information about the low-layer needs.

The output system receives the data, performs error checks, potentially requests retransmission of data that is not correctable, and then decompresses the data into a form that can be used. In this *postprocessing* stage, it may be possible for the information to be stored in long-term memory, displayed on a video screen, or printed on a high-resolution printer. In the latter two cases, interpolation of the data may be needed to increase the observed resolution if the receiving medium has a greater resolution that the received data.

Compression algorithms.   It can be noted from the above discussion that we are dealing with a simple original input data stream and a simple final data stream for output. Effectively, it is a series of bits. Error correction and retransmission can be handled by layers 2 and 3 of the OSI model—X.25 or X.75 is often used at these layers for data transport. However, the mass of data from a single page can be immense. At 200 pels/in, a U.S. standard page of 8.5 by 11 in could require 3,740,000 bits of information if it was not compressed—or about 1 min at full 64-kbps bearer channel capacity.

If necessary, this might be acceptable, but as discussed briefly above, this is not required. Two forms of compression are listed as supported for facsimile services. The Modified Huffman (MH) algorithm is the default for Group 3 facsimile services. The Modified Relative Element Address Designate (Modified READ, or MR) algorithm is defined for Group 4 services. The MR algorithm may also be used with Group 3 as an option. Each of these algorithms is considered to be an *information-preserving* technique—one which still contains the exact data that was originally processed. Approximation techniques are also sometimes used—particularly for video services. Approximation techniques may be sufficient for transient images. Another group of techniques is a differential technique. This is useful when series of data (such a screen images for live video) differ only in small amounts from the preceding image. For this discussion, we will concentrate on the MH and MR algorithms.

Modified Huffman compression.   The Huffman algorithms take advantage of something already noted. This is the fact that the data will be composed of series of similar information—a series of black or white pels. So, a series of eight white pels could be encoded by something which translates as an 8W code. Group 4 standards allow up to 1728 pels per line. This would indicate a need for 12 bits per code for all possibilities. There are two problems with this. The main one is that it would indicate a need for a fixed-length code. Twelve bits may be needed for a 1728 pel black line, but only 2 bits are needed for a single white pel. Thus, there would be 10 bits "wasted." The other matter for consideration is that the probability for each length of pels is not the same—particularly for text transmission.

The Huffman encoding techniques allow for varying probabilities for different codes. It also creates codes that are unique between adjacent codes. That is, no code can be a prefix of another code, so two adjacent codes cannot be confused with a different code. This allows variable-length codes to be used within a data stream.

A simple Huffman encoding, therefore, would encode a series of pels as a variable-length code representing the number of repetitions involved. The MH code takes advantage of the run-length requirements by having the codes of W0 through W63 and B0 through B63 represented by MH codewords and then supplementing them with other, longer run-lengths that can be used in combination (plus a special codeword for end of line). This modification cuts down the required codewords from 3557 (1728 pels per scanline times two values plus the end of line) to 184 codewords. An encoded line would be composed of a series of alternating black and white codewords followed by an end of line. By convention, white pel codewords are used first in a line. Thus, if the first pel was black, a codeword for a zero-length white pel would be inserted first.

Modified READ encoding. The MH encoding takes advantage of run-length data grouping. This may be considered a "one-dimensional" form of compression. Each line is compressed to a shorter stream of data. The MR code treats lines in combination with the previous line. Just as there is a probability that a pel will be followed by a pel of the same color, there is a probability associated with the differences between one line and the subsequent line. The MR code takes advantage of this to compress the image as scanned.

The exact details of this encoding will not be discussed further in this text; however, a brief discussion may be useful. Since the MR coding requires a series of delta codes based on the previous line, it is possible for errors to compound over the length of a page. Moreover, there must be an original line from which to produce a series of difference codes. Thus, the MR encoding makes use of MH encoding on a periodic line basis. This reduces the chances of errors compounding and gives periodic baselines for exact line matching. The ITU-T recommends that every other line use MH encoding for low-resolution scanning and every fourth line use it for high-resolution encoding.

X.25 Packet Layer Protocol

Out of the three bearer services declared as essential by ITU-T Recommendation I.230, two are associated with speech and audio signals. The third is that of unrestricted data. What is meant by unrestricted data? In an eventual evolution of ISDN, this may mean a situation where the network, itself, is fully transparent to the requirements of the data. Or it may mean that the out-of-band bearer service negotiations will become sufficiently defined—but expandable—to allow full knowledge of the protocols and purposes of the data channels. In any event, the eventual direction of the ISDN is unknown. What is known is that data must be transmitted reliably and efficiently. ITU-T Recommendation X.25 (or the more internationally oriented version, X.75) is a very useful medium for this at the present.

The data transported within the X.25 protocol can be of any nature. Thus, it can be used for video, facsimile, electronic mail, or any other variant. So, we see two reasons for ITU-T recommendations oriented toward X.25. The first reason is that it provides a sufficient mechanism to perform the lower-layer data transport function. The second is that the ISDNs can be interworked into many existing X.25 networks. Usability and practicality combine into a current reason.

It is also possible that X.25 will slowly fade as a primary vehicle for data transport. As mentioned before, X.25 has considerable redundancy built in between layers 2 and 3. It was designed during a period when data transmission was slow and unreliable. Other protocols may fill this need eventually—Frame Relay and cell-based broadband may come into full use. At this point, however, X.25 will continue to be an important mechanism.

Call setup revisited

In the previous chapter, we discussed the difference between VCs and PVCs. A PVC is an end-to-end data connection that is available at all times that the circuit is active. It may be reinitialized by use of a RESET packet, but the LCI is allocated by the network and used by the user equipment.

VCs, however, are transient in nature. They can be torn down by either side at any time—or they may be torn down implicitly by timer, or data link layer, indications that the channel is no longer available. A CALL REQUEST packet is sent by a DTE to initiate the channel. The LCI is allocated at this point of time.

Whether the channel is a VC or PVC, its purpose is to provide data transport capabilities. Once this data ready state (p4) is reached, the PLP portion of the ITU-T recommendation takes effect. The PLP provides sequencing, windowing, and optional retransmission facilities.

X.25 packet formats

Table 5.6 gives a full listing of the various packets supported by X.25. Of these, all may be used by PLP except for the category of call setup

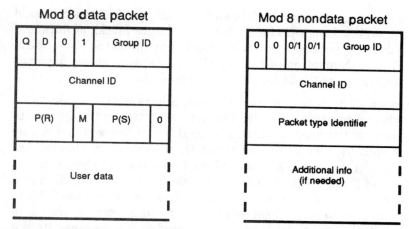

Figure 6.1   X.25 Modulo 8 packet formats. (*From ITU-T Recommendation X.25.*)

and clearing. In the case of PVCs, such packets are the only ones ever used. However, as noted in Chap. 5, the X.25 header is composed of 3 bytes. These are shown in Fig. 6.1 for the two general forms of data and nondata.

Within a data transfer packet, there are three fields that are unique to data transfer. These bit fields are located within the first and third bytes and are known as the Q-bit, D-bit, and M-bit. These may be used in various combinations and are explained in the following subsections.

D-bit usage.   The D-bit, or delivery confirmation bit, is originally negotiated (for VCs) as part of the GFI field (bit 7). If the terminating end is able to support this feature, the call is accepted. If not, the call is rejected. Delivery confirmation is a feature that provides application-to-application (or end-to-end) confirmation of data transmission.

In effect, use of the D-bit prevents data acknowledgment until the data has been passed to the upper layers. At this point, the upper layers may acknowledge the data explicitly, which will cause the network layer to send a layer 3 acknowledgment (and rotate its receive window). D-bit usage will slow data transport significantly, which is why the feature is an optional one for most user equipment. However, the bit of the GFI should be passed transparently by all networks.

M-bit usage.   The more bit (M-bit) is used to indicate that more data of the sequence is possible. The primary use of this is to provide for instances where maximum user data size differs between the originating and terminating ends. Thus, the network can make use of the M-bit to combine packets together from the side which uses smaller packet sizes into such as can be handled by the terminating side. In the case where both sides have identical packet sizes, the network will pass the packets unaltered. In such a case, the M-bit may be used when the higher layers prefer to deal with larger blocks of data and are not directly supportive of a protocol that indicates end-of-data blocks. The resegmenting of such blocks can be done at any layer as long as indication of the M-bit status is passed along.

Q-bit data.   The Q-bit, or qualifier bit, allows a distinction to be made between two types of data. The normal setting of the Q-bit is to a value of 0. A 1 is used to indicate special use of the contents of this frame. Naturally, this bit (or indication thereof) must be able to pass freely between the network layer and the upper layers to be useful.

One use of the Q-bit might be to distinguish between user data and control information. An application might, for example, use the Q-bit to pass information to an application on the far end indicating *what* is to be done with the user data. If the Q-bit is used in conjunction with the M-bit, all successive items of the packet sequence must have the same Q-bit value. The use of the Q-bit does not affect packet sequencing.

X.25 packet protocols

Some discussion of X.25 protocols took place in Chap. 5. These protocols concern the call setup and teardown of VCs and that of the restart sequence for an initialized physical channel. The remaining procedures concern data transfer and flow control, interrupts, resets, and optional packets such as diagnostic and registration packets.

Flow control.   Each layer 3 packet header of X.25 data contains both a sending and a received sequence number, designated by $P(S)$ and $P(R)$. Acknowledgment packets (RR and RNR) contain only receive sequence numbers [$P(R)$]. The default modulo operation of these sequence numbers is modulo 8. The window size, labeled W, is a default of 2. Within this situation, sequencing is performed at layer 3 in an identical fashion to that of layer 2—with the exception of the use of the D-bit as a possible higher-layer mechanism for acknowledgment and window rotation. A packet received out of sequence indicates that the data link layer has failed, and the channel undergoes a reset sequence.

Reset procedure.   A reset may be sent at any time. This procedure will effectively stop all data transfer, with the likelihood of lost data. Thus, if the higher layer is unable to recover, it is possible that the entire data transfer may have to be repeated. Resets occur under two primary conditions—failure of the data link layer and application requests for the resetting of the channel. A RESET REQUEST packet is sent, and timer T22 (180 s) is started. No data will be accepted until the RESET CONFIRMATION packet is received. If the confirmation is not received after R22 (default 1) retries, a VC will be torn down and a PVC will be indicated as out of order.

Interrupt procedure. Whereas a reset procedure (unless done while in an idle data transfer situation) may result in loss of data, an interrupt packet allows transmission of data without regard for flow control or windowing needs. This is very similar to that of the UNIT-DATA frame at layer 2. An interrupt packet may be used by the higher layer to send quick information to its peer for control purposes. It may be also used in a fashion similar to the "break" condition on modem-connected user equipment. The interrupt packet is acknowledged with an interrupt confirmation. If the interrupt is not confirmed within a period indicated by the T26 timer (default 180 s), a reset procedure will be initiated.

Optional reject usage. Layer 3 can use a REJ packet in the same manner that it is used by the data link layer. A REJ packet may be sent rather than a RESET when an out-of-sequence packet is received. The use of this optional procedure may be useful when the data link layer is not in multiframe procedure mode. This is an inverted situation from the "normal" use of the OSI layers but may be useful in certain interworking situations.

Optional diagnostic packet. The diagnostic packet may be used within some networks to indicate reasons for an error condition. In some situations, it is used (in possible conjunction with a REJ packet) as a layer 3 FRaMe Reject (FRMR) message. That is, the packet may contain the first 3 octets of the erroneous frame that was received. It may also be used in conjunction with timeout situations.

Optional user facilities

ITU-T Recommendation X.25 provides two sets of procedures (three if the call control is counted separately). One set, considered mandatory, has been discussed above and deals with proper data transmission and maintenance of the link. Another set, however, is optional and is associated with the registration and registration confirmation packets (see Figs. 6.2 and 6.3). A registration packet can be used to subscribe to facilities that are nonstandard for a fixed period of time. It may also be used to verify facility parameters that are known to the DCE.

The list of optional user facilities is extensive and some of them affect the basic operation of the layer 3 protocols. They correspond, in a loose fashion, to the supplementary services of the Q.932 and Q.95x call control procedures. These procedures will not be discussed in detail in this book. However, a brief overview of the types of facilities supported by X.25 may be useful.

Extended packet sequence numbering allows layer 3 to operate in modulo 128 instead of the default modulo 8. This is likely to be used in conjunction with the nonstandard default window size facility to

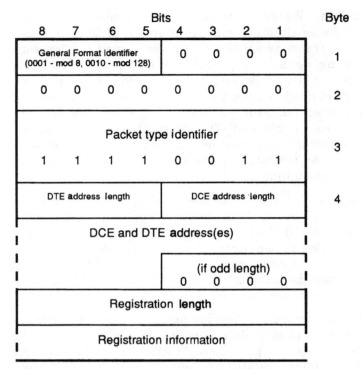

Figure 6.2 X.25 Registration request packet format. (*From ITU-T Recommendation X.25.*)

allow better usage of long-distance links such as satellite links.

The D-bit modification facility is an example of a method of retrofitting equipment that was implemented prior to D-bit standardization within X.25. This facility allows the user equipment to notify the network to set the D-bit of the GFI as part of any CALL REQUEST or CALL ACCEPT packets sent by the DTE (and clear such for incoming packets). In this manner, older equipment can interact with more recent equipment.

The packet retransmission facility allows use of the optional REJ packet. The incoming and outgoing calls barred facility types from shifting the equipment into originating or terminating only—a method of restriction, perhaps for security purposes. The LCI ranges may be changed (or requested) by the DTE. Nonstandard packet sizes, window sizes, and throughput classes may be requested.

A set of facilities operate on what is called a Closed User Group (CUG). This type of group can effectively set up a situation that can allow sets of DTEs to be treated as a single entity for purposes of facility negotiation or access. For example, if a set of DTEs is registered as being in a particular CUG, it is possible for the network to classify

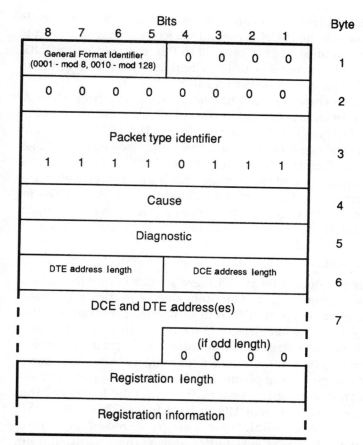

Figure 6.3 X.25 Registration confirmation packet format. (*From ITU-T Recommendation X.25.*)

the DTEs as being a subnetwork that allows access only between members of the CUG.

The next group of facilities is associated with network calling features and does not affect the layer 3 protocol. Reverse charging allows accounting information to be shifted. Hunt groups allow the network to reroute calls according to DTE usage. Call redirection allows the network to automatically forward calls to other DTEs. Many other facilities are currently available, but these give a good overview of the types that are available.

Interaction with ISDN

The examples of use of the low-layer IEs associated with the bearer capability IE showed methods of indicating the use of X.25 over a packet-mode connection. X.25 can also be used over a circuit-mode connection, although the ISDN is normally used only for connection to an X.25 packet handler in this case. ITU-T Recommendation X.31, which is discussed in greater detail in Chap. 9, allows several methods of negotiation for X.25 use over an ISDN.

X.31 details two cases for X.25 use over ISDN. The first, Case A, uses Q.931 only as an access mechanism to a Packet-Switched Public Data Network (PSPDN). This situation leaves the ISDN out of any possible parameter negotiations and allows the user equipment access to the specific facilities available from the PSPDN. In Case B, the ISDN provides packet-handling facilities itself and can provide varying services during negotiation of the call setup. Case A provides for B-channel access only. Case B can provide packet-handling support for D- or B-channels.

Software requirements

The software requirements for the PLP are similar to that of Q.931 or the call control portion of X.25. One item that is particularly important, however, is the support of facilities—particularly facilities that change window sizes, packet sizes, and modulo sequencing. The most efficient code uses fixed sizes for all parameters. If variable parameters are to be supported, a system of updating, checking, and use of those parameters must be implemented. In some instances, this will affect global parameters of the system. For example, let us say that packet sizes are increased from 128 to 1024 bytes. This means that the frame size at the data link layer must be increased in a corresponding fashion. It also affects buffer storage and manipulation. Window size may affect flow control of applications and so forth.

Terminal Adaption

User equipment that is designed explicitly to work with an ISDN does so at the S/T or the U interface reference point. One of the primary concerns contained within the various ITU-T documents is that ISDN takes an evolutionary role in the general development of networks. This means that existing networks and equipment must be allowed to be used within ISDN. To facilitate this, the ITU-T has designated the R interface reference point as a place where non-ISDN equipment can be connected into a TA which is able to make use of the ISDN.

Many different types of user equipment can be supported. They vary in physical interfaces, command use, flow control, synchronization, and other areas. Two general types of methods are used in the support of ISDN equipment. One way is to provide a special physical structure that can be used within the general 64-kbps framing structure of a B-channel. The other way is to "packetize" the data into HDLC frames along with additional needed terminal information (if any). ITU-T Recommendation V.110 follows the first philosophy. ITU-T Recommendation V.120 follows the second. Packet Assembly Disassembly (PAD) functionality of X.25 is also in the second category. The PAD methodology is covered in more detail in Chap. 9. V.110 and V.120 will be covered shortly in this chapter.

What are the needs of an interface between ISDN and non-ISDN equipment? One of the first is that of rate adaption. Most older equipment is used at speeds of much less than 64 kbps. A method is needed to convey that data over the greater bandwidth of ISDN. Rate adaption also includes issues of flow control. Another need is that of signalling. Old equipment has particular, perhaps widely accepted, conventions as to how the connections, speed, and facilities are to be supported. Another concern is that of multiplexing. If equipment makes use of only 12 kbps of the B-channel, the other 52 kbps is wasted. Multiplexing allows multiple terminals to make use of the same transmission facility.

Rate adaption

Most non-ISDN equipment (designated as TE2, by ITU-T documents) was designed to operate at speeds less than the normal B-channel supports. This difference in speeds causes two potential problems. These are the requirement to synchronize data flow between the two end-user pieces of equipment and the need to appropriately encapsulate the data within the higher-speed transmission circuit. Bearer Capability IEs and LLC IEs give the possibility of notifying the other end of the user rate of the equipment.

Rate adaption methods.   There are three primary forms of rate adaption. These are fixed-bit user data encoding, fixed-frame data encapsulation, and flag stuffing HDLC methods. As an example of the first type, consider a terminal running at a user rate of 8000 kbps. The TA could use 1 fixed bit of each byte and set the rest to another known value (or even let it vary). The receiving end then knows that only the fixed bit is actual user data and discards any other bits. This method, as well as the others following, requires that both ends know the rate adaption scheme in advance.

This bit use form may also be used for interworking of HDLC based protocols such as V.120. In such a case, the bytes of the protocol are shipped using only the lower 7 bits—requiring 8 octets to send 7 bytes. As long as both ends expect to use only the same bits within the data stream, this method may be used to adapt to lower data rates for any protocol.

The second method is that which is used by V.110 (as an example). A 64-kbps frame is created with the data encapsulated into this frame. This method has an advantage in that other, nonuser data, bits may be used for various control information. Its disadvantage is that it is likely to be a unique protocol for a particular use. It also does not directly fit the OSI model.

Flag-stuffing HDLC methods allow the full 64 kbps of the B-channel to be used. However, as an HDLC protocol family member, there is a degree of inefficiency due to the framing headers. It does offer the HDLC methods of error detection and automatic retransmission—very useful for high-speed data transfer.

Flow control.   Still, transient differences in transmission times between endpoints require some form of buffering—probably on the ISDN equipment (designated as TE1 by ITU-T documents) side of the R interface. Drastic speed differences between the two endpoints will, at some point, overwhelm any buffering capabilities without flow control mechanisms. However, if both sides are operating at approximately the same speed, a buffer mechanism will be sufficient to smooth out discrepancies.

Flow control is still useful because, even if the user rates of both pieces of equipment are the same, there will occasionally be bottlenecks within the system. Two categories of flow control exist—between the TE2 and the TA and between the two end TAs. Buffer management is a solution for transient flow control. Other mechanisms are needed in nontransient situations. X.25 has its system of windowing and local busy state (RNR situation). V.110 has the possibility of using control leads across the R interface to indicate a need for stopping transmission from the TE2. It may also use embedded XON/XOFF characters within the data stream, or it can use a special character within the V.110 frame, the X-bit, to indicate end-to-end flow control. V.120 can make use of the same flow control mechanisms as X.25 (dependent upon the layer 2 and layer 3 protocols supported) and also can pass hardware interface indications within V.120 header segments.

Signalling Conversion

A TE2 will support a particular signalling protocol, which is unlikely to be directly compatible with the Q.931 (or X.25 call control) packet syntax. Thus, some form of conversion is needed. A popular signalling protocol for asynchronous terminals is the de facto AT command protocol. For synchronous terminals, V.25 bis is popular as is X.21. PAD handlers make use of ITU-T Recommendations X.3, X.28, and X.29 to provide packet-handling functions for synchronous terminals into an X.25 network.

As an example, consider an asynchronous call control string from the AT command set. The sequence ATDT12345678901 is entered from the user interface. The characters are transmitted across the R interface. The TA then receives the characters, interprets them, and sends a formatted N_CONN_RQ primitive to the network layer. The network layer translates this primitive into the appropriate SETUP packet (for Q.931) and proceeds with the call. Later, an N_CONN_IN goes back up and is translated into the appropriate CONNECT message at the user interface. The same type of thing must occur for any user interface. At the R interface, the equipment sees a virtual copy of the interface for which it is designed.

Multiplexing

Multiplexing allows multiple data streams to be transported over the same physical transmission medium. All of the rate adaption methods listed above can be used for multiplexing. In the *fixed-format* method, different bits per sequence can be used for each data stream. Each end is responsible for routing of data. The fixed-frame method can be subdivided into smaller packages, with each package allocated to a particular data stream. V.120 and X.25 (X.31 use) allow layer 3 channel multiplexing. Generally, only the X.31 method allows the data streams to go to different physical addresses. This is a side effect of the packet handler system, which will either be part of, or interworked with, the ISDN. Any multiplexing system may, in theory, be used as a router mechanism for different addresses. ITU-T Recommendation V.110 does not have any implicit support of this, but the protocol can be overlaid as part of the data streams.

V.110

ITU-T Recommendation V.110 (I.463) is one of the four I.46x series of documents related to terminal adaption. V.120 (I.465) is discussed in the next section. X.31 (I.462) and X.30 (I.461) are discussed in Chap. 9. V.110 is a variable frame method of supporting synchronous or asynchronous DTEs on an ISDN. The method is highly conducive to hardware support, which allows a simpler software interface. It does not directly support multiplexing, however, and also circumvents most of the OSI layers. One particular advantage of V.110 is that the framing pattern is adaptable to 56-kbps interworking situations. Some ISDNs will insert this protocol into byte 5 of the low-layer information bytes to indicate interworking. (There is some question that this is good—but it is done.)

Rate adaption

V.110 has three separate data stream rates that are applicable to rate adaption. These are the user, intermediate, and bearer rates. The user rate is the speed at which data can be transmitted and received across the R interface. The intermediate rate is the effective speed of the data across the transmission circuit. The bearer rate is the final speed of the full transmission circuit.

The first thing that is important to recognize is that the speed is constant in a V.110 framing situation. Frames are always being sent, with or without user data. In the case of asynchronous data, it means that any periods without user data must be filled in with stop bits. In the case of synchronous data, the data stream will be constant but will contain either a synchronous frame or interframe flags. For this reason, V.110 breaks down the rate adaption into three sections. The first (RA0) is needed for asynchronous user data streams to fill the data stream. The second (RA1) is used to create intermediate frames to be transmitted at the intermediate data rate. The last (RA2) is used to adapt to the bearer rate.

Asynchronous fill rate adaption (RA0). This step, as noted above, is used only for asynchronous terminals. This is used to create a beginning rate that is compatible with synchronous rates. Synchronous rates are supported at multiples of 600 bps (600, 1200, 2400, 4800, 9600, and 19200). Thus, if an asynchronous terminal wants to have a user rate of 300 bps, the TA must (at the RA0/RA1 adaption point) interfill additional stop bits between data characters to bring it up to the necessary rate. V.110 also supports RA1 speeds of 3600, 7200, and 14400 bps, but these framing methods are not recommended for asynchronous data stream adaption.

80-bit intermediate frame formatting (RA1). The basic frame unit of V.110 is an 80-bit frame. These 80 bits are broken into 10 octets. Intermediate frame rates are 8, 16, and 32 kbps. An example of the frame formatting for a data rate of an adaption of 2400 bps to an intermediate rate of 8 kbps is given in Table 6.2.

Each frame is calculated according to the number of "real" data bits (they may actually be stop bits or interframe flags) in each 80-bit frame. Thus, an intermediate rate of 8000 bps means that one-hundred 80-bit frames will be transmitted per second. A data bit rate of 24 bits per 80-bit frame will therefore be transmitted at the user rate of 2400 bps. In the case of an intermediate rate of 16,000 bps, two-hundred 80-bit frames will be transmitted. Thus, if each frame contains 48 bits of real data, the user rate of 9600 bps is effected. The same frame can be used, therefore, for 4800 bps or 9600 bps with the intermediate data transfer rate making the difference.

TABLE 6.2   Intermediate V.110 Frame for 2400 bps

0	0	0	0	0	0	0	0
1	D1	D1	D2	D2	D3	D3	S1
1	D4	D4	D5	D5	D6	D6	X
1	D7	D7	D8	D8	D9	D9	S3
1	D10	D10	D11	D11	D12	D12	S4
1	1	1	0	E4	E5	E6	E7
1	D13	D13	D14	D14	D15	D15	S6
1	D16	D16	D17	D17	D18	D18	X
1	D19	D19	D20	D20	D21	D21	S8
1	D22	D22	D23	D23	D24	D24	S9

SOURCE: ITU-T Recommendation V.110.

Adaption to the bearer rate (RA2). ITU-T Recommendation I.460 indicates that the intermediate rates should be adapted to the 64-kbps bearer rate as follows. For an intermediate rate of 8000 bps, the low-order bit (first transmitted) of each octet is used to carry a bit. Thus, 80 octets will be transmitted at the bearer rate to transmit one 80-bit frame at the intermediate rate. The remaining bits of each octet are set to 1. An intermediate rate of 16 kbps uses bits 1 and 2 of each octet. The 32-kbps intermediate rate uses bits 1 through 4 of each octet/byte.

Rates greater than 19,200 bps do not make use of an intermediate frame in the strict sense of RA1. The frame is directly mapped to the bearer rate. V.110 gives two alternate framing methods for 56-kbps user rates. One method sends pure data, and the other sends flow control and interface lead mapping in addition to the data. The first form allows for direct interworking with 56-kbps bearer channels and should be used whenever the bearer rate is 56-kbps. The second allows flow control and In-band Parameter Exchange (IPE) between the terminals but should not be used unless the bearer rate is 64 kbps.

V.110 framing

As can be seen in Table 6.2, there are four different types of fields within a V.110 frame. These fields are composed of D-, S-, E-, and X-bits. The D-bits are the data bits. These can be actual data, stop bits, or interframe flag bits. For some 80-bit V.110 frames, these spots will be marked by F-bits. The F-bits indicate that this is a data fill character that is not needed to create the appropriate user data rate and may be filled with a 0 or 1 as desired (normally to be filled with a 1).

The E-bits are associated with three functions. These are to provide rate repetition information, to give network independent clock information for synchronization, and to provide multiframe information. Table 6.3 details the use of the E-bits.

TABLE 6.3   V.110 E-Bit Usage

Intermediate rate, kbps									
8	16	32	E1	E2	E3	E4	E5	E6	E7
600			1	0	0	C	C	C	1 or 0
1200			0	1	0	C	C	C	1
2400			1	1	0	C	C	C	1
		12,000	0	0	1	C	C	C	1
	7200	14,400	1	0	1	C	C	C	1
4800	9600	19,200	0	1	1	C	C	C	1

SOURCE: ITU-T Recommendation V.110.

E-bits E1, E2, and E3 are used as indications of synchronous inter-mediate rates. These bits are not needed for asynchronous use if low-layer IEs are present in the Q.931 SETUP message or the IPE is supported. E7 is set to 1 for all conditions except the fourth 80-bit frame for 600 bps. In this case, the bit is set to 0 to maintain the mul-tiframe synchronization. The bits marked as C (E4, E5, and E6) are used in network-independent blocking information.

The S-bits are grouped into two segments, the SA- and the SB-bits. The SA bits are those bits designated as S1, S3, S6, and S8. The SB bits are those bits designated as S4 and S9. The SA and SB bits can be seen conceptually as single bits. The groups allow repetition of the contents to fill in the 80-bit frame and to provide error detection. For example, in the 56-kbps frame used over a 64-kbps bearer channel, only bits S3 and S4 are used. This represents 1 bit for the SA group and 1 bit for the SB group.

The SA- and SB-bits are used for the V.24 interface circuit leads. The SA-bit group corresponds to 107 (Data Set Ready) and 108 (Data Terminal Ready) depending on direction of the frame signal. The SB-bit group corresponds to circuit leads 105 (Ready for Sending) and 109 (data channel received line signal detector). The X-bit is used in conjunction with the V.24 interface circuits also. It is used with circuit lead 106 (Ready for Sending) and is a primary method for applying flow control from the TA to the TE2. In all such instances, a value of 0 indicates that the circuit should be ON and a value of 1 indicates the circuit should be OFF. Thus, an X-bit of 0 means that flow control should be applied to the receiving TE2.

Frame synchronization

The 80-bit frame contains a built-in pattern for synchronization when using intermediate rate transmission. The first byte is all 0s, and the subsequent 7 bytes have the low-order bit set to 1. If this pattern is seen twice, synchronization may be assumed. For 48-kbps transmission rates a repeated pattern of 1011 in the low-order bit of subsequent bytes verifies synchronization. In the 64-kbps bearer rate version of 56-kbps data transfer, a pattern of 0YYY1111 in the high-order bits may be used for synchronization. The Y indicates that this bit may be either a 0 or 1 (corresponding to X- and S-bits in the frame). (It is recommended that this pattern be seen at least four times before synchronization is assumed.)

Network-independent clocks

In interworking situations, the ISDN 64-kbps channel may be interfaced with a DTE and modem on an analog network (or PSTN). In such a case, it is necessary to synchronize the ISDN bearer channel with the analog line. The network-independent clock bits allow for such synchronization. The support of such is indicated in the low-layer information parts of the SETUP message. Bits E4 through E6 are used to convey particular shift information. The specific algorithm is left to readers of ITU-T Recommendation V.110.

In-band parameter exchange

This optional procedure is described in Annex I of V.110. This procedure allows in-band negotiation of intermediate and user rates, stop bits, parity, character lengths, mode, and maintenance information. It is an involved process using the S-bits, X-bit, special codes, timers, and a complete state table sequence. This detailed algorithm will also be left to readers of ITU-T Recommendation V.110.

V.120

ITU-T Recommendation V.110 is oriented toward adaption of a single asynchronous or synchronous terminal onto an ISDN. The defined protocol is specifically designed for interworking. This gives some advantages to the protocol but also some restrictions. ITU-T Recommendation V.120 has much greater versatility over an ISDN but has greater interworking restrictions. Two features are particularly attractive in V.120. The first is that it is possible to set up a V.120 data channel via Q.931 out-of-band signalling procedures, in-band V.120 signalling procedures, or both. In this respect, it is very similar to that of X.31. The second advantage is that V.120 allows different protocols to be used over the bearer channel. There is a restriction if Q.921 is not used at the data link layer—the call setup must be done out of band. This is because Q.921 is needed for the multiple logical links necessary for in-band signalling.

Bits								Byte
8	7	6	5	4	3	2	1	
0/1 ext	0 Layer 1	1 ident.	User information layer 1 protocol					5*
0/1 ext	Sync/ asynch	Negot.	User rate					5a*
0/1 ext	Hdr/ no hdr	Multi frame	Mode	LLI negot.	Assignor /ee	In-band neg.	0 Spare	5b*
0/1 ext	Number of stop bits		Number of data bits		Parity			5c*
1 ext	Duplex mode		Modem type					5d*
1 ext	1 Layer 2	0 ident.	User information layer 2 protocol					6*
1 ext	1 Layer 3	1 ident.	User information layer 3 protocol					7*

Figure 6.4 V.120 LLC IE bytes. (*From ITU-T Recommendation V.120.*)

Protocol parameters

V.120 actively uses the LLC IE (or, for SETUP, the Bearer Capability IE) for both out-of-band and in-band signalling. This may be a problem for initial setup in ISDNs that do not transparently pass the low-layer information fields. It would be a good idea, at this point, to review bytes 5 through 7 of the LLC IE (as seen in Fig. 6.4) at this time. This time we will emphasize the fields used by V.120 and their effect upon the system.

Layer 1 protocol byte. Byte 5 of the LLC IE should indicate V.120. This is indicated by the value 00101000 (hex 28). In the case of interworking, some ISDNs will overwrite this byte to be ITU-T Recommendation V.110. Under such conditions, both ends must recognize that V.120 is to be used and the default parameters are to be associated with the link.

User rate byte. Byte 5a of the LLC IE has three important fields. These are the synchronous/asynchronous, negotiation, and user rate fields. The negotiation field should presently always be set to 0. A value of 0 in the synch/asynch field indicates synchronous use and a 1 indicates asynchronous use. For synchronous terminals, bytes 5c and 5d are unnecessary. The user rate is used similarly to that of V.110—indicating the data speed across the R interface reference point.

V.120 internal protocol byte. The fields in byte 5b of the LLC IE are concerned with V.120 internal protocol. The first, the terminal adaption header, indicates whether the TA header byte(s) is included. If the link is to be in "bit transparent" mode, the TA header byte will not be used. Thus, in such a case, the field should be set to 0 and ignored if set. In nontransparent cases, the first byte of the TA header is mandatory, and thus, this bit actually indicates whether the optional control state (CS) TA header byte is supported.

The second field of byte 5b indicates use of multiframe data link support. A 1 indicates that "normal" multiframe data link windowing and retransmission is to be used. A 0 indicates that only DL_U_DA_RQ and DL_U_DA_IN primitives are to be used with the data link layer. Effectively, this can provide the same streamlined service as Frame Relay (minus the maintenance and support functions).

The mode field uses a 0 to make use of bit-transparent operation and a 1 for protocol-sensitive operation. If this bit is set to 0, bytes 6 and 7 should not be present. The next field, called Logical Link Identifier (LLI) negotiation, effectively declares whether multiplexing can be present on the bearer channel. We will discuss LLI negotiation soon. Associated with this is the assignor/assignee bit. If the bit is set to 1, the LLI will be initially assigned by the originator (and, if 0, by the terminating side).

The last field of byte 5b is used to specify whether in-band negotiation is done (a value of 1 indicates that it is). If the LLI negotiation bit is set to 0, this is not very useful for most cases. It can still be useful if only in-band setup is done (over a permanent, or subscribed, transmission line). This allows a single logical link to be set up.

Interface use byte. Byte 5c is used to indicate the number of stop bits, data bits, and parity. Thus, it is useful only for asynchronous terminals. Unless bit-transparent mode is used, it is necessary to make sure that data bit lengths less than 8 are placed in the lower-order bits of the data bytes.

Modem use byte. Byte 5d indicates modem type and whether the modem is to be used in full- or half-duplex mode. This is useful only for the user equipment since the ISDN provides full-duplex transmission capabilities.

Layer 2 and layer 3 protocol bytes. Bytes 6 and 7 allow various protocols to be used via V.120. A very useful combination is to have the data link layer be Q.921 and the network layer to be null (not present). This provides a very efficient, reliable transmission link. Nine different protocols are explicitly supported at the data link layer and six at the network layer (plus null). The actual protocols supported depend on the implementation. Note that if Q.921 is not used at the data link layer, V.120 call multiplexing in not possible. (However, it is possible that another set of layer 2/3 protocols could be used to provide in-band signalling.)

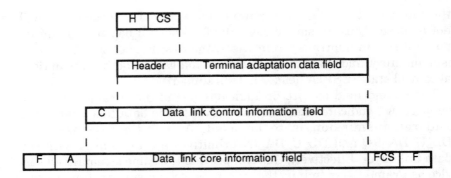

F   HDLC Flag
A   Address
FCS Frame Check Sequence
C   Control (HDLC format)
H   Header octet (optional for bit transparent mode)
CS   Optional header extension for control state information

Figure 6.5 V.120 layer relationships. (*From ITU-T Recommendation V.120.*)

V.120 protocol architecture

The September 1992 revision of ITU-T Recommendation V.120 added support of frame-mode services to that of circuit-mode services. Thus a total of four layers are specified in the recommendation. These are the physical layer, data link core sublayer, data link control sublayer, and the terminal adaption sublayer (see Fig. 6.5). If circuit-mode is used, the data link core sublayer and data link control sublayers merge into the data link layer. Note that the TA header field, composed of the H and optional Control State (CS) bytes, effectively is interposed between what would normally be the layer 3 data field and the layer 2 header.

Terminal Adaption Protocol

The 2 bytes of the TA (Fig. 6.6) header provide information associated with the terminals at the end of the connection. The first byte, called the H-header is mandatory for all protocol-sensitive operations. The second byte is optional and is called the CS-byte.

H-header byte. The fields are the extension bit (set to 0 if the CS byte follows and 1 otherwise), break bit, error control bits (C1 and C2), and segmentation bits (B and F). The break bit (BR) is used (value 1) to indicate use of a BREAK function in asynchronous terminals or an idle situation in synchronous terminals. The error control bits provide four possible values and are used in protocol-sensitive operation. A value of 00 indicates that no error has been detected (and should always be such for bit-transparent mode). A synchronous terminal may also indicate a FCS error, abort, or TA overrun (values 01,

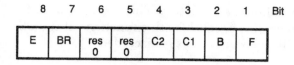

| 8 | 7 | 6 | 5 | 4 | 3 | 2 | 1 | Bit |

| E | BR | res 0 | res 0 | C2 | C1 | B | F |

Header (H-byte) octet

E   Extension bit
BR   Break/mark hold bit
C1,C2 Error control bits
B,F   Segmentation bits
res   Reserved for future standardization

| E | DR | SR | RR | res | res | res | res |

Control State Information octet

E   Extension bit
DR   Data Ready
SR   Send Ready
RR   Receive Ready
res   Reserved for future standardization

Figure 6.6 V.120 TA header bytes. (*From ITU-T Recommendation V.120.*)

10, and 11, respectively). An asynchronous terminal may indicate a stop-bit error, parity error, or both.

The segmentation bits allow segmenting and reassembly of user HDLC frames in synchronous mode. The Begin (B) bit is set to indicate the beginning of a series of frames and the Final (F) bit is set to 1 to indicate the end of a series of frames. Thus, the value 10 for the BF-bits is the beginning frame of a series. The value 00 is a middle frame of a series, and 01 marks the final frame of a series. A single frame should be marked as 11 (and this should be the setting for asynchronous and bit-transparent modes). An illegal sequence (two subsequent frames with 10 set, for example) may be the cause of an abort or other recovery situation.

Control state byte.   This optional byte allows for flow control by reflecting (and potentially effecting) control of the interface leads. The Data Ready (DR) bit corresponds to the DTR lead (or circuit 108/2) and says that the interface at the R reference point is activated. The Send Ready (SR) bit corresponds to the RTS (or circuit 105) and indicates that the TE is ready to send data, and the Receive Ready (RR) bit, corresponding to the CTS (or circuit 106), says that the TE is ready to receive data.

The CS header byte (if supported) is sent with the first TA header for the first I frame (or UI if flow control is needed). It is sent each time that the state changes. The SR- and RR-bits may be used in combination to provide a half-duplex function. The RR-bit is used for flow control (the DR-bit may also be used for this function). If the CS-byte is part of a received illegal sequence, it is discarded.

Data Link Control Protocol

After a new logical link has been set up (via out-of-band or in-band procedures), it is necessary to make sure that both ends are active and able to send and receive data. For multiframe mode, this is done with the SABME/UA exchange. For non-multi-frame, this is done with an XID command/response exchange. Furthermore, in the case of out-of-band signalling, a TE that receives a CONNECT from the network (terminating side) is required to transmit a frame or, if a CONNECT ACKNOWLEDGE is received (originating side), wait T200 or until a frame is received before transmitting.

In the case of frame-mode services, out-of-band setup is done via ITU-T Recommendation Q.933 protocol. The Data Link Core Protocol will be Q.922, Annex A, and bytes 5, 5a, and 5b (minimally) will be provided in the call setup information.

In-band Signalling Protocol

V.120 supports a subset of ITU-T Recommendation Q.931 for in-band call control. In order to do this, it uses a special logical link (similar to SAPI 0 on the D-channel Q.921) for signalling purposes. Thus, if *in-band signalling* is supported, two logical links are initialized upon out-of-band setup. These logical links are the signalling link and the first data bearer link. The default link is used if LLI negotiation is not supported.

Logical link identifier. First, let's discuss the use of the LLI. This LLI is used in place of the SAPI/TEI combination for Q.921. To provide Q.921 compatibility, the LLI is made of the same bit fields as the SAPI and TEI. The high-order 6 bits of the LLI are equivalent to the SAPI, and the low-order 7 bits are equivalent to the TEI. These 13 bits allow 8192 values for an LLI. LLI 0 is used for signalling. LLI 256 is used as the default LLI (if LLIs are not negotiable). The range of values from 257 through 2047 are available for LLI assignment and values 2048 through 8190 (and 1 through 255) are reserved. Value 8191 is set aside for in-channel layer management, but V.120 does not provide specific recommendations for its use.

Data link differences. The use of the LLI does not really change actions of the Q.921 since the fields are identical to those already used for a similar purpose. The C/R bit, however, is used differently since this is an end-to-end protocol. For V.120, a 0 *always* means a command, and a 1 *always* means a response. Additionally, an I-frame response is also permitted.

LLI assignor. The LLI may be assigned by either end, depending on the value assigned to the LLI assignor bit of the LLC IE. Thus, if the originating end is not the assignor, the LLI assignor bit is set to 0. The terminating end will then encode an LLI IE and set the LLI assignor bit to 1 in a responding message. Otherwise, the originator will send the LLI IE with the SETUP. If LLI negotiation is not supported, the data link will use LLI 256.

Call reference values. CRVs are used identically to those of Q.931. In fact, from a software point of view, the same call references may be used for outgoing connections as for Q.931. This may result in CRVs not in sequence for one entity, but this is acceptable. This reduces otherwise duplicated data space and code.

V.120 in-band establishment and release messages. V.120 supports the SETUP, CONNECT, RELEASE, and RELEASE COMPLETE message types. For in-band use, the protocol identifier is set to value 7. A SETUP message will contain (beyond the protocol identifier, call reference field, and message type) four optional IEs. These optional IEs are the LLI, Called Party Subaddress, Calling Party Subaddress, and LLC.

The CONNECT message contains optional LLI and LLC IEs. The LLI is mandatory if it is the LLI assignor. The RELEASE and RELEASE COMPLETE message types contain only an optional cause field. Only the values of 16 (for normal clearing) and 21 (for call rejected) are supported.

V.120 in-band protocol states

Since V.120 does not support the full Q.931 protocol set, the states used are not the same either. In fact, only states 0, 1, 7, 10, and 19 are supported. An incoming call makes a transition from state 0 (idle) to state 7 (call received) and will either return to state 0, if refused, or the active state (10) if accepted. An outgoing call makes a transition from state 0 to state 1 (incoming call received) and on to state 10 or back to 0 depending on acceptance. State 19 is used upon initiating a disconnection.

Software Issues

There are a number of software issues associated with bearer services. We have already discussed aspects of bearer service negotiation. This can happen either out of band, using Q.931 low layer information fields, or in band, using protocol-specific methods or encompassing various handshaking schemes. Once the bearer service is negotiated out of band, it is necessary to connect the bearer channel to the appropriate hardware device. This can be called *cut-through* of the channel.

Once the bearer channel has been connected to the hardware device, it is possible that nothing more is needed (if the hardware device handles all necessary data manipulation). In most cases, this is not the case. This means that the ISDN user equipment must be able to coordinate with other software layer entities to provide the necessary services.

TA is a special case of the bearer services. Some aspects of this are discussed in Chap. 9, which details use of X.25 services on a DTE connected into an ISDN. Other aspects, however, are concerned with how the data flow is accomplished between the TE2 and the TA. This section will discuss these issues.

Bearer channel setup

Once the out-of-band connection has been established, a channel ID has been obtained. This channel ID indicates what TDM digital channel is to be used for the service. In the case of voice (or speech) services, a CODEC device is used, which is a special form of analog/digital convertor. The CODEC can have the bearer channel routed to it over a data bus, with connections to a transceiver handset. Thus, the *bearer channel cut-through* removes all further responsibility from the ISDN—unless supplementary services are supported. If so, the supplementary service primitives must be routed in some manner. This may be from a digitized tone analyzer (making use of standard tone keys) or, in an integrated device, some type of user application. Once the call is terminated, the bearer channel is disconnected from the CODEC, and the system is back to the original condition. The situation described for voice may also be applied to 3.1-kHz audio. However, if data and voice are mixed, special devices will be needed.

Special hardware devices are needed, or useful, for many different services. A bit-oriented variable frame situation such as V.110 can make use of special hardware devices. Video chips may provide processing for video signals. Also, for combined bearer channels, various methods of *hardware bonding* can be used. The variety of potentially useful hardware support is as diverse as that of the bearer services.

Many of the categories of bearer services are associated with HDLC protocols. Thus, this is a particularly widely used device in an ISDN. The bearer channel is connected, via the data bus, to the HDLC device. At this point, a LLD is used to communicate data and commands between the hardware device and the data link layer. Detailed discussion of LLDs occurs in Chap. 11.

Data flow routing

After the bearer channel has been connected to a hardware device, the data must be routed through the system. The Q.931 data flow is discussed in detail in Chap. 5. Data flow for a bearer service is based on the needs of the service. As seen above, certain bearer services require no additional in-band support. (If Q.931 was used during call setup, it will also be needed for call teardown.) Others require software modules.

The first location of data flow routing is at the physical layer—more precisely, the LLD. The LLD will route data to and from the physical layer. For bit-transparent data streams, this may be directed to the upper layers. For other HDLC data streams, the data is carried between the physical layer and the data link layer. At this point, once again, choices must be made. For V.120 LLI 0 (signalling link), data is routed to the V.120 network layer. For V.120 LLI 256, or other data link LLI, the data must be routed between the data link layer and the appropriate network layer service.

An ISDN system that supports only one service may make *hard-coded* choices. The connection means that a particular hardware device is connected to the bearer channel. From there it is routed to another particular entity. For many architectures, the decision points are not so simple. For example, the V.120 link layer is *almost* the same as Q.921. The few differences involved force the data link layer to be aware of the protocol use of the logical link. Additionally, the LLC IE allows negotiation of different network layers—this must be made known to the data link layer for routing of data.

Coordination/management requirements

The coordination/management entity, which was discussed in earlier chapters, facilitates the ability to route data flow. Once a connection is requested, the connection is associated with a particular call handle ID until the connection is terminated. This means that, upon out-of-band call setup, the bearer service can be known in a central place. Different bearer service types can be implemented as software choices, with side effects particular to each service separated. Thus, an N_CONN_CF primitive for bearer service type V.120 may mean that a signalling logical link needs to be established, followed by a data link (default LLI 256 or otherwise). For X.31, the out-of-band setup means that a particular channel is available for in-band X.25 call control procedures. For V.110, it means that the hardware devices involved can be initialized for data flow.

We talked about the coordination/management entity providing a "software break." We now see that the coordination function provides the ability to allow different software protocols to coexist within ISDN equipment. This indicates a type of "glue" allowing out-of-band and in-band services to be coordinated.

Terminal adaptor functions

Remember from our initial discussions that the TA was separated from the non-ISDN equipment (TE2) by the R interface. Bearer services that interact with TE2s must have various TA functions performed by the TA. These functions can include handling of the required interface across the R reference point, translation of signalling functions, and packetizing of data. Some terminals may also have two separate modes. One mode is used for commands (such as call connection or setting up internal functions; the other is for data streams.

Modes. Many terminals have two separate modes. One such is the AT command set. If two modes are available, there must be specific protocols available to switch between the modes. In the AT command set, this is called an escape sequence.

During the command mode, commands are accepted from the TE2 to perform changes to the interface (such as changing speed, parity, local echoing of data, or command set behavior). Commands may also be associated with obtaining information about the TA and the transmission line. One command will provide a mechanism for making a transmission connection. From this point, the TA will be in data mode. An escape mechanism provides a method to change back into command mode without disconnecting the transmission circuit. Another command can then revert the TA into data mode.

Data mode can either be transparent to the TA or can mean collecting data to present to the lower-layer entities. A packetizing function is common on devices used on an ISDN. In such a case, the data is accumulated from the receiving side of the R reference point. When a local buffer is full, or after a predefined time period, the data is shipped to the lower layers in a primitive, with data put into the form expected. When received, the data is removed from the packet form and transmitted to the TE2 across the R reference point.

Bit-transparent operation. For some protocols, the coordination/ management entity sets up a data path directly from the R interface to the hardware device connected to the bearer channel. The TA functions only during the command mode. During the data mode, the data is transparent to the TA. V.110 is normally done in an operation of this nature. Given a sufficiently powerful processor, it is possible to do all of the V.110 framing within software. For practical reasons, it is usually quicker and less expensive to take advantage of specialized hardware.

R **interface handling.** The interface at the *R* reference point will use some form of hardware-oriented protocol. V.24, V.25 bis, X.21, and X.21 bis are all common interfaces. An LLD forms a software interface between a hardware device and a software layer. Thus, even though LLDs have primarily been discussed as performing a function associated with the physical layer as a lower layer from the data link layer, it is possible to have an LLD handling physical layer functions at a point that can be considered above the network layer.

As is true for most other software handling hardware chip needs, the LLD handling *R* reference point physical layer needs also needs to operate in an expedient fashion. This usually means that an operating system queue will be useful between the LLD and the software module incorporating TA functions. This minimizes the time spent in an interrupt handler.

TA functions. There are three primary TA functions. The first is to interpret the *R* interface command and circuit protocol. The next is to process data or to set up the appropriate hardware settings needed for pass-through data flow between a hardware device connected to the bearer channel and the *R* interface physical interface. The last is to manage data structures to provide information needed by other software layers.

Interpretation of the *R* interface command and circuit protocol may occur in the LLD or a separate TA entity. The functionality is that of the TA. The location depends on time needed to process the command. For example, the 105 circuit lead (RTS) may be toggled off indicating that the TE2 is no longer ready to accept data. It is possible that this only changes an internal data structure member—unnecessary (and extraneous overhead) to ship to a TA entity. However, if a series of data characters need to be parsed to interpret the function, the data should be handled separately from the LLD.

Data transport may be a null function in bit-transparent cases. In others, the data needs to be accumulated and then sent as a primitive to lower layers. For some protocols, considerable manipulation of the data may be needed (for example, a software solution for V.110).

Data structures must reflect the current status of the *R* reference point interface and the results of defaults and negotiated parameters associated with the protocol. This may be used by any protocol layer that requests it—but must be obtained through protocol-specific function calls to maintain interlayer separation for data. One use, for example, would be for routing data from the data link layer to higher layers. Another use would be to have appropriate information for a TA header in data packets issued through V.120. A third category of use is to allow proper responses to the interface protocol at the *R* reference point (such as knowing how the end of a command string is delimited).

Interconnection into the other software layers.   Unless a bit-transparent protocol is in use, it is best to route data between the R interface LLD and the coordination/management entity (with possible side routes into a TA software entity). This allows the data flow from higher layers to lower layers to be preserved independently from the source of the data.

Interlayer communication

If multiple protocols are supported at different layers, there are choices possible in the interlayer communication design. Previous chapters discussed physical layer, data link layer, and network layer primitives. If there are two different data link layers possible, data routing takes place between the physical layer and the data link layer. The point at which the routine takes place is a matter of software design for the ISDN. It would be possible to have two separate operating system tasks, one for the LLD to DL1 data flow and one for data between the LLD and DL2. It would also be possible to have data go between the LLD and DL and, within the data link layer, appropriate data flow routing takes place. In the first case, two separate operating system tasks (probably each with its own data stack) must be available. In the second case, only one operating system task is necessary, but the data link layer must know information that may not be relevant to the protocol.

In the downward direction, the situation is simpler. The data link layer should not know the meaning of the contents of the data. Thus, DL1 and DL2 may both make use of the same message queue and operating system task for primitives heading in a downward direction. Only in the case that two different protocols are used at the receiving direction are two separate operating system tasks likely to be needed.

In a similar fashion to the HDLC physical layer LLD and the data link layer, a call from the TA software entity to the R interface LLD may be a synchronous call. It is necessary to prepare the hardware for its task—but not to wait for it to complete. From the R interface LLD to the TA entity (or coordination/management entity), a software break is needed to keep interrupt handling to a minimum.

Summary of Chapter

This chapter has been concerned with bearer services. An ISDN without bearer services is of limited usefulness. Bearer services allow multiple data protocols and applications to be used over different bearer channels. A bearer service can be negotiated through Q.931 out-of-band protocols. Specifically, it can be part of a Bearer Capability IE or a LLC IE. Some switches do not pass the LLC IE, so it is necessary to design the ISDN user equipment to allow for various switch options.

The bearer service, and its parameters, can also be negotiated in band, after the bearer channel has been connected to the supporting hardware device. The in-band procedure can either be specific to a protocol or use a handshaking method of protocol recognition.

There are three "essential" circuit-mode bearer services defined in the ITU-T documents. These are for speech, unrestricted data, and 3.1-kHz audio. Other bearer services allow for two bearer channels to be set up as a pair or for multiple bearer channels to be used as a group to increase bandwidth.

Four terminal adaption protocols are mentioned specifically in ITU-T Recommendation I.460. These are X.30, X.31, V.110, and V.120. X.30 and X.31 are associated with packet-mode terminals to be used over X.25 networks. These protocols are discussed in greater detail in Chap. 9. V.110 and V.120 are oriented toward support of synchronous/asynchronous terminals. (V.120 also supports bit-transparent protocol service.) V.120 is used mainly in North America and Japan. V.110 is mainly used in Europe and other world locations. V.110 directly supports only one data stream but is well suited to use in interworking situations with non-ISDNs. V.120 supports a multitude of protocols and is much more versatile than V.110 within an ISDN. V.120 does have greater difficulties with interworking.

Software needs for bearer services mainly involve the possibilities of multiple protocols being supported. Multiple services indicate a need to be able to route data in different directions depending on the bearer service (and its parameters) in use.

Frame Relay

A frame is a group of data at the data link layer. A packet is a group of data at the network layer. Frame Relay is, thus, a system of transferring data at the data link layer level. In effect, it means that the network deals with only the physical and data link layers. The chained layers in such a network incorporate only the bottom two layers rather than the bottom three.

Frame Relay is designed to be an efficient data transport network based on reliable digital transmission lines. Most networks currently support only PVCs, but SVC support is likely to start being put into commercial use in 1995.

Philosophy, Standards, and Evolution

Frame Relay service architecture is based on that of X.31. The first stage of implementation has been based on permanent circuits—the same as PVCs in X.25. Frame Relay has two advantages over that of X.25. There is no built-in redundancy between the data link and network layers. This simplifies (and speeds) the network tasks involved. Second, all of the signalling is done at subscription time (for PVCs) or out of band (for SVCs). Thus, the protocol for the "core" part of Frame Relay is small, efficient, and simple. In fact, the protocol itself is so simple that a number of additions to the protocol are necessary to reinforce reliability. These additions are done in a periodic fashion to cut down on data transport overhead.

Frame Relay (or frame-mode bearer service) was initially proposed by the ITU-T in 1988 in ITU-T Recommendation I.122. ITU-T Recommendation Q.922 was published to define the data link layer requirements. A more important part, at the beginning of implementation, was the Annex A of the recommendation. This recommendation provided for a core layer to the data link layer. That is, Annex A provided for a stripped-down efficient protocol that could be used on permanent bearer channels. Still, at this point the ITU-T had not addressed SVCs.

In North America, frame-mode service was looked at as a method to improve speed and network utilization. (Narrowband ISDN was considered to be too slow for most business uses.) In keeping with this orientation, the ANSI T1 committee produced a series of documents to help support switched frame-mode bearer service. Recommendations T1.606, T1.617, and T1.618 helped to fill the existing gap of recommendations. Annex D of T1.617 provided an extension called the Local Management Interface (LMI) to provide maintenance and administrative capabilities. The ITU-T is continuing to produce new recommendations—some of which incorporate the basic substance of the ANSI documents. Thus, presently, Q.933 takes the place of T1.617, and the other I.2xx and I.3xx documents take over from the others. The Frame Relay Forum, a group of manufacturers, network providers, and end users, also continues to push standards in useful directions.

Frame Relay was also designed, from the beginning, to incorporate higher bandwidths. The philosophy in the recommendations has been that if a transmission medium is reliable, extra precautions for error correction and retransmission at the low layers are mainly added overhead. This does require that some mechanism at the higher layers be available to check for transmission problems. The data link core allows data transport at very close to the theoretical maximum possible for the transmission medium. If most packets come through in sequence, the delay needed for higher-layer retransmission protocols is acceptable. If not, lower-layer protocol retransmission support is better. Thus, error-free transmission is the foundation for Frame Relay (or frame-mode bearer services).

The small overhead of the data link core protocol requires that additional maintenance and administrative functions be used as supplementary procedures. The congestion control mechanisms are incorporated mainly within the network. LMI procedures act between the terminal equipment and the network.

Frame-Mode Data Link Layer Protocol

As is true with many of the HDLC family ISDN data link layer protocols, ITU-T Recommendation Q.922 is based directly on Q.921. This recommendation is known as LAPF. The data link core sublayer is specified in Annex A of the recommendation.

There are a number of differences between Q.922 and Q.921. The UNI is symmetric (no DTE versus DCE or originating versus terminating asymmetry). It is explicitly designed to be used on any form of bearer channel (B-, D-, or H-) and also to be able to be used concurrently with LAPD on the same channel. The Data Link Connection Identifier (DLCI) is similar to the LLI used for V.120—using the same fields as the SAPI/TEI of Q.921. A particular DLCI is allocated for layer management. Finally, Q.922 has been designed explicitly as a encapsulated transport method for other protocols. This is associated with the need only for the lowest two layers within the switched network. A "regular" data service, such as X.25 or V.120, may be carried over the frame-mode bearer service as an intermediary network.

Frame structure

The frame structure for Q.922 is based directly on Q.921. Only differences will be discussed in this section.

Address field. The address field (see Fig. 7.1) splits what would be the TEI field of Q.921 into three separate fields. These are the lower-order bits of the DLCI, a Forward Explicit Congestion Notification (FECN) bit field, a Backward Explicit Congestion Notification (BECN) bit field, and a Discard Eligibility (DE) bit field. The EA extension bit is explicitly defined for LAPF—0 indicating additional address bytes. If the address field is extended, a DLCI or DL-CORE control indicator (D/C) bit field is part of the last byte of the field. The 2-byte address field is to be available on all networks. The 3- and 4-byte definitions are not currently supported on a D-channel (the matter is for further study).

The C/R bit field, defined for symmetric use, differs from that of Q.921. A 0 indicates a command frame. A 1 indicates a response frame. The DLCI is used in the same manner as an LLI for V.120 or a SAPI/TEI group for Q.921. The FECN, BECN, and DE bit fields are discussed in the "Congestion Control" section later in this chapter. Table 7.1 lists the use of DLCIs for the 2- and 3-byte address fields. The D/C bit is used to distinguish between use of the byte as a DL-CORE control field or the lower-order bits of the DLCI. This is a method for not explicitly having a control field within the data link core sublayer.

Control field. The control field for Q.922 is identical to that of Q.921 with one exception. The use of a symmetric C/R bit means that it is possible to receive a Information response frame (similar to that in V.120). It also implies the possibility of the P/F flag being used as an active flag (sent in an initial frame rather than only as an echo of a received flag).

Invalid frames. There are six conditions considered invalid for Q.922. It must be bounded by flags. It must have at least 3 bytes between the address field and the closing flag. It must be an integral number of 8-bit bytes (octets). It must not contain an FCS error or single-byte address field. Finally, the DLCI must be a supported value. If any of these conditions fail, the frame should be discarded with no action taken.

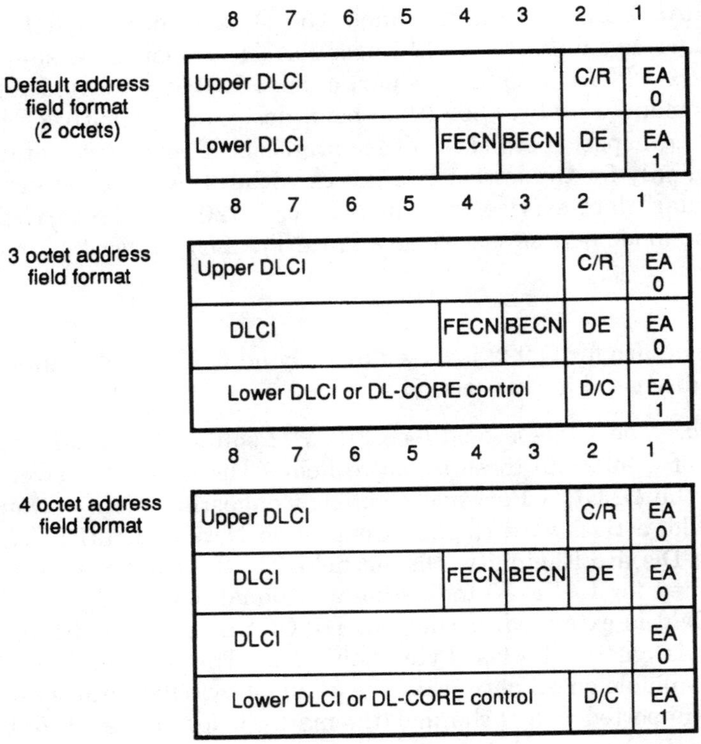

```
        8    7    6    5    4    3    2    1
```

Default address
field format
(2 octets)

| Upper DLCI | | | | | | C/R | EA 0 |
| Lower DLCI | | | | FECN | BECN | DE | EA 1 |

```
        8    7    6    5    4    3    2    1
```

3 octet address
field format

Upper DLCI						C/R	EA 0
DLCI				FECN	BECN	DE	EA 0
Lower DLCI or DL-CORE control						D/C	EA 1

```
        8    7    6    5    4    3    2    1
```

4 octet address
field format

Upper DLCI						C/R	EA 0
DLCI				FECN	BECN	DE	EA 0
DLCI							EA 0
Lower DLCI or DL-CORE control						D/C	EA 1

EA   Address field extension bit
C/R   Command response bit
FECN Forward Explicit Congestion Notification
BECN Backward Explicit Congestion Notification
DLCI Data Link Connection Identifier
DE   Discard Eligibility identifier
D/C   DLCI or DL-CORE control identifier

Figure 7.1   Frame-mode Address field formats. (*From ITU-T Recommendation Q.922.*)

Interlayer primitives

The primitives used between the data link layer and other layers are basically the same as for Q.921. There are three differences. One is that each place that Q.921 used the TEI and SAPI as parameters the DLCI is used. The second is that a Data Layer Connection Endpoint Identifier (DL-CEI) is used instead of a CES. The third is that a separate set of two primitives, M2N_ASSIGN and M2N_REMOVE, has been added to specifically address layer 3 to layer 2 management C-plane primitives (needed for splitting out data link core sublayers). Table 7.2 shows the use of these new primitives.

TABLE 7.1   Frame-Mode 2- and 3-byte DLCI Usage

10-bit DLCIs (2-byte address mode or 3-byte with D/C = 1)	
DLCI range	Function
0 (non-D-channel)	In-channel signalling, if required
1–15	Reserved
16–511	Network option: on non-D-channels, available for support of user information
512–991	Logical link identification for support of user information (PVCs may reduce availability.)
992–1007	Layer 2 management of frame-mode bearer service
1008–1022	Reserved
1023 (non-D-channel)	In-channel layer 2 management, if required

16-bit DLCIs (3-byte address mode with D/C = 0)	
DLCI range	Function
0 (non-D-channel)	In-channel signalling, if required
1–1023	Reserved
1024–32,767	Network option: on non-D-channels, available for support of user information
32,768–63,487	Logical link identification for support of user information (PVCs may reduce availability.)
63,488–64,511	Layer 2 management of frame-mode bearer service
64,512–65,534	Reserved
65,535 (non-D-channel)	In-channel layer 2 management, if required

SOURCE: ITU-T Recommendation Q.922.

Data link layer procedures

Most procedures of Q.922 are identical to Q.921 with the exception of the use of a DLCI for particular command routing rather than the SAPI. Exceptions will be discussed in the following text.

DLCI management can be assigned using C-plane primitives in conjunction with Q.933 or as permanently assigned (as for PVCs). For example, the M2N_ASSIGN_RQ primitive in the C-plane would then be routed to the U-plane as an MDL_ASSIGN_RQ primitive. The Q.921 state TEI ASSIGNED is retained for historical reasons—although it is really a DLCI ASSIGNED state.

Data link layer parameters may be negotiated as part of the connection establishment procedures defined in Q.933 or using XID frames. They may also be used as the default values.

An I-frame response with F-bit set to 0 is treated the same as an I-frame command with P-bit set to 0. If an I_R frame with the F-bit set to 1 is received, the protocol is a bit different. If the data link layer is not in a timer recovery state, the frame is noted as an error and then processed as if the F-bit was set to 0. If the data link layer *is* in a timer recovery state, the frame will act as a response to the outstanding frame and be treated as an I-frame with F-bit set to 0.

TABLE 7.2   LAPF Primitive Types

Generic name	Type*				Parameter*		Message unit content
	RQ	IN	RS	CF	PI	MU	
Layer 3 to layer 2 management							
M2N_ASSIGN	x					x	DL-CEI, DLCI
M2N_REMOVE	x					x	DLCI
Layer 3 to layer 2							
DL_ESTABLISH	x	x		x			
DL_RELEASE	x	x		x			
DL_DATA	x	x				x	Layer 3 peer-to-peer message
DL_UNIT_DATA	x	x				x	Layer 3 peer-to-peer message
Layer 2 to layer 2 management							
MDL_ASSIGN	x	x				x	DLCI, DL-CEI
MDL_REMOVE	x					x	DLCI
MDL_ERROR		x	x			x	Reason for error
MDL_UNIT_DATA	x	x				x	Management peer-to-peer message
MDL_XID	x	x	x	x		x	Connection and congestion management information
Layer 2 to layer 1							
PH_DATA	x	x			x (BRI)	x	Data link layer peer-to-peer message
PH_ACTIVATE	x	x					
PH_DEACTIVATE		x					

*RQ = request   CF = confirmation
 IN = indication   PI = priority indicator
 RS = response   MU = message unit
SOURCE: ITU-T Recommendation Q.922.

Frame rejection conditions are the same as in Q.921. However, in addition to reestablishing the link (as per Q.921), the equipment may transmit an FRMR response. (Q.921 user side only accepts FRMR responses, although LAPB can transmit such.) Unsolicited response frames are treated the same as in Q.921 with the exceptions of allowance of supervisory response frames with the F-bit set to 0, supervisory response frames with F-bit set to 1 in the timer recovery state, and for I-frame responses as detailed in the previous paragraph.

Timers in LAPF are used in an identical fashion to that of Q.921 (LAPD). Some of the default values are different. T200 has a default value of 1.5 s or, if known, the maximum of 3 times the round-trip transmission delay. The default window size, k, is 3 (the same as LAPD D-channel X.25 services) for a D-channel or 7 for a 64-kbps bearer channel. Q.922 adds additional sizes of 32 for a 384-kbps bearer channel and 40 for a 1.536 or 1.920-Mbps link.

Data Link Core Protocol

The core aspects of frame relaying bearer service are described in Annex A of ITU-T Recommendation Q.922. These are called the DL-CORE section. The core level is defined by use of ITU-T Recommendation I.122 and service description ITU-T Recommendation I.233. As a subset of LAPF, it provides a core to be used on any ISDN channel and may operate concurrently with LAPD. The core functions are considered to be frame delimiting, alignment and transparency, frame multiplexing/demultiplexing using the address field, detection of invalid frames, and congestion control functions.

The address field is the same as LAPF. The control field does not explicitly exist. (The format of the information field of the data link control sublayer may incorporate the equivalent of a UI control field.) The C/R bit is not used—but is passed transparently if set.

The primitive types supported in the data link core are shown in Table 7.3. Note that the PH_, MDL_, and M2N_ primitives are the same as for the full Q.922. The MC_ and DLC_ primitives allow direct use and control of the data link core sublayer. The management primitives are divided into layer boundaries. This allows easier segmentation of layers. For some implementations, this may be unnecessary. It is still useful in the case where various protocols are used within the same network or user equipment.

The MC_ASSIGN and MC_REMOVE primitives are a bit different in scope from the MDL_ASSIGN and MDL_REMOVE primitives. With the data link control management primitives, it is possible that the link will still need to be established. These primitives, for the data link core sublayer, imply readiness for data transfer.

ISDN Frame-Mode Signalling and Switching

The basic mechanisms for frame-mode signalling are given in ITU-T Recommendation Q.933. As is true of Q.922, this recommendation is also a differential document—this time based on that of Q.931. Annex A provides management extensions for control of PVCs that are similar in scope to that of ANSI T1.617, Annex D. These additions are discussed in greater depth in the section dealing with the LMI.

Q.933 breaks down the connection aspects of Frame Relay into two categories. Case A offers support for a circuit-mode switched circuit to

TABLE 7.3   Data Link Core Primitive Types

Generic name	Type*				Parameter*		Message unit content
	RQ	IN	RS	CF	PI	MU	
Layer 3 to layer 2 management							
M2N_ASSIGN	x					x	DL-CEI, DLCI
M2N_REMOVE	x					x	DLCI
DL-Core user to DL-Core							
DL_CORE_DATA	x	x				x	UI-carried data
Layer 2 to layer 2 management							
MDL_ASSIGN	x					x	DL_CORE CEI, DL-CEI
MDL_REMOVE	x					x	DL_CORE CEI
DL-Core to layer 2 management							
MC_ASSIGN	x	x				x	DLCI, DL_CORE CEI
MC_REMOVE	x					x	DLCI
Layer 2 to layer 1							
PH_DATA	x	x			x (BRI)	x	Data link layer peer-to-peer message

*RQ = request   CF = confirmation
 IN = indication   PI = priority indicator
 RS = response   MU = message unit
SOURCE: ITU-T Recommendation Q.922.

a remote frame handler that resides outside of the ISDN. Case B details procedures necessary for access to frame-mode services within an ISDN. As is true with X.31, Case A is supported only on the bearer channels. Case B may be provided on any bearer channel including the D-channel.

The general protocol is structured the same as in Q.931. Various message types are supported, each with its own mandatory and optional information elements. Supported message types are a subset of basic ISDN signalling, but additional IEs are defined to provide more specific frame-mode bearer service support.

States within the Q.933 protocol are the same as that of Q.931 for packet-mode network access. These vary from circuit-mode states in two ways. First, overlap sending and receiving are not supported. Next, the resume and suspend primitives do not exist.

Frame-mode connection control messages

Message types used within frame mode are a subset of the messages in Q.931. Table 7.4 lists these message types. The messages are used in the same manner as those for Q.931. There is a difference, however, in

TABLE 7.4   Q.933 Frame-Mode Connection Control Messages.

Message	Reference	Message	Reference
Call establishment messages:		Call Clearing messages:	
ALERTING	3.1.1	DISCONNECT	3.1.5
CALL PROCEEDING	3.1.2	RELEASE	3.1.7
CONNECT	3.1.3	RELEASE COMPLETE	3.1.8
CONNECT ACKNOWLEDGE	3.1.4	Miscellaneous messages:	
PROGRESS	3.1.6	STATUS	3.1.10
SETUP	3.1.9	STATUS ENQUIRY	3.1.11

SOURCE: ITU-T Recommendation Q.933.

the IEs that may be contained within the messages. Some IEs are not supported for frame-mode services, and some are special to frame mode.

There are two new IEs that are of particular significance for frame-mode bearer services. These are the DLCI IE and the Link Layer Core Parameters IE. Other new IEs used include a Packet Layer Binary Parameters IE, Link Layer Protocol Parameters IE, Connected Number and Connected Subaddress IEs, and X.213 Priority IE.

A set of IEs is also directly associated with use of the STATUS and STATUS_ENQUIRY message types for management of PVCs—and are not explicitly mentioned as IEs for any of the standard message types. These IEs are for report type, link integrity verification, and PVC status.

Bearer Capability IE. The Bearer Capability IE (as seen in Fig. 7.2) is used in a limited fashion to indicate frame-mode use of the network. It is designed to be used *only* by the network. Other information (Types II and III) must be contained in the link parameter IEs or the LLC IE. This requirement indicates a need for any ISDN support-

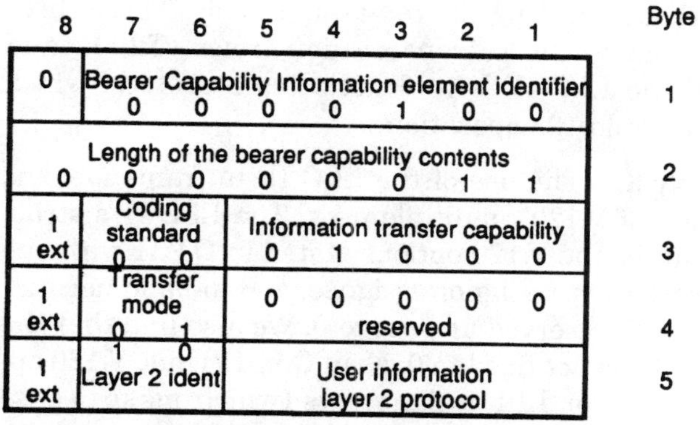

Figure 7.2 Q.933 Bearer Capability IE format. (*From ITU-T Recommendation Q.933.*)

ing frame-mode bearer services to transport these IEs in a transparent fashion (which does not currently exist on all ISDNs).

Although some of the fields in the bearer capability IE are the same as those listed in Q.931, the options for their use are strictly limited. The coding standard in byte 3 is to be 00 for ITU-T standardized coding, and the information transfer capability field in the same byte is 01000 for unrestricted digital information. The transfer mode field in byte 4 is 01 for frame mode. Only the layer 2 protocol information in byte 6 (byte 5 is omitted from the IE) has an option. The value of this field may be either 01110 for LAPF or 01111 for the data link core procedures.

Thus the encoding of the IE is almost fixed. Byte 1 has a value of 00000100 (hex 4) to indicate a bearer capability IE. The length, given in byte 2, is always 3. The next byte is 10001000 (hex 88) for unrestricted data, and byte 4 is 10100000 (hex A0) to indicate frame mode. Byte 6 can be either 11001110 (hex CE) or 11001111 (hex CF).

Data Link Connection Identifier IE.   The DLCI IE is used in the protocol in a similar way to that of a CHANNEL ID IE. The fields can support any of the DLCI lengths possible. The Pref/Excl field is an indication of whether a particular frame-mode channel is preferred or exclusive (required). The setting of this field may affect call setup procedures in case an "exclusive" DLCI is not available.

End-to-End Transit Delay IE and Link Layer Core Parameters IE.   These two IEs serve similar functions. The Transit Delay IE is used for indication of the delay inherent in a frame-mode connection. The Link Layer Core Parameters IE indicates requested quality values for the link. These values include outgoing and incoming maximum information field sizes, throughput, and burst sizes.

Link Layer Protocol parameters IE.   This IE (seen in Fig. 7.3) has three possible fields—each with its own identifier. Thus a field may be present, or not, without affecting the remaining fields. These three fields indicate the window size, retransmission timer (T200), and modulo sequencing (if the LLC IE byte 6 indicates a protocol which supports modulo 8 or modulo 128 operation).

Low-Layer Compatibility IE.   The use of the LLC IE in frame mode is an interesting extension of V.120's multiple uses. The LLC IE's structure is the same as that for the V.120 options stated in Q.931 with one exception. The LLI negotiation, assignor/assignee, and in-band negotiation fields of byte 5b are all reserved and set to 0. We also find that the negotiation field of byte 3a is *not* fixed to 0, as in Q.931. Thus, V.120 options are preserved except for LLI assignments (which makes sense since LLIs and DLCIs are in direct conflict within the address field).

Why V.120? V.120 allows three things that are extremely useful for

8	7	6	5	4	3	2	1	Byte
0	Link layer protocol parameters IE identifier							1
	1	0	0	1	0	0	1	
Length of link layer protocol parameters contents								2
0 ext	Transmit window size identifier							3*
	0	0	0	0	1	1	1	
1 ext	Transmit Window Value							3a*
0 ext	Retransmission Timer Value Identifier							4*
	0	0	0	1	0	0	1	
0 ext	Retransmission Timer Value							4a*
1 ext	Retransmission Timer Value (cont.)							4b*
0 ext	Mode of operation							5*
	0	0	0	1	1	1	1	
1 ext	Spare					Mode indication		5a*

* optional bytes

Figure 7.3 Q.933 Link Layer Protocol Parameters IE format. (*From ITU-T Recommendation Q.933.*)

specifying bearer services carried over Frame Relay. First, it supports terminal equipment. Second, it allows specifications for layers 2 and 3. This means that the V.120 format of the LLC IE can specify almost any ITU-T protocol. If you want to carry X.25 within the information field, specify LAPB at layer 2 and X.25 packet layer at layer 3. If you don't want to support a TE2, leave out the optional 5a through 5d bytes. If you want to do something special, use the optional 6a and 7a bytes with the layer 2 and layer 3 fields specified as user supplied.

There are limitations. This is frame mode. Thus, only frame-oriented HDLC protocols are supported. There is no possibility of using V.110 over the bearer channel. Also, signalling over V.120 is not possible (however, a layer 3 call control mechanism *is* possible).

Byte 3 specifies ITU-T unrestricted digital information. Transfer mode is set to frame mode (value 01). Layer 1 is specified to be V.120 (value 01000). Otherwise, the full variability (except as mentioned above) of the LLC is available.

X.213 priority IE.   This IE allows optional negotiation of priority for the frame-mode call. It is a maximum of 8 bytes long and is carried transparently by frame-mode networks.

Q.933 Signalling protocols

As stated at the beginning of the chapter, there are two categories of connections described as Case A and Case B. Case A is used for access to a remote frame handler. Case B is used for situations where the frame handler is part of the ISDN. This, however, is only part of the distinction. There are two parts to a call—gaining access and making an end connection.

Thus, Case A has two parts to it. The first is to use the Q.931 procedures on the D-channel to gain access to a remote frame handler. The next part is to use Q.933 procedures on DLCI 0 within the assigned bearer channel. Only after this second setup is accomplished is there an end-to-end frame-mode bearer channel available.

Case B is much simpler. A Q.933 frame-mode call setup is placed on the D-channel. If the ISDN channel is "semipermanent" (sometimes called nailed-up, or presubscribed), a non-frame-mode access (Case A) is created by using the in-band DLCI 0 Q.933 signalling protocol, or a frame-mode access (Case B) is established using Q.933 over the D-channel. It is also possible to have a semipermanent frame-mode connection—in which case no signalling is needed or allowed.

Data Encapsulation

Frame Relay provides an efficient method of transporting data—but, what kind of data? The protocol is actually indifferent to the type of data. On a PVC, Annex A talks about sending UI frames over the data link core sublayer. It also discusses various methods of adding, deleting, and marking various PVCs as inactive.

The LLC IE, as we have seen, provides enormous flexibility as to the content of the information field. That is unnecessary, actually, but it does provide useful signalling methods to determine protocols.

One of the best aspects of Frame Relay is that it can be used as an intermediary network. This means that a Frame Relay network can be set up between any two high-traffic areas. It does *not* have to be omnipresent. If you have a LAN in your company headquarters, a frame-mode router can transfer the data to another remote location to be used on that LAN. Other bearer services can be used to provide the interlinks, but Frame Relay is fast.

Another aspect of data encapsulation is that the data to be transmitted is likely to already have a protocol for error detection and retransmission. X.25, LAN protocols, and others have a set of protocols to deal with such problems. Why duplicate the effort? Frame Relay provides a method to encapsulate the protocol without redundancy. The negative aspects become positive ones. Yes, frames may be lost over the network. But, if it is a statistically insignificant number, the

upper layer protocols (that are present, even if the underlying protocol is more robust) provide a method of recovery.

Congestion Control

A frame-mode network has varying amounts of data traffic from various points. One of the advantages of a frame-mode (or packet-mode)
network is that the total capacity of the network does not fix the
amount of each channel. Yes, a maximum is available for each channel. A 64-kbps bearer channel cannot transport more than 64 kbps (actually less, with flags and overhead). However, a frame-mode network
with a 1.544-Mbps capacity *can* support 50 channels as long as, statistically, the total amount of traffic does not exceed the bandwidth.

This is fine, as long as the connected channels meet the statistical
requirement. A network cannot make the assumption that this will
always occur. The process of making sure that the network can handle the traffic, in a "fair" way, is called congestion control.

The first thing necessary for a network is to keep track of utilization.
The user equipment notifies the network of several parameters. These
parameters include necessary QOS, expected normal traffic load, and
potential "burst" traffic. This information, combined with ongoing
analysis of the actual data traffic load, allows the network to determine
how to allocate resources. Congestion may also occur as a side effect of
network resources that undergo temporary (or permanent) failure.

Congestion control is achieved by congestion avoidance or congestion recovery mechanisms or both. Congestion avoidance is a mechanism used upon noticing that overload traffic situations are imminent.
This relies on cooperation of the user equipment on the network.
Congestion recovery is used to keep the network operational when
overload occurs. This mechanism usually requires loss of data (and
recovery at the higher layers or the user equipment). Note that congestion recovery will not cure the problem in itself. That requires a
reduction in the overall traffic.

Congestion avoidance

ITU-T Recommendation I.370 discusses congestion avoidance mechanisms. Congestion avoidance takes place in the U-plane of ISDN.
That is, it requires the notification and cooperation of the user equipment protocol stacks. There are two *points* defined by I.370 as stages
of congestion. Point A causes a degradation of service—mainly indicated by increased transit delay. This point A is the final place on the
traffic-load curve where the QOS can be guaranteed. Point *B* is where
the network must discard frames. It is possible to go from point *A* to

point *B,* without an increase in user data traffic, if network resources fail or undergo reconfiguration.

Congestion management strategies seek to minimize loss of frames, maintain QOS, and allocate channel capacity fairly among user equipment while requiring a minimal overhead necessary to implement the strategies. They should also try to limit the congestion from spreading to other networks or subsystems within the network.

BECN and FECN. The BECN and FECN bits of the data link core sublayer are used by the network to notify, or request, user equipment to prepare for congestion. The BECN set to 1 indicates congestion in the direction *opposite* to that of the frame being received. The FECN set to 1 indicates congestion in the same direction as being received. Thus a BECN field is a request that the user equipment reduce the traffic that it is currently sending. The FECN field can be used by the user equipment in two ways. One, it can use the information as notification that frame loss may occur. Next, if its protocol requires responses, it may be able to reduce "backward" traffic by decreasing the "forward" traffic. Less data to acknowledge means less data coming back through the network.

If possible, the user equipment reduces its traffic load based on the ECN fields. It will either maintain this lowered level until further notification that the congestion is cleared, or it may progressively return to the originally negotiated information transfer rate. The user equipment must decide the best strategy for its own priority needs. Keeping the network down to point *A* (or before) is in the interest of all user equipment on the network. The user equipment decides whether the possible frame loss is acceptable. The network may apply congestion recovery mechanisms (frame discards) to a particular channel if the user equipment does not respond to ECNs.

CLLM. The Consolidated Link Layer Management (CLLM) message is defined in ITU-T Recommendation Q.922. This optionally supported message may be generated as an explicit notification of the causes of congestion, or current status, of network resources. It is sent as part of a Q.921/Q.922 XID response frame type. If the message is sent on a D-channel, the address will be a value of 62 for the SAPI and 127 for the TEI (broadcast link layer management). If the message is sent on a frame-mode connection, it will be sent with the maintenance DLCI. The message has a mostly fixed form, as detailed in ITU-T Recommendation Q.922.

Congestion Recovery

Once point *B* has been reached, the network must make sure that the traffic degrades the overall service as little as possible. It can only do this by either reducing the traffic or by allocating additional resources, if possible. At any rate, during this process, some frames will have to be discarded.

The congestion recovery mechanism has two general requirements. The primary task is to reduce the traffic load. The secondary task is to reduce the traffic in a manner which is as fair as possible for all the user equipment currently using the network. The concept of fairness is applied to mean that each piece of equipment is still allowed some access to the network. It is preferred that the user equipment be given continued access at the committed information rate that was negotiated at call setup time.

One thing that the network can, and should, do is to give priority to existing calls. Additional connections can be refused. As calls complete, the traffic load will inherently decrease. Another possibility is to allow only a reduced information rate at the time of connection. This will also reduce overall traffic over a period of time.

The network will carry out some type of algorithm in deciding which frames to discard. One item in the algorithm, as discussed, is how much over the Committed Information Rate (CIR) a given piece of equipment is currently transmitting. A fair algorithm would keep the traffic over the committed rate approximately equal among all user equipment. An exception to this might be if the connection was interworking into another congested network. Additional frame deletion, in this situation, may be warranted for a given piece of equipment.

The user equipment does have some degree of control over just *which* frames are to be discarded. This is provided by the Discard Eligibility (DE) bit of the data link core sublayer. The network should discard frames with this bit set to 1 before discarding other frames. If sufficient traffic reduction is achieved in this manner, no frames with the DE-bit set to 0 will be deliberately discarded.

Local Management Interface

The congestion control mechanisms are designed to allow the network a method of keeping traffic to levels that it can support. The LMI is designed to allow user equipment knowledge of the status of various PVCs to which the user equipment has subscribed. The ITU-T Recommendation Q.922, Annex A, does not specifically refer to these procedures by the LMI name. This Annex merely refers to them as "additional procedures for PVCs using unnumbered information frames." In this book, we will refer to this collection of procedures as the LMI.

The LMI procedures allow for notification of the addition of a PVC, detection of the deletion of a PVC, the availability of use of a configured PVC, and the link integrity. Note that these procedures only apply to PVCs (and, only for Case A configurations thereof). This means that the set of PVCs is actually a set of subscribed links. Addition means *only* notification that the PVC is now available in the network. Switched procedures are required for actual addition of links past the subscribed base.

STATUS and STATUS ENQUIRY message types

The main provision for the LMI procedures are the STATUS and STATUS ENQUIRY message types. The STATUS message type includes the Report Type IE, Link Integrity Verification IE, and the PVC Status IE. The STATUS ENQUIRY message type includes only the Report Type IE and the Link Integrity Verification IE. All IEs are potentially optional. Messages used for PVC status are sent using a dummy CRV on DLCI 0.

The Report Type IE may be a full status of all PVCs on the bearer channel, a link integrity verification only, or a single PVC asynchronous status report. The Link Integrity Verification IE allows a polled, sequenced, message to be exchanged between the user equipment and network. This is used to verify usability of the signalling channel (DLCI 0).

PVC Status IE. This IE (Fig. 7.4) may be repeated within a STATUS message to allow status information to be obtained on all PVCs available on a bearer channel. There is an exception in that, with 4 message header bytes and 5 bytes required per PVC, the 256-byte default maximum information field may not be able to contain all PVC information in one message. A limit of 50 PVCs in one STATUS message exists for the default message length. In such a case, multiple STATUS messages will be sent by the network.

8	7	6	5	4	3	2	1	Byte
0	PVC Status Information element identifier 1 0 1 0 1 1 1							1
Length of the PVC Status contents								2
1 ext	0 spare	Data Link Connection Identifier (Most significant 6 bits)						3
1 ext	Data Link Connection Identifier (2nd most significant 4 bits)			0	0 spare	0		3a
1 ext	0	0 spare	0	New	0 spare	Active	0 spare	4

Figure 7.4   Q.933 PVC Status IE format. (*From ITU-T Recommendation Q.933.*)

Each PVC Status IE contains the DLCI, and 2 information bits associated with the PVC. These information bits are declared to be the *new* bit and the *active* bit. The new bit is set to 1 for the first STATUS message in which a particular PVC is declared among the PVC Status IEs. It will be set to 0 for subsequent STATUS messages (once the receipt of the STATUS message is implicitly acknowledged by subsequent Link Integrity Verification IEs). The active bit indicates whether a PVC is active or inactive. An inactive PVC, although still configured, is to be considered unavailable for data transfer.

If the PVC becomes completely unavailable, the appropriate PVC Status IE will be deleted from the STATUS message. This indicates that the user equipment must maintain a list of PVCs and their status. Each time a full report STATUS message arrives, the list of PVCs may change.

Use of the STATUS and STATUS ENQUIRY message types.   The STATUS message type is always sent by the network. The STATUS ENQUIRY message type is sent by the user equipment. The STATUS message type is used as a response to a STATUS ENQUIRY message type or, optionally, sent on an asynchronous basis by the network.

The user equipment supporting LMI procedures keeps a running timer (restarted each time it expires) of T391 s (default, 10 s). Normally, when the timer expires, a STATUS ENQUIRY message requesting link verification is sent to the network. Every N391 (default 6) polling cycles a STATUS ENQUIRY message requesting a full report is sent. The STATUS message received back from the network is used to maintain PVC lists (as mentioned above).

Two other counters are maintained by the user equipment. One counter concerns the number of events that have happened. This counter starts at 0 and goes up to N393 (default 4). An event list is kept for all events up to N393. That is, only the last N393 events (or fewer in case of start-up) are kept. This event counter is used in conjunction with an error counter. If N392 (default 3) of the past N393 events are to be considered errors, the channel is to be considered temporarily unusable (except for continued STATUS ENQUIRY messages). If N392 events occur without error, the channel is to be considered usable once again.

From the user equipment side, an error is considered to be the non-receipt of a STATUS message of a full report type after such a STATUS ENQUIRY message type has been sent. A second error type is the receipt of a STATUS message with invalid sequencing in the Link Verification Information IE. Such an error indicates loss of frames on the channel.

From the network side, nonreceipt of a STATUS ENQUIRY message from the user equipment within T392 (default, 15 s) is considered an error. Also, receipt of a STATUS ENQUIRY message with invalid sequencing in the Link Verification Information IE would be considered an error. The same use of N392 and N393 applies to the network side as to the user equipment side.

LMI procedures for acknowledged operation

Unlike T1.617, Annex D, the Q.922 Annexes include provision of procedures for acknowledged operation mode. These are given in ITU-T Recommendation Q.922, Annex B. This annex is meant for Case A switched connections where both switched and permanent connections are available or for Case B (using signalling on the D-channel).

The procedures entail use of DLCI 0 STATUS and STATUS ENQUIRY messages. These are sent over DLCI 0 for Case A situations or over SAPI 0 for Case B situations. The contents of the message types are slightly different from those of unacknowledged PVC operation. The STATUS message includes a Cause, Call State, Channel ID, DLCI, and Display IEs. The STATUS ENQUIRY message type includes the Channel ID, DLC, and Display IEs.

The DLCI IE is used to specify a particular PVC. Two Cause IEs are included in the STATUS message. The first Cause IE will indicate value 30 (response to a STATUS ENQUIRY). The second Cause IE will contain either value 39 (permanent frame-mode connection out of service) or 40 (permanent frame-mode connection operational). The network should send a STATUS message for each PVC available as soon as the signalling channel is available. Timer T322 should be used in conjunction with sending a STATUS ENQUIRY message by the user equipment.

Other Frame Relay Network Options

Other options may be made available in Frame Relay in future releases of recommendations. One option, found in some proprietary networks, is to provide a multicast capability based on DLCIs to allow a limited broadcast capability. Global addressing capabilities may be useful for DLCI internetwork identification. Additionally, the use of 3- and 4-byte DLCIs needs further study.

Software Considerations

The software architecture needs for Q.922 are basically the same as for the Q.921 or V.120 data link layer. Additional management primitives are needed within the C-plane, but these are relatively transparent to the actual data transport protocols. The new bits and DLCI field must be allowed for within the address field. Most other matters are the same.

The difference lies in the implementation of the data link core sublayer. The functions needed are parallel to those of the data link control sublayer. Interactions between these two sublayers can be difficult to combine in an efficient manner. There are also needs for the data link core sublayer to act independently for support of PVCs using UI frames.

For these reasons, it is probably best to incorporate the SVC portion of Q.933 within the main data link layer (incorporating the data link control sublayer and the data link core sublayer). The data link core sublayer for PVCs should *also* be implemented as an entity. The additional task provides flexibility and efficiency. In this way, data link layer data fields may be relayed to the LLDs or to the data link core sublayer in a semitransparent manner.

So, a separate data link core sublayer exists. It provides for the LMI functions, PVC lists, and UI transfer over the PVCs. The data link layer entity handles SVCs and the Annex B LMI protocols.

Module entry points

The data link core sublayer and the data link layer act as parallel software modules. Each can have access to the LLD to control and make requests of the hardware. The data link layer (and, more importantly, the data link control sublayer) acts as a little higher layer in the protocol stack. Data may be sent from the data link layer down to the LLD or to the data link core sublayer. Data may be sent up from the data link sublayer to the higher layers or to the data link layer.

The data link layer has the same entry points as discussed in Chap. 4, with a decision as to whether data is sent "down" to the LLD or to the data link core sublayer. The data link core sublayer has the same type of entry points: one between it and the upper layers, one between it and the management control functions, and one between it and the LLD. The operation of the interface toward the LLD is synchronous. Interactions with the management functions are synchronous. Other interactions may use operating system asynchronous message queues. In particular, the transfer of primitives from the LLD to the data link core sublayer should have a *software break* so that interrupt handling is minimized.

Operating system use

It may be desirable to have a separate set of OS tasks for the core protocol. It is also possible to reroute to the data link core functions at the regular data link layer entry points. There are some structural advantages to having the data link core operate as separate tasks. Mainly, this allows versatility in reusing other modules as part of an embedded data transport protocol.

Coordination/management requirements

SVCs for frame-mode bearer services can be handled in the same manner as other switched services (with the addition of new management M2N and MC primitives). PVCs can be handled either within the data link core sublayer or within the coordination/management entity. For efficiency, it is probably best to leave the general LMI actions in the data link core sublayer. However, status of the active PVCs should be maintained in the coordination section of the system so that unnecessary primitives are not generated toward the core functions (such as a DC_DATA_RQ for a DLCI that is not available).

Summary of Chapter

Frame-mode bearer services are a bridge between N- and B-ISDN. They provide access to bearer channels of varying bandwidth. They also provide a mechanism for carrying embedded data protocols that can facilitate interworking. Since the frame-mode bearer service requires only the lower two OSI layers for switching functions in the network, it provides an efficient method of providing data transport. Efficiency depends, however, on the underlying network providing a relatively error-free framework.

Frame Relay is actually a network service that supports frame-mode bearer services. This can be support of PVCs only or of SVCs in combination with PVCs. *Frame switching* is sometimes used as a term to indicate that the information field is transported without the surrounding frame structure. This provides a means to use ATM or other B-ISDN mechanisms as a data transport mechanism between networks.

Frame-mode bearer services fall into two separate categories. One is the data link core sublayer, defined by ITU-T Recommendation Q.922, Annex A. The full data link layer, called LAPF, is described by Q.922. ITU-T Recommendation Q.933 specifies the methods to be used for either circuit-mode connection to a remote frame handler or in-band (DLCI 0) signalling access to a frame handler incorporated into the ISDN.

A Frame Relay network must handle data traffic flow. It is wasteful to allocate resources such that all data possible from each unit of the connected user equipment may be transported. Frame mode (and packet mode) implies a nonstatic data load. This variable data load is handled by use of congestion management procedures. Congestion avoidance implements a voluntary mechanism to allow the user equipment to reduce data traffic before it becomes a problem. Congestion recovery incorporates algorithms to discard frames that the network cannot support.

Frame Relay using PVCs with unacknowledged information transfer requires methods to determine the state of each PVC. The LMI procedures allow the user equipment to know the status of each subscribed PVC. Another set of abbreviated procedures can be used with PVCs supporting acknowledged transfer.

The next chapter discusses ISDN and ATM. Frame Relay can either be considered to be a form of B-ISDN or as a bridge between N-ISDN and B-ISDN.

8

Broadband-ISDN and Asynchronous Transfer Mode

B-ISDN, in a sense, is not a separate type of ISDN. Instead, what it refers to is the incorporation of wider bandwidths into the general ISDN. It is called broadband mainly to delineate it from the slower network (sometimes called N-ISDN). B-ISDN is designed to be used with different services—services that require large amounts of data to be transmitted quickly. Thus, there are particular bearer services that can be fulfilled by B-ISDN that cannot be fulfilled by N-ISDN.

A synchronous transfer mode will send data in a set sequence, delimited by some sort of data markers. A BRI falls into this category. The signal (as discussed in Chap. 3) is subdivided into different bearer channels. This organization stays constant for the entire connection—although it is possible, at network access points, that the information from one channel will be shifted from one physical line to another. In an asynchronous transfer mode, each Protocol Data Unit (PDU) carries its own identification and address with the data. This adds extra overhead to each unit of data but means that each PDU can be routed, encapsulated, or otherwise redirected in an independent fashion. The ATM, currently a primary architectural method for providing B-ISDN services, carries these PDUs within cells.

Philosophy of B-ISDN

The beginning B-ISDN recommendations of the ITU-T (then CCITT) in 1988 outlined a general philosophy for B-ISDN. These documents, ITU-T Recommendations I.113 and I.121, provided for the vocabulary to be used and the general aspects for broadband use of ISDN.

The ITU-T recommendations published in 1988 (the Blue Books) were the last to be issued as bound sets by the ITU-T for two reasons. One was that, with the increased number of revisions and new documents coming out, it was considered undesirable to have to keep releasing new compendiums of recommendations. This was for ecological as well as practical reasons. The other reason, however, was that the demand for standards covering higher-speed services was causing the old 4-year cycles to be unusable.

In response to these needs, the ITU-T has been issuing new documents concerning higher-speed services (such as Frame Relay and B-ISDN), interworking, and other critical standards at a much faster pace. So, by the time this book is published, many new recommendations will probably exist for B-ISDN. However, this chapter will provide a sound basis that can be used with those new recommendations. Most of the new documents, published since 1990, are concerned with the bearer services for B-ISDN. These recommendations are in the F.2xx series of documents.

ITU-T Recommendations I.113, I.121, I.150, I.211, I.311, I.321, and I.327 are concerned with vocabulary, functional aspects, and general architecture components of B-ISDN. ITU-T Recommendation I.432, discussed in Chap. 3, is concerned with the physical layer of B-ISDN. I.413 provides the reference models necessary for the UNI for B-ISDN. This will be discussed as part of the general text. ITU Recommendation I.460 gives the minimum functions needed to make use of the physical layer and ATM layers by customers and will be discussed more later in this chapter.

Most of this chapter involves discussion of three ITU-T documents that directly concern the implementor. These documents are called I.361, I.362, and I.363. These documents cover the B-ISDN ATM Layer Specification, B-ISDN ATM Adaptation Layer (AAL) Functional Description, and the B-ISDN ATM Adaptation Layer (AAL) Specification.

B-ISDN Services

B-ISDN does not have any more functional possibilities than any other form of ISDN. What it does offer is speed. Using ISDN for High-Definition TeleVision (HDTV), for example, would require about 20 Mbps for compressed real-time images. However, video phone service has been demonstrated, with reasonable quality, on BRI using the two B-channels in bonded service. Thus, B-ISDN is not a requirement for a particular type of feature—only for the amount of data in a given amount of time.

Australia is currently implementing ISDN on a widespread basis because the country is very large and the population is very dispersed. Remote medical applications become quite possible if the bandwidth is available. Place remote sensor equipment in the medical offices across the country and keep the expensive parts of the equipment centralized. Rather than needing duplicate equipment at many different points, use of a high-speed connection allows only the data to be transported. This may, indeed, help to keep medical costs down in many countries.

ITU-T Recommendation I.211 discusses service aspects of B-ISDN. It identifies two main service categories. These are interactive and distribution services. Interactive services have three subdivisions, which are conversational, messaging, and retrieval services. The distribution services are delineated by whether the user has any control over the information flow. Table 1 in ITU-T Recommendation I.211 lists divisions, and possible attributes, of broadband services.

Interactive services

The first subdivision of interactive services is called conversational services. This allows bidirectional communication with real-time information transfer. The fact that bidirectional transfer is possible does not mean that all such services will use the capability in that way. Video surveillance requires only unidirectional data transfer. I.211 lists videotelephony, video conference, and high-speed data transmission as examples of this type of service.

The next services, called messaging services, are no longer real time. Storage of data may take place at some point in the data flow. Thus, perhaps, data will be transferred at high speeds to a location and then, later, transferred to another location. Message-handling services and mail services for movies, high-resolution images, and audio information are listed as examples of this type of service.

The last subdivision of interactive services is called retrieval services. You might also call this a data library service. Film, high-resolution images, and audio and archival information (perhaps digitized images of original documents) can be stored at central locations and accessed.

A pattern can be noticed in the subdivision of interactive services. The first offers real-time services in a potentially bidirectional manner. The next two are not real time. The second expects data to go in both directions—but not at the same time. The third expects data to only be accessed in one direction. These aspects seem to fall into agreement with the ITU-T-defined transmission structures.

Distribution services

Distribution services imply that the information is always available. An interactive retrieval service sends the data only after they have been requested. Distribution services are always being transmitted. This is similar to cable broadcast services of today.

These services are divided into two categories. In the first case, the information flow is always fixed. Anyone who wants access to the data may tap into the data flow—but there is no control over the data presented. Examples include broadcast services for television and audio programs.

The second case allows the user individual presentation control. A fixed form of data is available, but the user may request the information be presented in a certain manner or starting at a certain point. In other words, the data are fixed, but the access is not completely fixed. A full-channel broadcast videography would fall into this category.

Multimedia

B-ISDN is very well suited to multimedia situations. Television has audio and video components. Closed-caption television has video, audio, and text components. A musical program on closed-caption television has video, audio, high-quality audio (music), text, and so forth. With remote equipment, it may be possible to watch an instructional program and have the program monitor your movements as you try to follow the instructions. This is bidirectional multimedia, which makes the physical classroom less important for instruction (but still important for learning social skills).

ITU-T broadband service specifications

One method of determining which services are generally seen as particularly important is from the types of services being addressed by standards. At the beginning of this section, we mentioned some of the F.*xxx* recommendations. F.310 discusses aspects of a broadband videotex service. F.732 is about broadband videoconference services. F.821 is about broadband TV distribution services, and F.822 discusses broadband HDTV distribution services. Note that all of these services interact in some manner with video. A video application is one of the more obvious bearer services that is ideally suited to B-ISDN because of the enormous (by today's standards) amount of data used in small amounts of time.

Transmission Aspects

The bearer services above tie directly into the types of transmission capabilities discussed in Chap. 3. There are three types of connections defined. These are a bidirectional 155.52-Mbps connection, a 155.52/622.08-Mbps asymmetric connection, and a bidirectional 622.08-Mbps connection. The bidirectional connections can be used for interactive conversational services, and asymmetric ones can be used for distribution services with user control. Other services are less well defined in their needs.

Maximum bit-rate support

Just because a connection supports 155.52-Mbps connection does not mean that the data transport rate is that high. The payload capacity of such a connection is only 149.76 Mbps, and, with the 5-byte header overhead of an ATM cell, the actual maximum throughput rate is 135.631 Mbps. In other words, data transport is about 87.2 percent of the full bandwidth. ITU-T Recommendation I.211 does not specify the maximum service bit rate of a 622.08-Mbps interface, but if the physical structure is the same, 87.2 percent of 622.08 Mbps produces a data throughput of 542.524 Mbps.

Transmission media

As mentioned in Chap. 3, ITU-T Recommendation I.432 does not specify the actual transmission media to be used for B-ISDN. The lower-rate 155.52-Mbps service might be handled over coaxial cable or fiber optics—although the coaxial solution does have length limitations. Current technology limits the 622.08-Mbps transmission to fiber optics. Of course, current technology soon becomes obsolete technology. The ITU-T documentation, therefore, emphasizes architecture and structure rather than mechanisms. It is hoped that ISDN architecture and services will be sufficient to allow a continual increase in services over the next few decades.

Cells and Cell Relay

Frames, frame mode, frame switching, Frame Relay. These terms give an orderly ranking of those involved in one type of data transport. A frame is a unit of data—a PDU, to use the later ITU-T nomenclature. Frame mode is associated with a service that makes use of a particular type of PDU (the frame). Frame switching is the process of routing the service using the PDU. *Frame Relay* is a term for the network which employs frame switching that routes the service using the PDU. This same hierarchy of names is used for other types of PDUs. It's true for packets, for example. It is even true for circuits—although the circuit mode was long in use before the concepts really got formalized. How is a circuit a PDU? After all, a circuit does not have a particular form of data—or does it? When we look at V.110, for example, we see that the entire circuit is, in itself, a PDU. It is possible to use it to carry other types of PDUs (such as frames or packets), but that is true of every other kind of PDU also.

8	7	6	5	4	3	2	1	Bit / Octet
GFC				VPI				1
VPI				VCI				2
VCI								3
VCI				PT		RES	CLP	4
HEC								5

CLP   Cell Loss Priority
GFC   Generic Flow Control
PT   Payload Type
RES   Reserved
HEC   Header Error Control
VPI   Virtual Path Identifier

Figure 8.1   B-ISDN UNI format. (*From ITU-T Recommendation I.361.*)

Cells fall into this same category. As we saw in Chap. 3, the general framework for physical transportation of cells is within a Synchronous Transfer Mode (STM-1 for 155.52 Mbps) container. The cells are the PDUs within this container. These cells travel within the cell-mode service and are routed by cell switches as part of a Cell Relay network.

The cells are more fully called ATM cells. ATM is basically the equivalent of the layer 2 for HDLC modes. Thus an ATM cell is in the same equivalent part of the B-ISDN architecture as the data link layer frame for N-ISDN. The ATM cell is composed of 53 bytes. The first 5 bytes are the header bytes. These 5 bytes are slightly different depending on whether the cells are passing the UNI, (Fig. 8.1) or the Network-Network Interface (NNI, Fig. 8.2). The fifth byte of the header is, in both cases, the Header Error Control (HEC) byte. The use of the HEC was discussed briefly in Chap. 3. It provides the equivalent of the Frame Check Sequence (FCS) of the data link layer frame.

ATM cell formats

A cell can be used by the physical layer for its needs, the ATM layer for its needs, or by the AAL for its needs. Table 8.1 shows the basic formats for the first 4 bytes of the ATM cell header. The lowest-order bit of byte 4 is called the Cell Loss Priority (CLP) bit. The setting of this bit indicates to the network that this cell may be discarded, if necessary. Note that the basic format of the first 4 bytes indicates

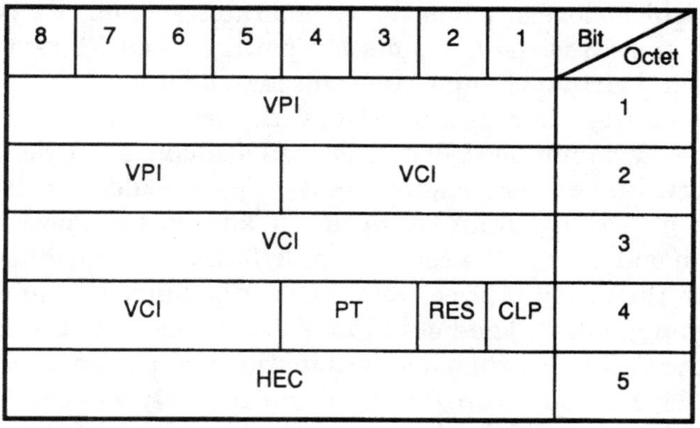

8	7	6	5	4	3	2	1	Bit / Octet
VPI								1
VPI				VCI				2
VCI								3
VCI			PT		RES	CLP		4
HEC								5

CLP   Cell Loss Priority
GFC   Generic Flow Control
PT   Payload Type
RES   Reserved
HEC   Header Error Control
VPI   Virtual Path Identifier

Figure 8.2   B-ISDN NNI format. (*From ITU-T Recommendation I.361.*)

TABLE 8.1   B-ISDN UNI Preassigned Cell Header Values

	Octet 1	Octet 2	Octet 3	Octet 4	Octet 5
Reserved for use of the physical layer	PPPP0000	00000000	00000000	0000PPP1	HEC
Unassigned cell identification	AAAA0000	00000000	00000000	0000AAA0	HEC

SOURCE: ITU-T Recommendation I.361.

that this same bit is indicative of a physical layer use of the cell. According to ITU-T Recommendation I.361, this bit is *not* used for the CLP mechanism as described in ITU-T Recommendation I.150. However, the effect is the same. A cell used by the physical layer is not passed to the ATM layer.

Physical layer cell use

The field marked *PPP* for the physical layer encoding identifies the particular Operation And Maintenance (OAM) use of the cell (as described in ITU-T Recommendation I.610). These uses are identified as F1 through F3 and are used for network line monitoring. F4 and F5 are used by the ATM layer for recognition of integrity at the end-to-end points. These are names of various connection points within a network, and such cells are used to determine network *alarm* (Alarm Indication Signals, or AIS) signals. These F (or *flow*) points are part of the OAM hierarchical levels. F1 is the regenerator section, F2 is the digital section, and F3 is the transmission path. F4 and F5 refer to the virtual path and virtual channel (to be discussed later).

An F1 path goes between a Broadband Network Termination 1 (B-NT1) point or a Line Termination (LT) and the cell switching function. An F2 path lies between two termination points. These could be a B-NT2 and a B-NT1, a B-NT1 and an LT, or an LT and an LT. The F3 path passes from an endpoint to the cell-switching function—providing an integrity path, at the physical level, from the user equipment to and from the cell-switching matrix. The F4 and F5 paths are used to check integrity *through* the Cell Relay from one termination point to another. The PPP values of F1, F2, and F3 are 001, 010, and 100, respectively.

Another physical layer use of the ATM cell is to provide the format for an *idle cell*. Such a cell means that it is not to be considered a PDU with useful information. The value of such a cell's first 4 header bytes is 00000000 00000000 00000000 00000001.

User-network interface and network-network interface

When we look at Fig. 8.1 on the UNI for B-ISDN, we find six fields. These six fields are the Generic Flow Control (GFC) field, Virtual Path Identifier (VPI), Virtual Circuit Identifier (VCI), Payload Type (PT), CLP, and HEC fields. The NNI, as shown in Fig. 8.2, has the same fields except for the GFC. Instead, the VPI has been expanded from 2 to 3 bytes.

The GFC is currently coded to be 0000. The purpose of this field is to allow flow control between the user equipment and the network (which is why it does *not* appear in the NNI). Its exact use is still for further study.

The VPI and VCI are considered together as a routing field. The UNI has 8 bits available for the VPI, and the NNI has 12 bits available. Both have 16 bits available for the VCI. The use of these routing fields will be discussed soon.

The PT field is a default value of 00. Other values are still under study. The purpose of this field is to describe the contents of the payload (or end-to-end data content). The CLP was described briefly. The value of 1 for the CLP bit gives the cell a lower priority and is to be considered a prime candidate for cell loss (similar to the DE bit of frame mode).

B-ISDN Protocol Architecture

We have now discussed (above and in Chap. 3) the basic underlying technology supporting Cell Relay. In the rest of this chapter, we will discuss how this technology can be used to provide high-bandwidth capabilities within the B-ISDN protocol architecture. ITU-T Recommendations I.113 and I.121 give the vocabulary used and general philosophy aspects of B-ISDN.

The general purpose of B-ISDN is to support a wide range of services. These include services that require large bandwidths, but they also include the lower-bandwidth services currently allocated to N-ISDN. Thus, B-ISDN is a superset of N-ISDN, and the term *broadband* is used mainly to distinguish technologies and capacities. As is true for N-ISDN, B-ISDN takes a layered approach to the needs of the network and should support both switched and nonswitched (semipermanent) connections.

The ATM is seen as the basic underlying mechanism to be used for B-ISDN. This provides a limit (though much higher than current networks), and thus, it may be expected that some other mechanism will be defined in the future to incorporate even higher bandwidths as the need surfaces (perhaps Ultra Broadband ISDN, or UB-ISDN). At any rate, the ITU-T documentation attempts to set together an architectural philosophy that will allow growth while maintaining support of other levels. Thus, B-ISDN is to support the N-ISDN functions and services, and other protocols to come must support B-ISDN functions and services to maintain integrity within the architectural basework of ISDN.

Virtual Channels and Virtual Paths

In the discussion on cells and Cell Relay, we discussed the VCI and VPI fields very briefly. These two fields correspond roughly to a logical channel of X.25 and the address field of Q.931. The VCI, like V.120, may also be treated as a DLCI. That is, *unlike* X.25, a particular VCI may be identified as a signalling channel or a maintenance channel or for some other particular purpose. All of the Virtual Channel Connections (VCCs), however, are associated with a particular Virtual Path Connection (VPC). The VPI, however, may only be a partial address (many signalling aspects for B-ISDN are yet to be defined).

Use of VPI

The point of a VPI is to identify how a stream of cells is to be switched from one network node to another. The UNI version, which contains only 8 bits for the VPI, acts as an identifier to the next point in the Cell Relay switch. This network node, in turn, will have the cell stream routed to another point, perhaps using a totally different VPI. The signalling system identifies endpoints; the VPIs identify routing.

Let's say that, via some yet not fully defined signalling mechanism, a connection is made between two endpoints. The network will return a VPI to the calling equipment. This VPI will be associated with the connection by the network. Therefore, from the point of view of the user equipment, the VPI *is* the identifier associated with the connection to the endpoint. The network, however, knows that this is only the connection between the first network node and the user equipment. This means that the total number of possible connections is limited to 64 (28 possible VPI values) for a particular B-ISDN UNI. It is possible that a particular VCC will be switched at the end of the VPC.

The network knows the full address of the connection. Each UNI will have a fixed VPI for the connection (though probably not the same VPI). In between, however, the connection may be routed in any manner deemed convenient by the network. Thus a VPI of value hex 62 might be routed on by the NNI VPI of 113, another of 173, one of 893, and then a UNI VPI of 54. If a particular network node became unavailable, the network might reconfigure the path as UNI VPI 62, NNI 809, NNI 1073, NNI 15, and UNI VPI 54. To the user, the connection would be the same. It would even be possible to change the routing path for each cell, although this might cause transit delay problems and even loss of sequencing.

Cell switching

The Cell Relay switch may look very similar to that of an old analog step-by-step. One such switch configuration is called a Banyan switch (seen in Fig. 8.3). Many other topologies are possible, but the Banyan switch works well as an example of the possibilities. Each node connects to two other nodes. Thus, the first address bit performs a switch between nodes. The next performs a switch between two more nodes, and so forth. A set of n stages allows for a cross-matrix switch of 2^n addresses.

Cell switching has some restrictions imposed upon it by the requirements of B-ISDN. The first is that sequencing of cells must be maintained. The second category is that a particular QOS be maintained. Keeping the various switch connections at an equal length helps to keep delay factors in line—helping to maintain sequencing and other factors.

OSI Layers Applied to B-ISDN

Fig. 8.4 (adapted from ITU-T Recommendation I.121) shows a model of the B-ISDN protocol for the ATM. The ATM layer takes the place of the layer 2 and the AAL takes the place of layer 3. Note the division of higher-layer protocols and functions. They are divided into the U- and C-planes. The C-plane can be expected to be used for control functions—determining the function and connection of the service to be used by the U-plane. The Plane management functions are repre-

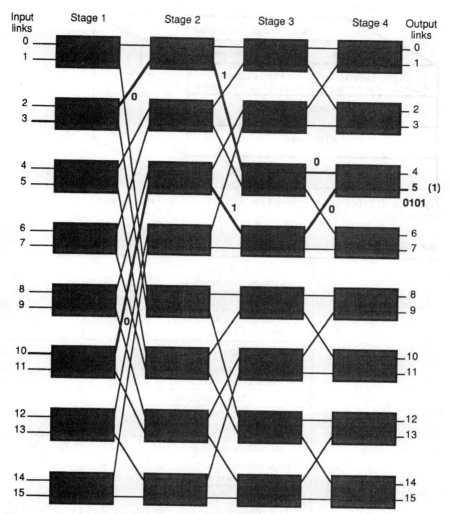

Figure 8.3   Banyan-type cell relay switch example.

sented by a flat plane butting up against all the OSI layers. These are functions used in common by various layers. Sometimes the C-, U-, and S-plane (Supervisory-plane) functions merge in an application with some of the Plane management functions. This is an aspect of implementation rather than architecture.

In the 1990 version of ITU-T Recommendation I.121, the functions supported at the various levels are presented in more detail (see Table 8.2). The three layers (as already mentioned) are the physical layer, ATM layer, and AAL.

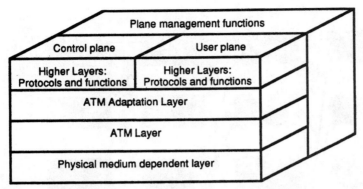

Figure 8.4 B-ISDN protocol model for ATM. (*From ITU-T Recommendation I.121.*)

TABLE 8.2   B-ISDN Layer Functions

	Higher-layer management functions	Higher layers	
Layer Management	Convergence	CS	AAL
	Segmentation and reassembly	SAR	
	Generic flow control	ATM	
	Cell header generation/ extraction		
	Cell VPI/VCI translation		
	Cell multiplex and demultiplex		
	Cell rate decoupling	TC	Physical Layer
	HEC header sequence generation/verification		
	Cell delineation		
	Transmission frame adaptation		
	Transmission frame generation/recovery		
	Bit timing	PM	
	Physical medium		

SOURCE: ITU-T Recommendation I.121.

Physical layer

The physical layer is split into two sublayers. These are called the Physical Medium (PM) sublayer and the TC sublayer. Some aspects of the TC were discussed in Chap. 3. The PM sublayer deals with the actual physical transmission medium (unknown, at this time, although it is likely to be coaxial or fiber optics). Whatever the medium, the PM sublayer must provide a certain set of functions. One is called bit timing. This is a means of synchronizing the transmission and reception of a stream of bits. Timing information may be included within the transmission medium or may follow some form of format pattern resynchronization (such as for V.110). The other is called the physical medium functions—that is, the actual manipulation of the signal. This will probably really be done by hardware support.

The TC sublayer is responsible for cell rate decoupling, HEC header sequence generation/verification, cell delineation, transmission frame adaptation, and transmission frame generation and recovery. Cell rate decoupling involves insertion of idle cells into the cell stream to maintain a constant data flow within the data "envelope" structures. This is similar to the HDLC flag idle insertion algorithm.

The HEC must be generated on transmission and checked on reception. This is true even for idle cells. (All cells, including OAM cells, must have valid HECs.) Cell delineation functions are concerned with making sure that the cell boundaries can be known at the endpoints. Scrambling techniques may be used to maintain line balancing. If so, the cell boundary will not actually be received in the tidy form presented in Chap. 3. However the TC sublayer has the responsibility to scramble (if desired) the cells to maintain cell integrity. This is discussed further in ITU-T Recommendation I.432.

Transmission frame adaptation is involved with placing the cell into the appropriate payload form. Although the SDH STM-1 payload structure is suggested as one means of transporting cells at the 155.52-Mbps rate, it would still be possible to send the cells in a stream. The transmission method might be different at different sections of the network. Adaptation means to adapt the cells to be carried in the appropriate manner. Transmission frame generation is the process of telling the PM sublayer to transmit the now existing frame (and recovery is equivalent to receiving).

ATM layer

The ATM layer deals with the contents of the ATM cell header—both creation and interpretation. The first function listed is that of cell multiplexing and demultiplexing. Each ATM cell has a VPI/VCI combination detailing the precise virtual channel and path that the cell is associated with. Assuming one B-ISDN physical connection to the network, it is expected that various VPI/VCIs will be used by the same user equipment. In HDLC, this function is done by the data link layer, the network layer, or both. The ATM layer has the responsibility of putting all the different cell stream paths together and breaking them out upon receipt.

The VPI and VCI have local significance. Each set is associated with a particular logical connection and, possibly, a particular function. It is the responsibility of the ATM layer to map the VPI/VCI into the appropriate local architecture identifier (sometimes called a Service Access Point, or SAP, by some documents) and vice versa during transmission.

The cell header generation and extraction function is the process of adding the ATM header onto the AAL data package. The fields of the first 4 bytes are filled in appropriately according to the needs of the system. The HEC byte is not actually filled in by the ATM layer (this is done by the TC sublayer of the physical layer), but the fifth byte must be present as part of the ATM cell header as presented. Cell header extraction takes an ATM cell header and translates the first 4 bytes into information used by the system.

The final function of the ATM layer is that of GFC (the upper nibble of the first UNI ATM cell header byte). The exact algorithm is yet unknown. In general, this function is to provide a method of maintaining the subscribed transmission rate between the user equipment and the network.

ATM adaptation layer

The AAL consists of two sublayers. These are the segmentation and reassembly (SAR) and the convergence (CS) sublayers. The SAR functions fall into the category of needing to adapt a variable-length amount of data into being transported by fixed-sized cells. Thus, the SAR function must segment the data into 48-byte units and pass this data (with appropriate addressing information) to the ATM layer. Upon receipt, the various 48-byte units need to be put back together into the appropriate higher-layer data unit. The CS sublayer is application specific and defines the services provided by the AAL to higher layers.

Interlayer primitives

Only four interlayer primitives are currently defined for B-ISDN. There are PH_DATA_REQUEST, PH_DATA_INDICATION, ATM_DATA_REQUEST, and ATM_DATA_INDICATION. These primitives flow between the ATM and the physical layers and between the

ATM layer and the AAL. The PH_ primitives are basically the same as those for flow from the data link layer to physical layer (except for possible parameter differences). The ATM_ primitives correspond to the DL_ data primitives. ITU-T Recommendation I.413 discusses the information flows associated with the physical layer. These are between sublayers in the physical layer, between the physical layer and the management plane, and between the physical layer and the ATM layer. Management interaction primitives are not yet specified.

ATM Layer

Much of this section is a summary of information already discussed. The ATM layer corresponds to that of the data link layer in N-ISDN. It provides a mechanism for taking data transmission requests from the AAL and routing the data to the physical layer. It is also responsible for flow control and the ATM cell header (except for the HEC byte). Note that the ATM layer is *not* responsible for retransmission—although HEC violations may be passed to the upper layers.

The ATM cell is similar to a frame. It consists of a header and a body of data. The header, which is 5 bytes, contains an HEC which corresponds roughly to the HDLC FCS bytes. It also contains address information called the VPI and the VCI. Like the frame-mode DE bit, the ATM cell header contains the CLP bit.

Use of virtual circuits and virtual paths

ITU-T I.150 discusses characteristics of virtual paths and circuits. Both have characteristics described as QOS, ability to have both switched and semipermanent connections, cell sequence integrity, and traffic parameter negotiation and usage monitoring. The QOS parameters concern the transit delay, percentage of cell loss allowable, and so forth. Just as is true for frame mode, packet mode, or circuit mode, a connection may be transient or semipermanent. A switched channel or path exists only for the duration of the call. A semipermanent (or subscribed) path or channel is expected to always be present and available for use.

Traffic parameter negotiation and usage monitoring is closely related to QOS. Once negotiated, the network can monitor the actual usage to determine that the load is within the limits agreed to.

Virtual paths have one additional restriction that does not apply to virtual circuits. Certain VCIs may be precluded from use on a virtual path. That is, VCIs may not be available for use by the user. This restriction may reserve certain VCIs for network-only use or as a reserved identifier for services not yet implemented.

Establishment of VPCs and VCCs

A VPC is defined in ITU-T Recommendation I.113 as "a concatenation of virtual path links that extends between the point where the virtual channel identifier values are assigned and the point where those values are translated or removed." The exact methods of establishing or releasing a VPC are not fully defined but may be expected to fall into the "normal" three classes. One, it may be semipermanent, or subscribed, and available at all times without any signalling process. The next two classes are the equivalent of an outgoing and incoming call. That is, a VPC may be established by the user equipment using a signalling method, or the network may use signalling methods to indicate a VPC being initialized toward the user equipment.

During the process of establishing a VPC, various parameters are negotiated. (In the case of a semipermanent VPC, these parameters are part of the service as the subscription indicates.) Traffic parameters may be renegotiated during the life of the connection. QOS parameters exist for the life of the connection, and therefore, any QOS parameters for VCCs used over the PVC can be only equal to, or less than, those QOS parameters which are part of the VPC.

A VCC is defined in ITU-T Recommendation I.113 as "a concatenation of virtual channel links that extends between two points where the adaptation layer is accessed." In other words, the VPC provides a conduit, and the channel provides a logical data stream that is used by the applications at each end.

A VCC at the B-ISDN UNI can be established in one of four ways. First, it can be semipermanent, or subscribed. No signalling is necessary, but the VCC can only be used if an underlying VPC has been established. The VCC can be established or released using *meta-signalling* procedures. This is mainly used to set up a signalling VCC within the VPC.

The last two methods employ the signalling channel VCC that has already been established. An end-to-end VCC can be established by using signalling procedures from the user-to-network interface (or vice versa, for incoming VCCs). A signalling VCC can also be used, end to end, to establish (or release) additional VCCs. The VCI will be determined by subscription or as an end result of signalling procedures. In general, the VCI may be any value, but some values may be reserved by the network for special purposes (such as meta-signalling) or reserved for future standardization.

ATM Adaptation Layer and Service Classes

ITU-T Recommendation I.362 presents the B-ISDN AAL functional description. In the 1993 version, it mentions that the AAL provides an enhancement of the features available in the ATM layer. The AAL does functions required by the U-, C-, and M-planes and supports the mapping between the ATM layer and higher layers. Thus, the AAL is not exactly a layer 3 protocol—rather, it provides support for layer 3 protocols. The AAL is dependent upon the service desired.

The AAL can handle transmission errors, perform segmentation and reassembly, and handle errors within the PDUs and flow and timing control. However, not all of these functions may exist in all AALs. The AAL basically provides an interface between the higher layers and the ATM layer by mapping PDUs into the ATM structure. It provides service by interacting with its peer AAL entity. The AAL, as discussed earlier, provides SAR and CS sublayers. The CS sublayer is highly service dependent.

The services supported by the AAL are classified according to the timing relation between source and destination, bit rate, and connection mode. Fig. 8.5 shows the basic service classifications for the AAL. The various combinations provide four classes of service. Note that 2^3 (8) combinations are possible, but only four classes are defined by the ITU-T.

The ones not defined are as follows. Required bit timing is not supported along with connectionless service (removing two possibilities), and a constant bit rate is not supported without bit timing (removing the other two possibilities). These absences make a lot of sense. If a service is connectionless, how can bit timing be maintained between two undefined endpoints? Second, how can a constant bit rate be supported without bit timing?

ITU-T Recommendation I.363 discusses how these four classes of service correlate to particular AAL protocols. Constant Bit Rate (CBR) services, which are defined by Class A, are supported by AAL Type 1. Connectionless (CL) services utilize AAL Types 3 and 4. Class C services also can use AAL Types 2 and 3. It is also possible that the AAL does not provide an additional service but acts only as a function

	Class A	Class B	Class C	Class D
Timing relation between source and destination	Required		Not required	
Bit rate	Constant	Variable		
Connection mode	Connection-oriented			Connectionless

Figure 8.5 Service classification for AAL. (*From ITU-T Recommendation I.362.*)

to send and receive ATM Service Data Units (SDUs). In general, the four classes correspond to the four defined AAL types. A particular service, however, might be able to make use of more than one class and therefore more than one AAL type. An *AAL Type 5* was initially proposed by the ANSI T1 committee in 1992 and may end up as part of the official ITU-T recommendations. AAL Types 3 and 4, as defined in the 1991 recommendation, provide very similar functions in a very similar way and were merged into one type (designated Type 3/4 to prevent confusion to those with older recommendations). It is important to keep track of changing standards—particularly in the case of higher-speed architectures such as B-ISDN.

AAL Types

The AAL types, as defined in the 1991 ITU-T Recommendation I.363, give some functional descriptions of the SAR and CS sublayers for use with each of the classes. The CS structure and coding is mostly "for further study," although some of the suggested CS functions are listed. Fig. 8.6 shows the AAL structure for the original four types as car-

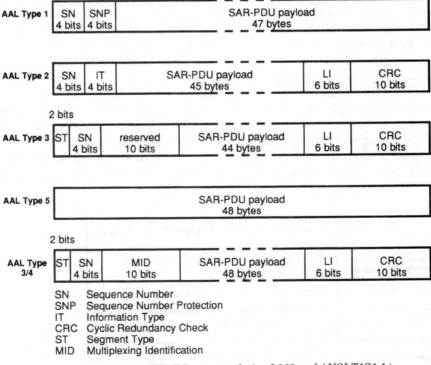

Figure 8.6 AAL types. (*From ITU-T Recommendation I.363 and ANSI T1S1.1.*)

ried by the ATM cell plus the AAL Type 5 as defined by ANSI. Only the SAR structure elements are defined for AAL Types 1 through 4, since the CS sublayer elements are strictly application specific. For AAL Type 5, no SAR functions are defined. The subsection on AAL Type 5 will go into its use in more detail.

SAR bit fields in the AAL

There are eight fields defined as part of the four (or five) AAL types. These are the Sequence Number (SN) field, Sequence Number Protection (SNP) field, Information Type (IT) field, Segment Type (ST) field, and Multiplexing IDentification (MID) field for the header. (The RES field is just a reserved field that is not used for the old AAL Type 3.) Two fields exist in the trailer. These are the Length Identifier (LI) and CRC fields.

The SN field is used to provide a modulo 16 sequencing for the PDUs. This is available on all of the original four ITU-T AAL types. The SNP field, used in AAL Type 1 (Class A) PDUs, is available for an undefined error-detection and correction capability. The IT field is used to indicate segmentation information (beginning, continuation, end of message) or can be used as an indication of a single segment message or special service. The IT field is used only on Type 2 (Class B) PDUs. The ST field is used in AAL Types 3/4 and is the basic equivalent of the IT field. The MID field (whose position is "reserved" in the original AAL Type 3 SAR-PDU) provides a method of multiplexing and demultiplexing CS PDUs at the *layer 3* level.

AAL Type 1

This type supports Class A (bit-timed, constant-rate, connection-oriented) services. It provides a transfer service, in accordance with Class A requirements, and an indication of lost or erroneous information which is not able to be recovered. The AAL Type 1 service should be able to pass information to the management layer concerning transmission errors, lost or misinserted cells, errors concerning AAL protocol control information, or loss of timing and synchronization.

AAL Type 1 provides a number of functions. Most would be provided by the SAR sublayer, although I.363 does not so specify. These functions include the SAR functions, handling of cell delay variation, handling lost and misinserted cells, and recovery of the source clock frequency by the receiver. The monitoring and handling of Protocol Control Information (PCI) and other user information field bit errors is also a likely function.

A few specific functions are considered for the CS. These indicate a use of AAL Type 1 for primarily bit-pattern-oriented protocols such as voice or video. It mentions that some error correction, or possible bit interleaving, may be performed by the CS. The CS sublayer may also provide clock recovery capability for the receiver (perhaps by monitoring the buffer filling) or time stamp patterns inserted into the CS-PDU for explicit time indication. Some form of handling lost and misinserted cells may also be performed.

The SN (and SNP) field allows lost cells to be noticed. This provides a minimal error-detection mechanism. The CS-PDU is needed to provide any real correction mechanism.

AAL Type 2

This type supports Class B (bit-timed, variable-rate, connection-oriented) services. It provides a transfer service for a variable bit rate and an indication of lost or erroneous information which cannot be recovered. The AAL Type 2 service should be able to pass information to the management layer concerning transmission errors, lost or misinserted cells, errors concerning AAL protocol control information, or loss of timing and synchronization.

AAL Type 2 provides a number of functions that are very similar to those of AAL Type 1. Most would be provided by the SAR sublayer, although I.363 does not so specify. These functions include the SAR functions, handling of cell delay variation, handling lost and misinserted cells, and recovery of the source clock frequency by the receiver. The monitoring and handling of PCI and other user information field bit errors are also likely functions.

A few specific functions are considered for the CS. These indicate a use of AAL Type 2 for primarily bit-pattern-oriented protocols such as voice or video that have a variable bit rate. This type of service may use forward error correction for audio and video services. The CS sublayer may provide clock recovery capability for the receiver (perhaps by the use of time stamps or another real-time synchronization word). Sequence numbering may be used to detect, and handle, lost or misinserted SDUs.

AAL Type 2 provides for three additional fields over that of AAL Type 1. The IT field can be used to indicate the beginning, continuation, or end of a message (BOM, COM, or EOM). This field can also be used as a component of a video or audio signal. Since the bit rate is variable, the LI field is necessary to determine how many of the bits within the payload are useful information. The CRC section takes the place of the SNP of AAL Type 1 and provides a method for error detection. The ITU-T definitions of the various types require that they be of the same form. So, it is quite possible that the CRC field may not be used until after a PDU with the IT field indicating EOM. However, the field will exist in each ATM payload. (This is part of the reason for the ANSI AAL Type 5 recommendation.)

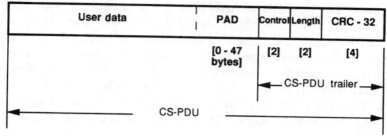

Figure 8.7 AAL Type 5 details. (*From ANSI T1S1.1.*)

AAL Type 5

This type, defined by the ANSI T1S1.1 committee and being studied by the ITU-T, passes the responsibilities of the SAR sublayer to the CS. It is used for Class C services—variable bit rate service over a connection-oriented link. As can be seen in Fig. 8.7, only the CS-PDU has any fields. This type of the AAL is called the Simple and Efficient Adaptation Layer (SEAL), and it is designed to provide a lower overhead, with better error detection and correction, than the other AAL types.

AAL Type 5 provides for greater efficiency by not containing any overhead bytes except for the last PDU of a sequence. There is no header (either SAR or CS), and only the last PDU of a sequence contains the CS-PDU trailer indicated in Fig. 8.7. This allows the full 48-byte payload to be used for data, with only the PAD bytes (to bring the payload to the full 48 bytes) and the CRC-PDU trailer (8 bytes) as overhead.

This is a considerable improvement over the fixed 4-byte overhead (plus any unused "padding" bytes needed to fill out the payload) per SAR-PDU of AAL Type 3/4. It does give up a few things from the AAL Type 3/4. For one, only the entire sequence can be checked for errors (no SN or CRC fields per payload). Another is that the MID field is not available to allow SAR multiplexing and demultiplexing. (However, this capability may exist within the CS-PDU payload. An example of this would be the carrying of X.25 protocol within the AAL SDU.)

This is compensated for by the fact that the CRC-32 is better able to check for errors on the entire resegmented data stream. A CRC-32 polynomial check allows 4 bytes for the check sequence (rather than the 10 bits of the SAR-PDU CRC of AAL Type 3/4). Note that if, because of variable bit occurrence, a particular AAL SDU does not have a full 48 bytes of data, either a special pattern of padding data must be used or the CS-PDU trailer should be included. This means that some indication must be available to indicate the last AAL payload of a series. It has been suggested that the PT field of the ATM layer be used to do this.

The fields of the CS-PDU of AAL Type 5 are as follows. The user data goes to and from the higher-layer protocol. It may be up to 65,536 bytes in length (as limited by the size of the length field). The padding characters make sure that the payload is constant at 48 bytes in length. There may thus be up to 47 such characters. Note, however, that these bytes must be able to be delimited from the real user data if they are not used in conjunction with the last payload that contains the CS-PDU trailer. The length field is 2 bytes long and the CRC-32 field is 4 bytes long.

In conjunction with AAL Type 5, a couple of sublayers are proposed for use with this service. The first of these is called the Frame Relaying Specific Convergence Sublayer (FRCS), which is designed to provide interworking between Frame Relay and ATM. The second is called the Service-Specific Connection-Oriented Protocol (SSCOP) for B-ISDN signalling. This CS will provide error control using retransmission, flow control mechanisms, ability to adjust frame size, optimization of SAR functions in accordance with the characteristics of the channel, pipelining capability, and an ability to detect lower-layer failures.

AAL Types 3 and 4

The original ITU-T AAL Types 3 and 4 varied from each other only by one field (the MID) not being actively used in AAL Type 3. Thus, it was decided to not call them separate types—although, connection-oriented application channels can be considered AAL Type 3 (or Class C), and connectionless application channels can be considered AAL Type 4 (Class D).

The main purpose of Type 4 is to be able to provide the multiplexing/demultiplexing function. Two modes of service are defined for AAL Type 3/4. These are the message mode and the streaming mode services. Both types of service may provide either assured or nonassured operations. In assured operations, the AAL-SDUs are all delivered with exactly the data content that was transmitted. This is provided by retransmission of missing or corrupted PDUs. Flow control is necessary.

In nonassured operations, individual AAL-SDUs may be lost or corrupted without an attempt to retransmit or correct. These SDUs may either be discarded or forwarded to the user. Flow control is still possible for the system on point-to-point ATM connections.

The use of the multiplexing field (MID) allows the AAL to choose between different channels of varying QOS parameters. These are identified by the AAL-Service Access Point (AAL-SAP) and are used by sending primitives through the AAL to the ATM layer. The primitive to the AAL is called an AAL_UNIT_DATA_REQUEST and may contain a SAP-ID plus the data. This is translated into a particular ATM_DATA_RQ. Coming up, an ATM_DATA_IN is translated into a AAL_UNIT_DATA_INDICATION primitive with an (optional) SAP-ID and the data and reception status. The reception status parameter is an indication of the failure or success of reception.

The SAR sublayer uses the ST and SN fields to indicate the type, and proper sequencing, of an SAR-PDU. The LI field indicates how many bytes of the SAR-PDU payload are to be passed as the CS-PDU. The CRC field allows a minimal check of single bit errors of the payload. The MID allows multiplexing/demultiplexing of multiple CS-PDUs over a single ATM connection. The SAR-PDU header and SAR-PDU trailer provide a constant overhead of 4 bytes.

The CS functions are performed on the CS-PDU, after analysis by the SAR sublayer. The CS functions may include preservation of the AAL-SDU, mapping between AAL-SAPs and ATM layer connections, error detection and handling, message segmentation and reassembly, identification of information, and buffer allocation size.

Summary of AAL types

The AAL types, as defined by the ITU-T and ANSI, allow the various classes of services to be supported. These classes are made of various combinations of bit timing, constant versus variable bit rates, and connection-oriented versus connectionless service. The original AAL Types 3 and 4 have been merged, due to the relatively minor differences between them. The ANSI-proposed AAL Type 5 supports Class C services but removes the SAR sublayer functionality into the CS. AAL Type 5 also adds new sublayers that are application specific. Table 8.3 summarizes the various AAL types and their definitions.

Higher Layers

The ITU-T recommendations, so far, only cover the equivalent of layer 1 through part of layer 3. The CS functions are partially defined, but the coding and structure are still "for further study." The ANSI T1S1.1 committee proposals start on some definitions for the CS coding and use—however, these are still not incorporated into the main standards.

The types of functions that need to be defined more clearly are how error recovery is to be done, signalling processes, and switched channel techniques. These will be done relatively soon since they are badly needed.

TABLE 8.3 ATM AAL Protocol Types

	Service provided	Functions in type	SAR sublayer	CS functions
Type 1	Transfer of SDUs with constant bit rate. Timing information. Indication of lost or errored unrecoverable information.	SAR. Cell delay. Handling lost and misinserted cells. Source clock recovery. Monitoring and handling of AAL-PCI errors. Monitoring and possible corrective action for user information bit errors.	For further study.	High-quality audio and video may have FEC. Clock recovery capability. Time-stamp services. Lost and misinserted cells handled.
Type 2	Transfer of SDUs with variable bit rate. Timing information. Indication of lost or errored unrecoverable information.	SAR. Cell delay. Handling lost and misinserted cells. Source clock recovery. Monitoring and handling of AAL-PCI errors. Monitoring and possible corrective action for user information bit errors.	For further study.	High-quality audio and video may have FEC. Clock recovery capability. Lost and misinserted cells handled.
Type 5	Efficient and versatile CS handling.	CRC checking. Ability to handle variable data rates. Control field for future use.	None.	Error checking. Retransmission. Flow control. SAR. Pipelining. Data link failure detection. Signalling (SSCOP).
Type 3/4	Message-mode service. Streaming-mode service.		Preservation of CS-PDU. Error detection. Multiplexing and demultiplexing.	Preservation of AAL-SDU. Mapping between AAL-SAPs and ATM. Error detection and handling. SAR. Information identification. Buffer allocation size.

SOURCE: ITU-T Recommendation I.363 and ANSI T1S1.1.

Software Considerations

B-ISDN is an aspect of the general ISDN architecture. It is necessary to be able to integrate the N-ISDN protocols into the B-ISDN methodology. For this, it is likely to be useful to redirect primitives at the bottom of the N-ISDN protocol stack to the functions of B-ISDN (since B-ISDN will carry N-ISDN and not vice versa). In our discussions about Frame Relay and the data link core functions, we talked about some aspects of redirecting primitives based on general use. The same philosophy applies to B-ISDN

Hardware-intense protocol

Some hardware silicon chips support the data link layer, as well as the physical layer, of N-ISDN. B-ISDN has even greater need of hardware support, due to the greater speeds involved. Some chips exist for ATM layer support. At the AAL segment of the protocol, specialized hardware support will be needed. As an example, let us consider a video application. The AAL class would be Class A (or B). We know the general SAR functions needed for AAL Type 1. However, the CS functions will be specific to the video application. The ATM layer will pass the SAR-PDU to another hardware device, where the SN and SNP fields will be checked. The SAR-PDU payload will then be processed by a video processor. Although it is not impossible, it is highly unlikely that software will be directly involved once the VCC has been set up for the application.

Another possibility of use would be for carrying N-ISDN protocols over the B-ISDN. The ATM layer chip set would pass the data up to the AAL (probably AAL Type 5). At this point, the CS would reassemble the frame and, after checking for errors, pass the frame to the data link layer as a PH_DA_IN primitive. The data link layer would then process the data as if it had arrived via the TDM channel in the HDLC frame. The rest of the protocol stack would be as previously described for N-ISDN.

On the downward direction, the network layer and data link layers would process the data. At the data link layer to physical layer interface, either the primitive would be changed to an AAL_UNIT_DATA_REQUEST primitive (rather than a PH_DA_RQ primitive) or the physical layer handler would process the frame (adding CRC, for example) and then pass to the AAL for segmentation and other CS functions.

An option would be to have the physical layer create a full HDLC bit stream (flags, data, CRC, and flags). This bit stream would then be processed by the AAL, packaging the data into an appropriate ATM cell package. In this manner, the entire N-ISDN bit stream would be encapsulated into the B-ISDN. Either method is possible. The first is more efficient. The second allows direct interworking between B-ISDN and N-ISDN without having to involve the chained layers at the B-ISDN network node termination.

The difference in hardware needs between N-ISDN and B-ISDN is not a matter of protocols. It is a matter of speed. A microprocessor operating at a 33-MHz clock rate cannot handle data at a 155.52-MHz data speed. The interface between N-ISDN can occur at the protocol level or at the hardware level. This is an architectural decision to be made according to application needs.

Bottlenecks and throughput

A bottleneck occurs in software whenever one portion of the architecture is slower than the capacity of the other portions. Such a bottleneck might occur in a disk system, microprocessor, software task within the processor, or output line of the hardware. These are only examples. Bottlenecks are just defined as the place within the architecture where other portions are forced to wait. That doesn't inherently mean that the architecture is bad, because something will always be the slowest portion of an architecture. If the transmission speed were faster or the microprocessor clock was faster or a particular area of code was faster, the overall capacity of the system could be faster. This is how system improvements, or *tuning,* is done. Analysis of the system is done and then efforts are made to improve that area. Once that area is improved, analysis is done again, and emphasis is given to the new slow area.

A system will thus have a criteria against which to base judgment. If you have a 64-kbps line capacity and the rest of the system is able to keep this capacity in use, the system is adequately designed. If you speed up the line capacity, the rest of the system may need to be speeded up also to take advantage of this new criteria. A throughput of 59 kbps on a 64-kbps transmission line is very good (considering physical protocol framing overhead). A throughput of 59 kbps on a 155.52-Mbps transmission line would be considered very bad. Most of the components are the same, but the criteria have changed.

In the case of B-ISDN, the criteria will be quite stringent. Parallel processing and primary-secondary architectures may be necessary to take advantage of the bandwidth. Data buses may need to be designed to allow multiple devices and processors to multiplex onto the transmission line. Other possibilities include special hardware to support the bandwidth.

Example scenarios

As an example, a video service for HDTV (around 20 Mbps compressed) will probably be set up by software, but the data will be processed completely by hardware. Let's assume a 155.52-Mbps coaxial transmission line into a business or residence (Class A service probably). The payload possibilities are about 135.631 Mbps for such a connection. AAL Type 1 has an overhead of 1 byte out of 48. Assume that the rest of the SAR-PDU payload is usable data. There are 132.805 Mbps of video (and audio) data available over the transmission line. This would allow six HDTV channels to be broadcast over the transmission line.

The *tuner* for an HDTV television would literally set the channel (VCC) at the ATM layer. The VPC would probably be a semipermanent connection. So, the ATM layer would pass up all of the AAL-PDUs for this channel. The AAL would check for appropriate sequencing and (in this application) probably toss away any erroneous PDUs. The payload would be passed to a hardware chip that did the HDTV processing, and the result would be video and audio from the set. With the appropriate hardware support, the existing standards are almost sufficient for such an application.

Now say that you wanted *channel-in-channel* possibilities on an HDTV set. This means that two different VCCs would be processed. There are a couple of different possible scenarios. One would have the same ATM layer hardware chip set for two different VCCs but would be capable of routing the different VCCs to two separate AAL hardware chips and two separate (almost functionally equal) video/audio processors. Another possibility would be to have the ATM cell stream pass through two hardware chips supporting the ATM layer. Each is set up to capture one VCC, and each has a supporting AAL hardware chip and video/audio processor chip. This second scenario has a simpler architecture but requires additional hardware support. The choice would depend on the overall system requirements, including cost.

Let's try a third scenario. This time, we have a 155.52/622.08-Mbps transmission line. A CD-quality audio transmission line requires about 1.4 Mbps. For an AAL Type 1 service, the SAR-PDU payload would be about 135.631/542.524 Mbps. In other words, about 387 separate CD-quality input channels would be possible. An audio library system could be designed to broadcast to various stations. A user would ask for a particular selection (using the in-bound transmission side). If a broadcast channel was available, the music would be sent on a particular VCC/VPC which would then be processed for the user to receive. Privacy could be ensured for each listener by setting up the hardware such that the VCC is not set directly by the user—but, rather, by the software protocol after receipt of instructions have been received.

These are all feasible scenarios. Will this be what is actually implemented? It is unknown. However, applications will certainly develop as the standards become more widespread.

Bearer service support

What do bearer services mean in relation to ATM Cell Relay? Well, in one respect, they mean the same as they do for all other ISDN service support. For what is the data channel to be used? In other respects, B-ISDN use of the bearer services is a little different from those of N-ISDN. The main difference is the class of service. This affects the underlying SAR sublayer at the AAL. Plus, there is the fact that the signalling methodology for the ITU-T specifications has not yet been finalized.

Let us assume that some form of Q.931 signalling will be done for B-ISDN (in conjunction with some specific AAL CS functions). This indicates that the same possibilities for bearer services exist for B-ISDN as for N-ISDN. New IEs may be added for additional requirements, but the Bearer Capability and LLC IEs will be used fully. Perhaps there will be an AAL IE to indicate AAL SAR structure type and CS functions. Out-of-band negotiation will be possible, as will some form of in-band protocol negotiation.

Bearer services for B-ISDN will probably be in two categories: transportation and protocol services. Transportation services use the B-ISDN to act as a high-speed intermediary network for interconnection of other protocols or LANs. That is, the data transported are transparent to the B-ISDN protocol stack. The protocol services will require the network or user equipment to interpret and respond to the data that exists on the transmission channels. If a transportation service is done, will it be necessary to pass along its low-layer information fields to the other end, or will this be done over the meta-signalling channel? Many questions remain for the ITU-T to answer.

Summary of Chapter

B-ISDN is a continuation of the overall ITU-T architecture for ISDN. As such, it is designed to be compatible for interworking with other digital services such as voice and N-ISDN. It also, due to its higher capacities and speed, supports particular applications that are not possible at lower speeds. These include many video technologies.

The basic physical medium for B-ISDN is not a requirement. Currently coaxial cable is possible for short distance with a reliance on fiber-optic technologies for longer distances and for the highest bandwidths. The basic underlying mechanism for physical transport of data for B-ISDN is the cell. A cell is a self-contained data unit that is put into a digital framing envelope. The ATM is the protocol scenario used for such cell transport.

There are three layers defined for B-ISDN. The physical layer makes use of the SDH as defined by the ITU-T. The next layer is the

ATM layer followed by the AAL. The AAL supports four classes of service which are defined according to use of bit timing, the variability of the bit rate, and whether it is to be transported on a connectionless or connection-oriented transmission path. In support of these classes, the ITU-T originally defined four AAL types. The ANSI T1S1.1 committee further defined another type which more efficiently supports higher-layer protocols. The AAL has two defined sublayers: the SAR and the CS. The SAR functions depend on the AAL type. The CS functions are application specific.

Many areas of the B-ISDN are yet to be defined. A basis for general interconnection as a system is now known. Further work needs to be done on application support and filling out the standards to provide a basis for current interworking.

9

DTE Protocols
in Relation to ISDN

A DTE is a piece of user equipment which terminates a network connection. This is distinguished from that of CPE in two ways. One, CPE can be either network (such as Private Branch Exchange, PBX, equipment) or user oriented. Second, DTE implies not only a particular type of usage but also a protocol role to be fulfilled. A particular shift in ITU-T nomenclature over the years within their documentation has occurred. Originally, data networks were presented as if there were two firmly divided providers. The network equipment was presented only from the point of view of the UNI. Emphasis was always made on how the network and user equipment were to interact. The details of internetwork details were often left unclear.

There were two major reasons for this. During the earlier years of the ITU-T (at that time, CCITT) standards, almost all countries had a single network resource provider for within their borders. Second, there was a movement to open up to give various manufacturers the ability to connect their equipment to the network. Thus, it was important to standardize the UNI but not as important to publish the NNI. It also meant that the network providers were able to continuously change the interior of the network without informing users and manufacturers.

The division of equipment into the DTE and DCE meant that the network providers concentrated on the services provided to the user equipment. A service also implied a tariff. Most early use of the PSTN thus was for only one service—voice. It was possible (if a person *wanted* to spend more money or truly needed a high-grade, low-error line) to obtain a special shielded line from the local company. However, this was still a matter of a single service. Whenever new services were made available, they were associated with a tariff. Tone signalling has a tariff associated with it in the United States, although the actual handling of tone signalling usually requires fewer network resources than does pulse signalling.

DTEs and DCEs

So, based on history of resource and standards use, most user equipment designed to make use of the public network used voice lines and were required only to be compatible end to end. There were also private networks that were used with equipment. Most such private networks handled proprietary protocols—protocols that were not public knowledge and which were supported only by the company which had defined the protocol. One exception to this was ITU-T Recommendation X.25. X.25 was probably the first widespread data communication protocol implemented based upon a public international standard.

Other protocols used over the existing networks were concerned with methods of communication—encoding methods for transmission of digital information over the analog network. The current V.42 bis high-speed standard is an ITU-T recommendation. The purpose of these standards was to provide mechanisms for multiple vendors to have their equipment work together. It was also to be sure that the standards developed would be able to work properly, without interference, over the analog networks.

For example, many network uses of the analog network took advantage of the 3.7-kHz audible bandwidth of analog speech connections by using frequencies outside the normal bandwidth for signalling information. It was important that any digital-to-analog conversion (MOdulator-DEModulator, or modem) standards not make use of this signalling portion of the spectrum. The technology needed for long distance networks was also involved in standards. A long distance transmission line makes use of echo cancelers and repeaters to ensure line quality. These devices were originally designed for voice traffic. Additional use of the voice lines for other types of data (unrestricted digital, fax, etc.) meant that the standards must allow for the physical needs of the network.

A caveat concerning the definition of DTEs and DCEs needs to be made: They are used for unbalanced protocols (protocols where one side has particular responsibilities, and the other side has a different set). Thus, a piece of user equipment can be considered a DTE and the modem, or TA, considered a DCE. That same modem, or TA, can be considered a DTE to the network. Two or more sets of standards apply. Across the R interface reference point, the TA operates as a DCE to the TE2. Across the S/T or U interface reference point, the same TA operates as a DTE.

Some protocols (such as ISO 8208) apply for DTE-to-DTE applications. These are balanced protocols—each piece of equipment has the same responsibilities in the same circumstances. (For example, in the case of originating a call, the originator always follows the same procedure—no matter which piece of equipment it may happen to be.)

Types of DTEs

A good method of understanding the classifications for DTEs is to examine the low-layer information fields of the Bearer Capability IE or the LLC IE. The bytes (or octets) of field 5 are particularly important. The first byte (octet 5) is concerned with the user information at layer 1. It is possible to give a value for a particular rate adaptation scheme (such as V.110 or V.120), a generic HDLC rate adaption (adaptation and adaption are used interchangeably within ITU-T documents), or a speech encoding method. Any of these can be used with a DTE, although speech encoding methods are not likely to provide sufficient flexibility for most data transfer needs.

Byte 5a gives two parameters that are of specific use. These are the *synchronous/asynchronous* bit field and the user rate. The user rate has been discussed before. This is involved with the data transmission rate across the R interface reference point.

Asynchronous versus synchronous is concerned with how the data is presented to the TA, across the R interface reference point, and to the network (and peer entity). Asynchronous terminals are often referred to, within the ITU-T recommendations, as *start-stop mode* terminal equipment. This is because data may transfer across the interface leads at any time. Special indications are used to indicate the *start* and the *stop* of valid data. These can be set in such terminals by the number of *start* and *stop bits*.

A synchronous terminal will package data into frames. Often these frames will contain FCSs and be delimited by flags. They may actually use an HDLC family protocol. Each frame is presented as a complete package across the R interface. Depending on the resource capabilities of the TA and network, such frames may need to be segmented before transmission (and reassembled upon receipt).

An in-between terminal adaptor is called a PAD facility TA. This PAD allows for the collection of data by the TA. The TA is then responsible for putting the data into a packet format before transmission. In the direction of reception, a PAD must disassemble the packets that are received and process them in the form needed across the R interface reference point.

ITU-T Recommendation V.120 points out a third broad type of terminal equipment to add to that of asynchronous and synchronous types. This type is called *bit transparent*. A bit-transparent interface takes the data as presented and ships it onto the S/T interface in the same approximate form. Bit-transparent protocols are highly individualized and will not be covered in this book.

Requirements for DTE Use of ISDN

There are a number of requirements for a DTE to make use of ISDN. They are implemented as functions within the TA. They fall into the general categories of signalling conversion and data formatting. Signalling conversion is involved with the interactions of the TA with the DTE (or TE2) before a connection is made. Data formatting is required after the connection is made and data are ready for transfer. A third category involves setting up the TA for particular behavior. This does not directly involve the ISDN but may indirectly affect some aspects. One such function is the user rate across the R interface. The data are transferred only across the R interface point—but the user rate may affect the way that the ISDN bearer channel is used.

Signalling conversion

Signalling conversion operates across the R interface reference point. There are two general classes of signalling at this point. One is an interface lead protocol—the toggling of the states of the physical interface leads changes the state of the TA. The other is a data-oriented protocol. Particular sequences of data are interpreted as special commands during the *command mode* operation of the TA. The AT command set is an example of this type of protocol.

A data-oriented protocol must always have two modes since the method of transference of data and signalling control information occurs over the same physical interface. These are normally called the *command mode* and the *data mode*. Mode-oriented interfaces require *escape commands*. Such an escape command allows transition from data mode to command mode (and may provide for transition back to data mode without interruption of the connection).

An interface-lead protocol does not explicitly require modes. Instead, it usually implements a state machine for interpretation of physical interface handling. As is true for all state machines (see Chap. 10), operation will begin in a default beginning state. This state is defined as a particular configuration of physical interface leads that mean an idle condition.

TA setup

We discussed a third category of interactions that may occur between a TE2 and a TA. This category involves the setting up of the TA to behave in particular ways that are acceptable to the TE2. A TA will normally start off in a *default* configuration, which will include a user rate (but the hardware may support automatic baud detection) and a number of other operational characteristics. The TA interactions with the TE2 may occur before call setup or during a data connection. If the interactions occur during an active connection, some means must be allowed for further command interaction.

One important type of command is that of setting the user rate (although some hardware may support automatic baud recognition). Another is setting up the escape sequence. Many other features may be possible—a conversion of line feed characters into carriage returns, for example. Other sequences may provide ways to change the responses given by the TA to particular commands from the TE2. It may even be possible to configure the TA to perform special functions during data transfer. (One common function is that of *data echoing* where all data transmitted across the R interface reference point are transmitted back to the TE2 by the TA.)

TA setup commands take place either within a TA state that allows such commands or within the command mode for mode-specific TAs. The escape sequence allows transfer between the data and command modes (another function may allow transfer back to data mode without interruption of the data connection). All these commands are useful to allow for variable behavior on the part of the TA.

Data formatting

Formatting of data is the process of putting data received from (or going to) the TE2 to the TA into a form necessary for the ISDN bearer channel. We looked at such formatting for V.110 and V.120 in Chap. 6. V.110 requires specific frame construction based on the user and bearer rates. V.120 employs the HDLC framing and flag insertion methods to provide a proper data sequencing across the bearer channel. V.120 requires a type of PAD functionality for asynchronous terminal usage.

In the case of synchronous terminal equipment used with a TA, it will probably be necessary to check, and strip, any FCS information passing across the R interface reference point. This may be optional if the frame size is appropriate for direct transmission on the bearer channel. In most cases, the data contents of the frame are split out from the synchronous interface and then placed within the appropriate protocol format.

ITU-T Recommendation X.31 is concerned with synchronous terminals using X.25 across the ISDN to an X.25 packet handler. The protocol parts can be on either side of the R interface reference point. An efficient method of TA PAD functionality, however, would pass only the data contents of the layer 3 X.25 packet across the R interface. After checking for proper reception, the TA would pass the data to the network layer for X.25 header additions—which would then proceed through layers 2 and 1.

ITU-T DTE Recommendations

Most DTE-applicable UNI references for the ITU-T are found in the X.*xx* series of documents. We have already discussed ITU-T Recommendation X.25 in considerable detail. ITU-T Recommendation X.75 is a superset of X.25 that contains additional details for interworking of X.25 networks in an international situation. We have also briefly discussed X.31, which is involved with how the X.25 packet handler and ISDN may be used together. In the rest of this chapter, we will discuss some of the various ITU-T X.*xxx* documents in greater detail.

ITU-T Recommendations X.3 and X.29 discuss PAD aspects of a DTE. ITU-T Recommendation X.28 defines the DTE/DCE interface for a start-stop mode DTE that is accessing the PAD. These three documents provide an overall perspective of a method for using the start-stop DTE on a HDLC form of data transfer over ISDN.

ITU-T Recommendations X.3, X.28, and X.29 are concerned with a start-stop mode DTE. ITU-T Recommendation X.21 defines an interface between the DTE and DCE for synchronous operation. ITU-T Recommendation X.30 adds to this definition by giving the correspondence between X.21 and Q.931 commands. This defines a full set of interactions for terminals using the X.21, X.21 bis, or X.20 bis interface standards being used on an ISDN.

ITU-T Recommendations X.31 and X.32 are concerned with usage of synchronous protocols (X.25, specifically) over the ISDN. ITU-T Recommendation X.32 is concerned with the R interface reference point, and X.31 is concerned with the *S/T* interface reference point (see Fig. 9.1). Some of the details in the various ITU-T X.*xxx* recommendations overlap. Together, however, they provide a good definition of how to use older terminal equipment on an ISDN.

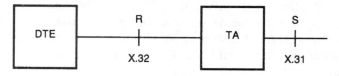

Note - The DTE and TA functionalities may be implemented in the same piece of equipment in the case of a TE1 terminal. In this case this Recommendation (X.32) covers layer 2 and 3 operation in the B-channel while the S reference point procedures are described in Recommendation X.31.

Figure 9.1 ISDN reference points. (*From ITU-T Recommendation X.32.*)

PAD Functionality

The primary purpose of a generic PAD function is to collect and assemble data into packets and to take received data and split it into appropriate data streams across the R interface reference point. This gives the PAD its name and purpose. More specifically, within the ITU-T recommendations, a PAD is used to provide access of a start-stop mode DTE to the capabilities of an X.25 network. However, as discussed above, any DTE TA function must also provide signalling and TA setup functions. In the ITU-T Recommendations X.3, X.28, and X.29, which deal with start-stop mode terminals, these functions are provided via the use of *parameters*. The parameters configure the operational mode of the TA. Command sequences are in-line data streams.

A PAD is normally expected to be used in conjunction with X.25 over a network. ITU-T Recommendation X.29 details how peer PAD communication can take place. Since the PAD is used with X.25, it may also be involved with X.31 and/or X.32 in its use over an ISDN.

ITU-T Recommendation X.28

This recommendation provides definitions for the "DTE/DCE interface for a start-stop mode DTE accessing the PAD in a public data network situated in the same country." The document begins with a discussion of the physical layer interface across the R interface reference point. The interface depends upon the exact user rate supported. ITU-T Recommendations V.21, V.22, V.23, and V.24 are all referenced as interface specifications (depending on the user rate). ITU-T Recommendation V.28 provides the electrical characteristics, currently with a pin assignment as specified by ISO 2110. Table 9.1 shows the various interchange circuits as described in X.20 bis.

X.28 makes use of an "access information path" between the DTE and the PAD for communication of parameters and commands. This

TABLE 9.1   X.20 bis Interchange Circuits

Number	Designation
102	Transmitted data
104	Received data
106	Ready for sending
107	Data set ready
108/1	Connect data set to line (used for autocontrol of direct call facility)
108/2	Data terminal ready (used for switched data network service)
109	Data channel received line signal detector
125	Calling indicator (not provided in leased circuit service)
141	Local loop back (not provided in those networks which do not provide automatic activation of the test loops)
142	Test indicator

SOURCE: ITU-T Recommendation X.20 bis.

path is set up by the DTE or PAD by signalling a 1 on pins 103 and 104 (transmit data and receive data). The DTE can disconnect the path by reverting the data circuit to the voice mode or turning off the 108 circuit (DTR) for a period of time. The PAD disconnects the path (while 108 is ON) by turning off circuits 106 and 109 (ready for sending and data channel received line signal detector). When the PAD disconnects the access path, the DTE should respond by turning off circuit 108.

Once the access information path is available between the DTE and the PAD, parameters and commands may be exchanged. These are exchanged via use of International Alphabet No. 5 (IA5) characters as described in ITU-T Recommendation T.50. For most purposes, the IA5 characters correspond to the standard ASCII character set. In this chapter, the printable equivalents of the binary code values will be used as sequences. These sequences of characters are used with similar functionality to that of the asynchronous AT command set functions.

X.28 describes the use of the data stream to use only 8-bit data characters. Parameter number 21 (P21) affects the actual use of these data. For example, if P21 is set to 0, the PAD will ignore the eighth bit received (just before the stop bit) and treat the data as 7-bit data. If it is set to 1, the eighth bit is treated as a parity bit and set in accordance to the type of parity. A number of parameters and their values affect the actual treatment of the data.

X.28 states.   ITU-T Recommendation X.28 defines the state description of the protocol used for exchange of information between the DTE and PAD. There are two general paths for state transitions in X.28. These are shown in Fig. 9.2 and are taken from ITU-T Recommendation X.28. State 1 is called the active link state and is the initial state after the access information path has been established. State 2, the service request state, is entered when the DTE sends a service request signal to the PAD. This signal gives data rate, code, and parity (if any) used by the DTE and used to select an initial profile.

A profile is a set of standard parameter values to be used for a terminal. X.28 defines two such profiles. They are the transparent standard and simple standard profiles. These two standards vary in the setting of seven parameters (P1, P2, P3, P4, P5, P6, and P12). Parameter usage will be discussed shortly.

States 3a and 3b are both considered to be DTE waiting states. State 3a exists when the DTE transmits binary 1 following the service request signal. State 3b is entered after the DTE responds to a PAD command request signal with the transmission of a binary 1. Transition to state 4 (service ready) is accomplished after the PAD responds to the binary 1, indicating DTE waiting with a PAD identification PAD service signal.

The PAD will enter state 5 (PAD waiting) following the transmission of a PAD service signal and transmit binary 1. The next transition (to

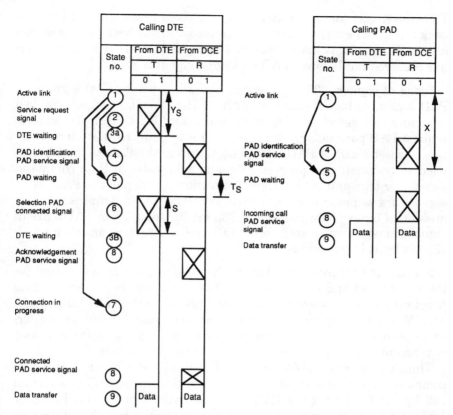

Figure 9.2 Call establishment event sequence. (*From ITU-T Recommendation X.28.*)

state 6, PAD command) is entered by having the DTE transmit a PAD command signal. It may also enter state 6 by escaping from the data transfer state (state 9). A selection PAD command from the DTE (followed by binary 1) causes the PAD to go to state 7 (connection in progress). State 8 (incoming call PAD service) is used for calls being originated by the network. State 9 (the data transfer state) has been mentioned before. This state is the equivalent of data mode in mode-oriented protocols.

Use of timers. A variety of timers are used by the PAD. These are designated by T10 through T15, T20, T21, T30, and T31. T10 is used in state 1 as a timer to disconnect the information path. State 5 uses T11 (60 s) to wait for a PAD command signal. State 6 uses the T12, T13, and T16 timers (>60 s) to receive a complete PAD command signal. T20 is used in state 4 for acknowledgment of a DTE initiated service request signal by a PAD identification service signal. T21 is a clear waiting timer in state 6. An incoming call is safeguarded by the

T30 timer, and the T31 timer waits for the 108 circuit to be turned back on. If T31 expires before the circuit is turned back on, the access information path is disconnected. Finally, T14 is used after a recall has occurred but before a PAD command signal is sent.

Types of signals. There are two general types of PAD signals. A PAD command signal (from the DTE to the PAD) is used to establish or clear a call, select profiles, select individual PAD parameter values, request PAD parameter values, send an interrupt, request circuit status, or reset a call. A PAD service signal (from the PAD to the DTE) is used to transmit call progress signals, acknowledge PAD command signals, or transmit information about the operation of the PAD. Most signals are sequences of IA5 characters. The break signal is a transmission of binary 0 for more than 135 ms (by the DTE or PAD). Break signals must be separated from start-stop characters or other break signals by at least 100 ms.

Formats of PAD command signals. The IA5 character set is used for PAD command and service signals. The space, delete, +, and editing function characters are not used as part of PAD command signals. The PAD will recognize a <CR> or + as a command delimiter. Upper- or lowercase characters may be treated as equivalent—although older equipment may require use of only uppercase characters.

Thus, a sequence of STAT (S, T, A, T, ended by a <CR> or +) is a PAD command signal to request the status information regarding a virtual call by the DTE from the PAD. The PAD will respond with FREE or ENGAGED. A CLR is used to clear an established call. A PROFx sets up a standard profile to be used. Other command signals exist.

Examples of PAD service signals include RESET to indicate peer resetting of a link (or local reset due to PAD protocol problems). A CLR can be used to indicate remote clearing of the call—or a CONF in response to a PAD command signal of CLR. These examples are for an English PAD setup. A PAD command signal of LANG can be used to set other language protocols.

ITU-T Recommendation X.3 and PAD parameters

ITU-T Recommendation X.3 gives the set of possible values for PAD parameters. There are 29 standard parameter reference numbers. Parameter reference numbers 1 through 12 are considered essential for all PADs. P13 through P29 are considered as additional parameters possibly supported by PADs or international networks.

P1 is called the PAD recall character. This is equivalent to an escape character in mode-oriented protocols. A value of 0 means that it is not possible. A value of 1 indicates that a standard <DEL> character can cause recall. Some PADs may optionally support values 32 through 126 to indicate a specific character to be used.

P2 is the echo parameter reference number. 0 indicates no echo. A 1 indicates echo of all characters based on the echo mask being used (P20). A 2 means that all characters will be echoed dependent on the values of P25 (and, if implemented, P29). It is also possible to set the P2 up as a *scrambled echo character*. The value assigned to P2 (decimal 32 through 126) is echoed each time a character is entered (rather than the character itself). This is useful during password entry sequences, for example.

A data-forwarding character is used as an indication that the PAD should now transmit the collected packet data. P3 is used to select the appropriate sets. The value of 0 is used to indicate that this is not possible. Alphanumeric characters are indicated by a value for P3 of 1. A <CR> uses value 2. The <ESC>, <BEL>, <ENQ>, and <ACK> characters are indicated by value 4. <DEL>, <CAN>, and <DC2> are value 8. The <EXT> and <EOT> characters are used for data forwarding if P3 is set to 16. Characters <NT>, <LF>, <VT>, and <FF> are represented by decimal 32, and other characters in columns 0 and 1 of IA5 are represented by decimal 64. Note that all of these values are bit positions in a binary field. Thus, combinations of sets can be used by setting multiple character sets.

P4 sets up the idle timer delay in increments of 1/20 s (possible values 0 to 255). The idle timer acts as an automatic command to send packetized data. P5 is used for ancillary device control (i.e., flow control from the PAD to the DTE). A value of 0 indicates no flow control. A value of 1 allows use of <DC1> and <DC3> within the data transfer state, and a value of 2 allows flow control (using <DC1> and <DC3>) in data transfer or command states.

P6 controls the PAD service and command signals. For example, a value of 0 means that no PAD service signals are transmitted to the DTE. A value of 1 says that service signals other than the prompt PAD service (an indicator to the DTE that input is desired) are in standard format. A 2 means that editing PAD service signals are transmitted according to P19. A value of 4 indicates the prompt PAD service signal is transmitted in standard format, and values 8 through 15 indicate a network-dependent format. Values above 15 are indications of extended dialog mode format. A 16 says they are in English; 32 indicates French. Higher multiples of 16 can be used in conjunction with 0, 1, 2, or 5 (4 + 1) to allow combinations of modes.

P7 is used for behavior control of the PAD upon receipt of a break signal from the DTE. A value of 0 says for the PAD to ignore a break signal. A 1 means to send an interrupt packet (X.25), and a 2 means to send a reset. The value of 4 says to send a special *indication of break*

PAD message. Escape from data transfer state is indicated by a value of 8, and a 16 indicates that output should be discarded in the DTE direction. As binary bit values, these can be used in combination.

P8 can be used to force the PAD to discard output to the DTE (a value of 1; a value of 0 indicates normal delivery). This would presumably be used on either a half-duplex system or on a temporary basis. P9 can be used on a printing DTE to indicate the number of padding characters to be inserted by the PAD after a carriage return. P10 is used for *line folding.* A 0 indicates this function is not used. A value between 1 and 255 gives the number of characters before the PAD inserts formatting characters.

P11 is used as read-only access to the binary speed of the other DTE. Values of 0 through 19 indicate different speeds from 50 bps to 64 kbps (although not in ascending order). Finally, P12 is used in the same manner as P5 but for flow control of the PAD itself (rather than an ancillary transmission device).

The preceding 12 parameter reference numbers are the essential parameters to be supported by a PAD. The remaining 17 will not be covered in detail here (it does not seem to be a useful item) but can be examined in X.28 or X.3. P13 deals with line-feed insertion after carriage return. P14 is concerned with line-feed padding. P15 works with editing. P16 specifies the character delete character, and P17 specifies a line delete symbol. P18 is a line display value, and P19 deals with the editing of PAD service signals. P20 is an *echo mask,* which was referred to previously. P21 describes parity treatment. The *page wait* condition is controlled by P22, with the size of input field handled by P23 and end-of-frame by P24. P25 deals with extended data forwarding signals, and P26 deals with the display interrupt. Confirmation of display interrupts are handled by P27, diacritic character editing by P28, and an extended echo mask by P29.

These parameter reference numbers and their values determine the operating characteristics of a PAD in accordance with the DTE. This has been presented in detail to give the implementor a more precise concept of the types of functions that are to be controlled. Implementation, as always, should refer to the direct ITU-T recommendations.

ITU-T Recommendation X.29

The above discussion has detailed the functionality and use of a PAD in conjunction with a start-stop mode DTE. ITU-T Recommendation X.29 describes how the PAD is to be used with respect to a peer PAD, that is, how PAD parameters can be transmitted to a remote PAD so that two DTEs can operate with each other. It also describes the general use of X.25 packets in accordance with PAD functionality.

Call request and incoming call packets use the protocol identifier and the call data fields. Any incoming call data will be sent to the DTE as part of the call data block of an incoming call PAD service signal. Data is transferred using the Q-bit set to 0 and the D-bit set to 0. The D-bit in incoming data may be set to 1, but this is an optional support matter for a PAD and DTE. The Q-bit data field is set to 1 to transmit PAD messages.

User data is transferred when a data-forwarding situation has occurred or upon receipt of a SET, PAR?, or SET? (SET and READ) command (or an RSET, RPAR?, or RSET? for remote PAD access). The PAD will not transmit empty packets.

If a PAD is interacting with a packet-mode DTE as its peer, PAD messages (Q-bit set to 1) may occur by invitation of the peer. If a PAD is interacting with another PAD, remote PAD control signals are sent using the Q-bit method, and local PAD control signals are handled directly by the local PAD. Thus, start-stop mode DTE with PAD functionality would query the parameters of the remote PAD by use of a RPAR? PAD command signal. A local query would use a PAR? command signal.

Table 9.2 shows a list of the types, and codings, of the first byte of user data for PAD messages (Q-bit set to 1). Nine values are currently defined. These are for set (RSET), read (RPAR?), set and read (RSET?), parameter indication (remotely set), invitation to clear (ICLR), indication of break (break signal at remote end), reselection PAD message, error PAD message (generated by remote PAD) and reselection with TOA/NPI.

Most of the possible messages correspond directly to what we have already discussed concerning the PADs. The reselection messages are an extra matter. These messages are used by a remote packet-mode DTE. They indicate that the PAD should transmit remaining data, clear the call, and then set up a new call with the indicated address. As an exam-

TABLE 9.2   Coding of Octet 1 of PAD Messages

Type	Message code			
	Bit 4	Bit 3	Bit 2	Bit 1
Set PAD message	0	0	1	0
Read PAD message	0	1	0	0
Set and read PAD message	0	1	1	0
Parameter indication PAD message	0	0	0	0
Invitation to clear PAD message	0	0	0	1
Indication of break PAD message	0	0	1	1
Reselection PAD message	0	1	1	1
Error PAD message	0	1	0	1
Reselection with TOA/NPI	1	0	0	0

SOURCE: ITU-T Recommendation X.29.

ple, this type of facility might be used when a start-stop mode DTE needs to obtain information from one, or more, of a set of terminals. Each packet-mode DTE would check its ability to fulfill the request. If all information had been sent, it could notify the start-stop mode DTE to get more information from another packet-mode DTE handler.

Adaptation of Start-Stop Mode DTEs

Adaptation of start-stop mode DTEs for use on digital networks has a few additional requirements over that of other types of terminal equipment. For one, there is a need to collect the data and place it into packets. Another important aspect is the setting up of the TA into the configuration required for the application and interface. Packet-mode DTEs also must be configured, but the method of doing such differs from that of start-stop mode DTEs.

The above description of PAD adaptation for X.25 networks is an example of a mode-oriented interface to the TA. In ITU-T Recommendation X.28, the data mode is described by state 9. Other states are used as part of the command mode. Other mode-oriented protocol interfaces (such as the AT command set) do not break the command mode into separate TA states, but the effect is the same.

The next section will cover the aspects of a circuit-oriented interface. The specific recommendation that we will use as an example is ITU-T Recommendation X.21. A circuit-oriented interface uses the physical interface leads to signal different states rather than using character streams on the transmission path.

Most earlier DTE-oriented ITU-T Recommendations use X.25 as an example of a packet method for transmission and control onto a digital interface. Later documents allow for the interworking of X.25 into an ISDN or provide for separate digital transmission methods (such as V.110). ITU-T Recommendation X.30 describes a framing method (very similar to that of V.110) for use of X.21, X.21 bis, and X.20 bis DTEs on an ISDN.

Thus, we have V.110 (and V.120) for use of adaptation of start-stop V.xxx series (such as V.24) terminals onto an ISDN, PAD (X.28, X.29, X.3) TAs onto an X.25 network, and X.30 mapping X.21, X.21 bis, and X.20 bis terminals onto an ISDN. Later in this chapter, the final interconnection of bringing X.25 onto the ISDN will be covered. All of these separate recommendations allow use of older terminal equipment on the new networks.

Circuit-Oriented DTE Adaptation

ITU-T Recommendation X.21 is titled the "interface between data terminal equipment and data circuit-terminating equipment for synchronous operation on public data networks." The circuit-terminating equipment is an emphasis on the circuit-oriented control of the state of the DTE and their use for control mechanisms. A list of nine interchange circuits is given for X.21. Five of these circuits are used for general electrical requirements. There are a signal ground, a DTE common return, and three timing circuits defined. Four circuits encompass the functions of transmit, receive, control, and indication. These are referred to in the document as T, R, C, and I. T and C are circuits from the DTE to the DCE. R and I are circuits from the DCE to the DTE. The T, R, C, and I circuits correspond to the V.24 103, 104, 109, and 105 circuits as given above in Table 9.1.

The DTE and DCE use these circuits in combination to indicate changes in state of the system. They must be able to send binary 0 or binary 1 on R or T with special conditions on C or I for at least 24-bit intervals (as supported by the timing signals). Such a state for 16 contiguous bit intervals will be considered to be a steady-state condition.

A quiescent phase is defined as the readiness to enter operational states such as the call control phase or data transfer phase. Fig. 9.3 shows the basic quiescent states. The quiescent states are a subset of the total set of available states, which number from 1 through 35. States 21 through 29 are loop states designated by Lnn (where nn is the state number). States 31 through 35 are used as test states within the loop mechanisms. Thus, states 1 through 20 are the active operational states. Emphasis will be made to define these states as an example of a circuit-oriented control mechanism for a DTE.

The actual circuit signals for the various states are given in Table 9.3, as used from X.21. The quiescent states are states 1, 14, 18, 22, 23, and 24. These are referred to by the conditions of readiness of both the DTE and the DCE. Thus, state 1 is a special quiescent state where the DTE and the DCE simultaneously signal DTE ready (T = 1, C = OFF) and DCE ready (R = 1, C = OFF). The other quiescent states are when one or both of the DTE and DCE are *not* ready (as shown in Fig. 9.3). A notready condition can either be controlled (operational but not currently able to handle new signals) or inoperational (T or R = 0; C or I = OFF).

States 2 through 12 are associated with call progress situations. State 2 is entered when the DTE wants to place a call while the DTE and DCE are in the ready state. States 3 through 7/10A are also part of an outgoing call scenario. State 8 is entered from the ready state when an incoming call arrives at the DCE to notify the DTE. State 15 is a call collision state, and state 12 is a *ready for data* preliminary state.

State 13 (with substates 13S and 13R) is a data transfer state. States 16 through 21 are used during the process of clearing a call, with the final resting state back at the ready state (state 1). The exact

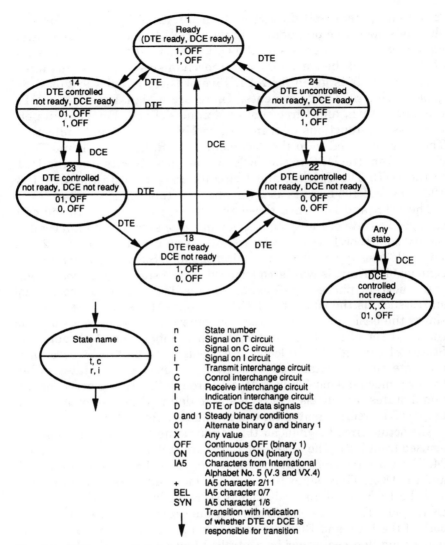

Figure 9.3 X.21 quiescent states. (*From ITU-T Recommendation X.21.*)

transitions are given in SDL descriptions in X.28. They will not be discussed in greater detail here.

States 7 and 10 are special states in that additional specific information is relayed across the circuits. This information is concerned with call progress signals (state 7) and DCE-provided information (states 10A and 10B). DCE-provided information includes line identification and charging information. This information is provided by the DCE to the DTE over the R circuit while the I circuit is set to OFF. Facility information may also be provided—called *selection* signals.

TABLE 9.3   X.21 Interchange Circuit Signals

T,	C	R,	I	State no.	Reference in the recommendation
1,	OFF	1,	OFF	1	2.5.3.1
1,	OFF	0,	OFF	18	2.5.3.3
0,	OFF	1,	OFF	21, 24	2.5.3.6
0,	OFF	0,	OFF	17, 20, 22	2.5.3.4
1,	OFF	BEL,	OFF	8	4.1.5
01,	OFF	1,	OFF	14	2.5.3.2
01,	OFF	0,	OFF	23	2.5.3.5
X,	X	01,	OFF	L27	7.3.2.8
0011,	OFF	D,	ON	L21	7.3.2.1
011,	OFF	0011,	OFF	L22	7.3.2.2
*,	OFF	BEL,	OFF	9B	4.1.6.2.2.1
*,	OFF	IA5,	OFF	10C	4.1.6.2.2.2
IA5,	OFF	SYN,	OFF	25	4.1.6.2.2.4
*,	OFF	SYN,	OFF	6C	4.1.6.2.2.3
1,	OFF	D,	ON	13R	5.2.2, 5.3.1.2, 5.3.2.2
0,	OFF	1,	ON	16	6.1
0,	OFF	0,	ON	16	6.1
0,	FF	D,	ON	16	6.1
1,	ON	1,	OFF	11	4.1.10
1,	ON	O,	OFF	19	6.2
0,	ON	1,	OFF	2	4.1.1
0,	ON	0,	OFF	19	6.2
1,	ON	BEL,	OFF	9	4.1.6
1,	ON	+,	OFF	5	4.1.4
1,	ON	SYN,	OFF	6A, 6B, 9C	4.1.7
1,	ON	IA5*,	OFF	7, 10A, 10B	4.1.8, 4.1.9
0,	ON	BEL,	OFF	15	4.3
0,	ON	+,	OFF	3	4.1.2
IA5,	ON	+,	OFF	4	4.1.3
D,	ON	1,	OFF	13S	5.2.1, 5.3.1.1, 5.3.2.1
D,	ON	0,	OFF	19	6.2 Fig. A-3/X.21
1,	ON	1,	ON	12	4.1.1.1
D,	ON	D,	ON	13	5.1, 5.2.3, 5.3.3

SOURCE: ITU-T Recommendation X.21.

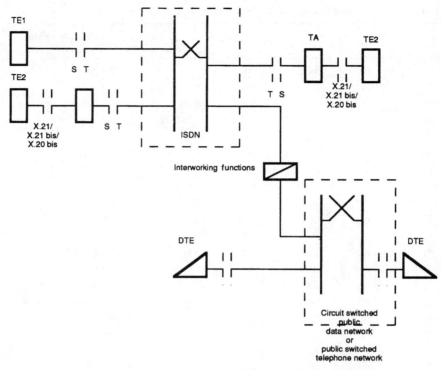

Figure 9.4   X.30 interworking configuration. (*From ITU-T Recommendation X.30.*)

The formats of these information blocks are composed of various IA5 characters and are defined in a Backus-Naur format in ITU-T Recommendation X.28.

ITU-T Recommendation X.30

This recommendation is used to describe the methods for connecting X.21, X.21 bis, and X.20 bis terminals to an ISDN. Figure 9.4 shows a basic network interworking configuration. The TA functions are divided into three areas. These areas are the rate adaption functions, X.21 to Q.931 mapping functions for call control, and data alignment.

Data alignment falls into two cases. The first case is where each TA has the same user rate. In such a case the bearer channels are used directly for mapping to the *R* interface reference point. The second case has one TA adapting to the rate of the other terminal. The calling terminal's rate will be used for the bearer channel and for data transfer. Thus, if calling TA *A* is at 2400 bps and the called TA *B* is at 1200 bps, the frame rate will be set for 2400 bps, and TA *B* is responsible for the flow control and buffering necessary to allow proper data communication.

		Bit number							
		1	2	3	4	5	6	7	8
Octet 0	Odd frames -	0	0	0	0	0	0	0	0
	Even frames	1	E1	E2	E3	E4	E5	E6	E7
Octet 1		1	P1	P2	P3	P4	P5	P6	SQ
Octet 2		1	P7	P8	Q1	Q2	Q3	Q4	X
Octet 3		1	Q5	Q6	Q7	Q8	R1	R2	SR
Octet 4		1	R3	R4	R5	R6	R7	R8	SP

Figure 9.5   X.30 frame structure. (*From ITU-T Recommendation X.30.*)

Frame structure

Figure 9.5 shows the basic frame structure used for frame adaption within X.30. The frame is shown, as in ITU-T Recommendation X.30, in the transmission order (which is reversed from the logical order). You will notice a basic correlation to the frame structure of V.110—although some control elements are different to support the X.21 control needs. In particular the SP-, SQ-, SR-, and X-bits are set according to the monitoring of the C and I circuit interface leads.

X.30 uses the same rate adaptation steps as for V.110. However, X.30 only supports baud rates of up to 9600 bps. This means that the intermediate rate can be either 8000 or 16000 bps. The E-bits are used to identify the user rate at the intermediate rate of 8000 bps (only 9.6 kbps is transported on the 16-kbps intermediate rate). The values of 100 for E1 through E3 are used to indicate 600 bps. A 110 indicates 2400 bps, and 011 is used for 4800 and 9600 bps (dependent on intermediate rate).

Call signalling mapping

Each of the X.21 call signalling requests maps to a Q.931 packet. The SETUP, CALL PROCEEDING, ALERT, CONNECT, DISCONNECT, RELEASE, and RELEASE COMPLETE messages are used. Fig. 9.6 shows an example of DTE call establishment and clearing. Mapping is straightforward in general. There is a message for a call that wants to be set up—a call request which maps to a SETUP packet. There are call progress indicators that map into CALL PROCEEDING and ALERTING messages (although, as may be noted from the figure, the CALL PROCEEDING is not explicitly passed to the TE). Finally, there are the clear requests for tearing down a call that translate into the DISCONNECT, RELEASE, and RELEASE COMPLETE messages.

General DTE to DCE Interface Requirements

ITU-T Recommendation X.32 is used as a description of the generic interface needs between a DTE and the access to a network. In a PSTN, this is equivalent to the interface between the DTE and the

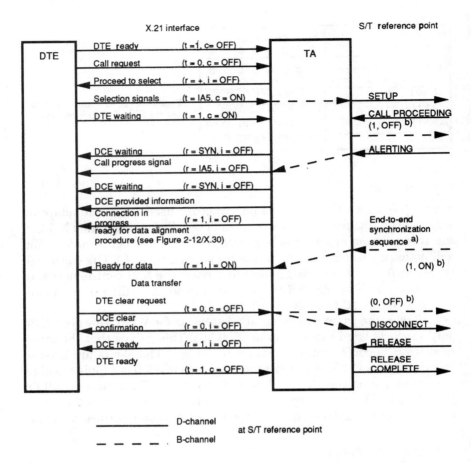

Figure 9.6 X.21 example of DTE call establishment and clearing. (*From ITU-T Recommendation X.30.*)

modem. For ISDN, this is the *R* interface reference point, and for a Circuit-Switched Public Data Network (CSPDN) it coincides with the X.21 (or X.21 bis) interface. The purpose of the network is to establish the access path and to provide the transmission circuit. It may also service addressing needs.

DTE service types and identification

There are three general types of services provided for DTEs. These are based upon the DTE being unidentified, identified, or customized. Unidentified DTEs are provided with a standard set of services—special, and optional, services are not provided, and standard parameters take effect without any special subscription requirements. An identified DTE allows the possibility of providing additional call charging services. A customized DTE has access to special nondefault parameters. Customization implies subscription for services from the network.

For the latter two categories of DTEs, a method must be followed to obtain the identification of the DTE. ITU-T Recommendation X.32 describes three methods of identification. The first is that identification is provided by the Public Switched Network. In other words, the terminal identification is associated with the terminal UNI. The second is by use of data link layer XID procedures. The third method uses the NUI selection facility in call setup packets. Identification can be prior to any virtual call establishment or per virtual call.

DTE service descriptions

ITU-T Recommendation X.32 lists a considerable number of DTE services—considerable, that is, if the number of options available per service are taken into account. The actual services are a list of items that are primarily associated with identified and customized DTEs. Unidentified DTEs basically are given a set of default and network-specified services. Identified terminals have some of the services modifiable according to the terminal. Customized DTEs have access to all of the services.

The first group of services is associated with the identity and address of the DTE. Both identified DTEs and customized DTEs have a DTE identity. They can also both use any form of identification method, although an identified DTE is limited to the network default method (customized DTEs also have this limitation prior to VC establishment). Identified DTEs can also have an address and a registered address, as a network option—this is always available for a customized DTE. A customized DTE also has the possibility of a user selectable Public Switched Network number.

An X.25 subscription set may be available to an identified DTE and *is* part of the user selection process for a customized DTE. X.25 subscription sets include data link layer parameters (including timers), call restrictions, throughput negotiation, and network layer parameters. Call restrictions and throughput negotiation affect the application capabilities, but they do not directly affect protocol support. The ability to negotiate data link layer and network layer parameters requires that this be allowed for in the design of the DTE (or TA).

The DCE identity presentation is set by the network for all types of DTEs. The use of optional X.32 user facilities is supported only for customized DTEs. All other services are supported as network defaults for unidentified and identified DTEs. These services are called logical channel assignment, dial-out access type, and data link layer address assignment. They are user selectable for customized DTEs. The final service of dial-out-by-the-PSPDN availability may be supported for identified DTEs but is always available for customized DTEs.

Implementation concerns for DTEs

There are many options available for the designer of a DTE (currently referring to the DTE as the entity that works with the network). Options are expensive—in terms of code space, efficiency, and maintenance. Thus, the designer of a DTE needs to decide exactly what options are to be supported. Will data link layer parameter negotiation be supported? If so, this affects the design. Does a particular DTE identification method need to be implemented—or possibly more than one method? How is this associated with the TE? Any optional messages need methods of communicating needs through the layers.

Connecting to a Packet Handler

ITU-T Recommendation X.31 has been mentioned several times in this book so far. X.31 describes the architectural, and protocol, needs for connecting a packet-mode device to a packet handler. The same type of architecture is being used for Frame Relay SVCs (and is likely to be the basis for B-ISDN connected into Q.931 signalling methods).

There are two basic categories of connection. The first, called Case A in the recommendation, makes use of the ISDN to make a connection to a PSPDN. In other words, ISDN is used to set up a conduit to a packet handler. The second case is called Case B. Case B integrates the Packet-Handling (PH) functions directly into the ISDN. A Case A situation only allows use of B-channels as data conduits. A packet handler integrated into the ISDN (Case B) may provide service over a D- or a B-channel.

Configuration of DTEs

From discussions toward the beginning of this book, we can recall the definitions of a TE1 and a TE2. The TE2 is used, across the R interface reference point, in conjunction with a TA to provide ISDN access. The TE2, itself, is *not* ISDN compatible. Many of the recommendations we have discussed in this chapter have been concerned with TE2s in combination with their TAs. A TE2 is directly ISDN compatible. Thus, for purposes of connection into the ISDN, a TE2-TA pair and a TE1 may be considered as equivalent entities. Each operates over the S/T interface toward the network.

The same types of circuit connections are available for Case A situations as for any other type of ISDN bearer service. These fall into subscribed, or semipermanent, connections and switched connections. If a semipermanent connection is used, the gateway Access Unit (AU) device will be an X.25 router directly connected to the PSPDN. A switched, or demand, circuit will use an address that is associated with the AU on the PSPDN. Case B provides a direct link via a packet handler (using the Bearer Capability IE elements mentioned in Chap. 6).

For Case A situations, two numbers are required if an on-demand circuit connection is made through the ISDN. Thus, an application has two separate connect requests to make. The first, using the out-of-band addresses, is routed as a SETUP message to the ISDN signalling handler. After the connection has been confirmed and the bearer channel has been connected, a physical link is available to the X.25 router, but no end-to-end data connection is yet available. This requires a second connect request, this time using an in-band X.25 CALL_REQUEST message, to the X.25 network. After the logical channel has been obtained, data can pass transparently through the ISDN bearer channel to the X.25 network and then can be routed to the addressed endpoint.

For Case B scenarios, the PH function is provided within the ISDN. Either a B- or a D-channel may be used as the bearer channel. In the case of a D-channel, SAPI 16 is used over LAPD for the data link layer. Many networks that provide PH capabilities allow only semipermanent D-channel connections. The packet handler may be provided as a packet-switching module within the ISDN, or it may be provided in the local exchange. In other words, the precise location of the packet handler is transparent to the use of the connection. We saw an example of the Bearer Capability IE needs for establishment of a packet-mode bearer service in Chap. 6.

If Case B applies, the SETUP message used to provide the packet-mode service may not require a Called Number (CDN). This depends on the implementation of the network. If any packet handler can provide the in-band connections desired, the address is unimportant. However, if multiple packet handlers are supported by the ISDN, each of which has access to different network nodes, the address may be required. Use of particular fixed (nonautomatic) TEI values may be sufficient to identify the packet handler.

Access to packet communications

For Case A DTEs (connecting via an ISDN to a PSPDN), an on-demand connection is established by sending a SETUP message with a Bearer Capability IE of unrestricted (or restricted) digital information link at 64 kbps in circuit mode. The DTE may also provide additional information in the LLC IE (if supported by the network). The DTE may ask for an exclusive, preferred, or any B-channel—depending on the capabilities provided by the network. Normal call connection scenarios apply for Q.931 at this point. (The network sends a CONNECT message which may be responded to with a CONNECT ACKNOWLEDGE message.)

As mentioned before, Case B DTEs send a SETUP message with unrestricted digital information, packet mode, with layer 2 and 3 specifying X.25 in the Bearer Capability IE. If the D-channel is selected, layer 2 is specified as using Q.921 (LAPD). The calling number (and subaddress) may be required by the network for purposes of terminal identification.

Incoming connections work in the same manner, except that the bearer channel will usually be identified in the SETUP message. A Case B situation may also ask for D-channel access in the CHAN ID IE. An ALERTING message from the DTE will not be forwarded through the network to the calling endpoint. In cases of call collision, a connection requested by a DTE will always have priority over a connection received through the network.

In either Case A or B, the bearer channel provided in Q.931 on-demand signalling will give access to the packet handler. The originating terminal will have the responsibility for initiating (establishing) the data link layer with the DL_EST_RQ primitive. A RESTART message should be sent after the establishment of the channel has completed.

Connection clearing

For on-demand connections (either Case A or B), there are two categories of connections involved. The first is the connection through the ISDN (circuit mode for Case A, packet mode for Case B). The rest are in-band through the connected bearer channel. In the case of the data link layer losing integrity (DL_REL_IN or PH_DI), all virtual connections should be released immediately. Similarly, if the Q.931 ISDN connection is released, all X.25 virtual calls should be released appropriately. An X.25 virtual call may be released independently at any time without affecting other virtual calls on the bearer channel or the ISDN connection. Once all virtual calls are released, the behavior is dependent on the needs of the TE. The data link layer on the bearer channel may be released (DL_REL_RQ), the ISDN circuit may be disconnected, or both. The on-demand circuit should be treated as a separate logical link by the application independent of the X.25 virtual circuits in use.

Semipermanent connections

For most purposes, semipermanent connections can be treated as if the equipment is directly tied to the packet-mode network. Thus, a channel will have the data link layer established (usually by the network). A RESTART packet may not be required on a semipermanent connection. One difference for the implementor is that X.25 PVCs are available on semipermanent (or subscribed) lines. This is largely due to the ability to be fully subscribed as a particular node on the network. For most on-demand connections (Case A or B), PVCs are not supported.

A PVC does not require a CALL REQUEST packet for setting up of the line. A dedicated LCI is used for identification of the link. It is probably always good (and may be required) to send a RESET packet at the time the equipment is powered up. This establishes availability of the circuit.

Terminal adaptation concerns

Generally, terminal adaptation on an X.31 connection will make use of HDLC flag stuffing rate adaption methods. Some PSPDNs may provide appropriate X.21 framing methods for access to the network. This can be determined by use of the layer 1 identification within the low-layer information fields.

One aspect of terminal adaptation, not yet discussed, is that of the mapping of Q.931 cause and result codes to the appropriate X.25 (or other DTE protocol) codes. This is necessary whenever the codes are to be passed on to a DTE that does not understand the Q.931 cause codes (thus, any DTE2 equipment that requires access to cause values). ITU-T Recommendation X.31 provides several tables for the mapping of cause values and other result codes.

Software Considerations

The main question for an implementor is always going to be, "For what is the equipment going to be used?" A DTE that will support only one type of service is always going to have a simpler architecture and a more efficient code structure. The simplest such architecture for a DTE adaptation will be a DTE attached via a semipermanent connection. In such a case, the TE2-to-TA functions will be implemented along with the standard packet network protocol (probably either X.25 or the more interworking-capable ITU-T Recommendation X.75).

In most cases, only one type of terminal (TE2) will be attached to a given TA. This means that the R interface must be supported. This will be done by direct hardware integration or by use of a LLD that is able to communicate appropriately with the TE2. The next step is to implement the appropriate TA entity. Similar to the design requirements of an I.430 (ISDN BRI physical layer) interacting with the data link layer, there will be a likely need for a software break between the

LLD and the TA. This is required each time that multiple possible connections can be made. In the case of a DTE interacting with ISDN, it will almost always be needed (because the D- and the B-channel protocol will require independent task scheduling).

The TA should be designed to interact directly with the "top layer" of the implemented protocol stack. A command comes into the DTE via the LLD connected to the R interface reference point. The information is passed to the TA entity. At this point, the command is translated into the appropriate ITU-T primitive (for example, an N_CONNECT_REQUEST primitive). If the system allows flexibility of service, the N_CONNECT_REQUEST primitive is routed by the coordination function of the equipment.

Coordination functions

A protocol stack can be implemented for a particular piece of equipment, or it can be implemented for use within different pieces of equipment (or for various services on the same piece of equipment). In the first case, all routing of data can be done automatically. If various services are to be supported or the software is to be used within different pieces of equipment, it is very useful to have primitives routed into the protocol stack by a separate coordination entity.

The coordination entity allows variable routing of data depending on the service required. As an example, say that the equipment can make use of SVCs or PVCs. It is possible that the TE2 will have knowledge of the type of connection that is desired. The command can then pass this information along to the coordination entity such that it is known that the data is routed directly to the X.25 network layer software using a particular LCI. The coordination entity can know the state of the PVC and perform a RESET, if necessary, on the PVC before starting data transfer.

The TE2 may *not* know whether a particular connection is a PVC or SVC. (The user only knows that a particular address is correct for a particular service.) The coordinating entity can translate this address into an LCI and then act upon connection requests in the same manner as described above. In the first case, the knowledge of the type of connection resides in the TE2 (and the user). In the second case, the connection knowledge is kept transparent to the user.

The same type of coordination can be applied in the case of an SVC. If the TE2 knows that the address is an SVC, the information can be passed as part of the connection request. If not, it would be the responsibility of the coordinating entity (or, perhaps, the TA entity that forwards the primitive) to recognize that a N_CONNECT_REQUEST primitive needs to be forwarded—with acknowledgment (or denial) passed back to the TA.

The above two examples are for a situation where there is a semi-permanent connection available to the packet network. If the use of an on-demand connection is required, two separate connection establishments are required for the first data circuit. The TE2 can know this and request two (or more) separate connections. The coordinating entity can also handle this task. The first connection request arrives (via the LLD and TA entity) at the coordinating entity. The coordinating entity knows that an on-demand connection to a packet handler is required. It formats an N_CONNECT_REQUEST to be forwarded to the network layer of Q.931. If the request is denied, a response is given to the TE2 (possibly with mapping of Q.931 cause codes to the appropriate format expected by the TE2). If the request is accepted, the coordinating entity proceeds to forward the original connection request to the X.25 network layer. Results of the connection attempt are passed back to the TE2.

Initialization of the TA

The preceding section details various options for data flow from the TE2 to the network. The exact functions needed within the TA depend on the location of data about the network. If the TE2 has the information, special addressing or commands may be needed at the interface boundary. If the TA has the information, it must be informed about conditions of the network. Part of this decision is one of implementation. Part is dependent on the flexibility of the existing interface. It is important to remember that a TE2 is, by definition, a piece of non-ISDN equipment that has been developed for use on previously existing networks. Thus, for most TE2 situations, it will be necessary to have the TA be aware of all ISDN requirements.

The information needed must be accessible by the TA. This may be done by special initialization methods, front panel setup programming, or any other method. The TA will either have been implemented for a particular service type, or it must be told the current service type to be used. A TE1 must be designed to allow the passing of this type of information.

Management functions

Management functions for DTEs fall into two categories. One is the normal set of buffer and timer management functions needed for most protocols. The other is concerned with data management for TA functionality. The TA can store, within its own data space, any information needed only by the TA. Data that are needed by the coordinating entity (or other parts of the software) must be communicated to those entities from a coordinated data structure. The ITU-T primitives are designed only for the main services needed within a protocol stack. Such functions include connect requests, clearing requests, assignment and establishment requests, and so forth. Other primitives are needed on a per-implementation basis. Thus, management primitives are needed for initialization of hardware, changes in state of interface requirements (such as changes in user rates), and communication of data to other software tasks.

Bearer service negotiation

Bearer service negotiation is a little different for TE2-TA pairs than for TE1 equipment. The mechanisms are the same, but the contents may vary. Knowledge of requirements is needed within the implementation. It is possible to use the LLC IE (or Bearer Capability IE) for some of this information. However, if multiple services are to be provided, methods of communicating the agreed-upon service must be implemented. Some of this may be part of special addressing or pre-programming of a TA.

Connection to a PSPDN requires a different set of fields within the Bearer Capability IE than does a direct ISDN packet handler. The equipment must be able to format, and parse, the correct IEs. If this is part of a varying service offered to a TE2, mapping is needed at one part of the system or another.

Hardware versus software reference point handling

Much of the work needed to properly work with the interface at the R interface reference point will be done by hardware. Such devices fall into two categories. One is called a *dumb* chip—which is not talking about the hardware device's abilities but rather the ability to program the device. Many very inexpensive hardware devices are available for serial interface interactions. These have the advantages of low cost and simple interfaces. If you are working with a direct interface, this may be the way that you decide to go in the hardware architecture.

Other systems require the possibility of changing the mode of the device. A TA might support asynchronous or synchronous terminals, for example. Other devices are designed to support particular protocols—such as ITU-T Recommendation V.110. Each device will have its own requirements for programming and use.

Extensibility

An architecture that allows for change is usually more complex than one that is designed for a specific task and requirements. However, an architecture that is *not* designed to allow changes in the future will be difficult to change in a maintainable and efficient fashion. Once again, the primary requirement for the system designer is to determine the uses of the product and the methods by which those uses can be supported.

Will different services be allowed? Will different TE2s be supported? What options are available to the user? How is information stored or programmed? These types of issues mold the shape of the design.

User interfaces

A user interface is what is presented to the user and is how the user tells the equipment what is needed. A TE2-TA pair will be somewhat limited in its options. This is because the TE2, by definition, is a pre-existing (or predefined) piece of equipment. The interface is already defined, and modifications can be made only within the existing interface specifications. Special character delimiters, for example, may be able to be used to indicate out-of-band versus in-band addressing. It is also possible that *register sets* that are used for customization of the functions performed by the TA can be extended for other requirements. The designer must be very careful in such cases to make sure that such extensions do not cause future problems.

In many cases, the TA designer will not have access to the TE2 interface at all. The TA supports a particular type of equipment and interface (or a set thereof). It may be possible for the TA designer to give interface requirements that any TE2 must be able to follow. Additional requirements may cut down on marketability of the product.

The designer of TE1 equipment has many more options for the user interface. A specific interface may be designed as part of the equipment. This interface should be able to meet the various needs of the system. The amount that the equipment needs to be programmed, or customized, should be minimized to present a robust, and easy, piece of equipment to the user.

Finally, many DTEs (TE1s and TE2-TA pairs) will work with existing communication programs. These communication programs act as filters between the users and the communication devices. It is highly desired to support existing popular interfaces. More versatile but less widespread interfaces can be supported in addition—but support of popular existing interfaces can be of great market support.

Summary of Chapter

Discussion of two types of DTEs was made in Chap. 6 about ITU-T Recommendations V.110 and V.120. These are standards that have been devised to take full advantage of the possibilities of an ISDN. However, other DTEs already exist. Part of the primary architectural reasons for the ITU-T's recommendations is to provide a migration path from existing equipment to new equipment and networks. Thus, it is very important to support existing equipment.

PAD is a generic name for a device that is able to accumulate data and then present them to other software entities in a condensed form. In the ITU-T recommendations, a PAD is more directly associated with interfacing DTEs into an X.25/X.75 packet network. Characters from a start-stop mode terminal may be accumulated until a data packet can be formed and forwarded to the protocol stack. Such accumulation and forwarding requires particular signalling interface requirements and methods of timing and mapping.

ITU-T Recommendations X.3, X.28, and X.29 provide the general interface needs for a start-stop mode terminal to work with a packet network. Timers, interfaces, and mapping of results are detailed in the recommendations.

Other types of terminals may also be supported as a TE2 to be used on an ISDN. Asynchronous terminals may have special protocols defined (such as V.110 and V.120). They may also be used in conjunction with PAD functionality to present a synchronous stream of data to the network.

Equipment may use a character-stream-oriented method or a circuit-lead-oriented method for communication across the R interface reference point. ITU-T Recommendation X.21 provides a standard for character-stream-oriented interface methods. In this case, an access interface is established. This access interface is then used with character streams to request functions (or receive responses).

ITU-T Recommendation X.30 provides a standard for circuit-lead-oriented interfaces. In this case, character data are used only to provide additional information needed for particular commands. The commands are interpreted by the TA changing states in response to the change of state of a set of four circuit leads. Two leads are controlled by the TE2, and two are controlled by the TA.

ITU-T Recommendations V.110 and V.120 were specifically designed for use with ISDN. Most of the recommendations covered in this chapter have dealt with how to adapt non-ISDN equipment to the needs of a packet network or ISDN. A final link needs to be made for such terminals—to actively make use of the equipment on an ISDN.

ITU-T Recommendation X.32 provides general needs for a terminal making use of a network. ITU-T Recommendation X.31 presents two cases of interworking. The first, Case A, uses the ISDN as a transparent conduit for access to another type of existing network (such as X.25/X.75). The second, Case B, uses the ISDN signalling protocol to provide access to special services. X.31 refers primarily to an X.25/X.75 network or packet handler. However, the general architectural difference between Cases A and B may be used with other protocols (such as switched Frame Relay, as seen in Chap. 7).

This concludes the discussion of the basic protocols used in many ISDNs. It does not cover all such protocols or standards but tries to give a sufficient overview to allow the implementor (or user of equipment) awareness of the needs of a system. Specific implementation will always require direct access to, and use of, the direct specifications and standards. The next three chapters go into the implementation concerns for a real-time system—more precisely, an ISDN system.

Implementation Concerns

Part

8

Implementation Concerns

10

State Machine Designs

A state machine is the basis for most protocol designs and is made up of three basics. These are the state (which reflects the history of past events), the event, and the responses to the events according to the state. Protocol state machines can be implemented in hardware, software, or combinations of both. They are used in programming appliances, communication equipment, and any other device that requires variable responses to events based on previous events.

The design of a state machine is probably the single most important part of the architecture for protocol systems. It affects efficiency, maintenance concerns, modularity and reusability, and the ability to debug and modify. This chapter will describe, in detail, many of the aspects of design of state machines as well as the different methods of implementation.

Description of a State Machine

An example of an extremely simple state machine is a door. The door has two states. These states are open and closed. Of course, one can also extend the definition to include aspects such as being absent, under construction, broken, and so forth. To make it more interesting, let's include substates. A closed door can have substates of being locked or unlocked. An open door can have substates of the lock being set or not being set. The primary event that happens with a door is that someone walks through the doorway that the door covers. Another event might be that the wind blows the door shut.

What happens if an event occurs depends directly on the state of the door. If the door is closed, the person is unable to walk through. If the door is open, the person can walk through. If the door is closed, the person must open it to walk through.

In this example, we have the essential elements of a state machine. The state implies current conditions (and may imply history). The set

TABLE 10.1   A Simple State Table

	State			
	Closed (1)		Open (2)	
Event description	Locked (1.1)	Unlocked (1.2)	Lock set (2.1)	Lock not set (2.2)
Door approached (with key)	Open door (2.1)	Open door (2.2)	Enter doorway	Enter doorway
Door approached (without key)	No action	Open door (2.2)	Enter doorway	Enter doorway
Wind blows	No action	No action	Door shuts (1.1)	Door shuts (1.2)

of events is what is possible to do within the system. The state implies what events can be done and possible further reactions. Table 10.1 gives an example state table for such a situation.

Another less essential but often present aspect of a state protocol definition is a set of alternative actions. If the door is closed, the person is unable to go through the doorway. Options include going to another doorway where the door is open. Another, more likely, possibility is to open the door. A requested event requires an action which then changes the state and allows the requested event to be performed. In description form or SDL notation, options can be listed. State tables (or state/event matrix tables) normally list a preferred combination of state/event/response.

A response can also inform and elicit more information. I approach a door that is closed and I call out "Is anyone there?" Depending on the response, I may leave, open the door (if unlocked—an aspect of the state of the door), unlock the door (if I have the key—part of my state), or ask someone to open the door. Figure 10.1 gives an SDL form of the state table that allows additional options in response. The various descriptions in the state table exist—but only as one variant of what is presented in the SDLs.

A state machine is the implemented algorithm of a state/event description. The state machine is responsible for maintaining the current state (electrically, magnetically, or even physically), interpreting events, and responding to the events according to the current state. Note that responses may include the changing of the current state.

Parts of a State Machine

A state machine incorporates the various parts needed to implement the responses to events according to the current state. Thus, there are a variety of requirements placed upon a state machine. First, it must be able to keep knowledge of the current state of the system. Next, it must be able to analyze events. Last, it needs to be able to respond to

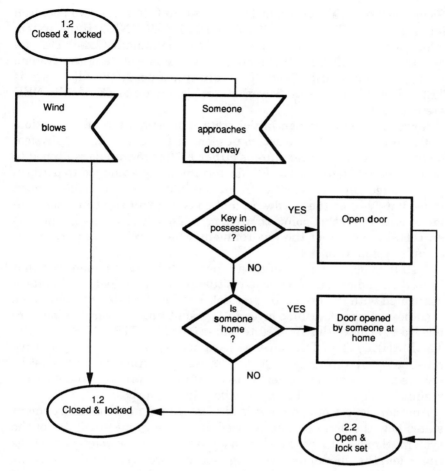

Figure 10.1   A simple SDL representation.

the events and update the current state, if necessary. A final requirement is not directly part of a state machine but is necessary for any
useful device. This is input and output (I/O). Data must be able to
enter and exit the system. Otherwise, the system is completely useless.

For the rest of this chapter, we will concentrate on communications
protocols (specifically, those that are useful for ISDN). The door example is useful as a simple explanation of what a state system is, but it
is too simple to allow for the various possibilities that are encountered in a complex system.

State data structures

The state of a system reflects the current situation. In LAPD (Q.921),
there are eight possible states. Two states are potential *starting states*.

Sometimes these are also called *idle states*. In Q.921, the two potential idle states are TEI_UNASSIGNED and TEI_ASSIGNED. If the system allows for the use of either automatic or nonautomatic (fixed) TEI values, the TEI_UNASSIGNED state will be the idle state of the system. If only nonautomatic (fixed) TEI values are used, the idle state is TEI_ASSIGNED (and the "lower" states leading up to TEI_ASSIGNED are not used).

A system always has an initial state. This state, the idle situation, is set up with default characteristics for the system. In I.430 (physical layer for BRI), the initial state is either F1 (for terminals equipped to perform power detection) or F3 (for terminals not equipped to perform power detection). A transition from the idle state is possible almost immediately—but the hardware and software will start in a particular condition. At this point, there is no history. There may be, however, certain aspects of the environment which are set as part of the default system use.

A state is actually a set of data. There will be a state identifier that is used for indication of the macro situation. Such an identified state is associated with the use of the protocol. The LAPD states are tied to the readiness of the data link layer to transmit data. Several things are needed to achieve this final state (state 7, or MULTIPLE_FRAME_ESTABLISHED). The TEI must be assigned. A DL_EST_RQ primitive must be received from layer 3. A *valid* response from the peer must be received. The state identifier is used within the protocol to define the readiness of the protocol to achieve its primary purpose.

Another aspect of the state is the environment. The environment states how the protocol is to be used and the current attributes of the system. Some of these data are associated with parameters such as timer lengths, modulo sequencing, specific protocol variant (such as V.120, LAPF, LAPD, or LAPB), There are also data structures that act as resources. A set of data space can be reserved for I-frame queues. These resources, at the initial state, are unused and should be initialized to reflect this condition. Sequence numbers are set to initial values.

The environment will, thus, have an initial situation that is associated with the idle state. Data structures are initialized as not being in use. Timers are marked as not being set. Default parameter values are stored. The environment changes as the system is used. For example, sending an I frame will cause Q.921 to increment (based on the modulo sequencing) a send sequence variable. This is a change of the environment. A timer will also be set—this is another change of the environment.

The last item associated with state is optional. This is a history of how the state was achieved. Substates may be used for this, or it may be stored in environment variables that are used in multiple states. Q.921

has three substates for AWAITING_ESTABLISHMENT. These are Establish, Reestablish, and Pending Release (noted as states 5.0, 5.1, and 5.2). These substates reflect how the major state (AWAITING_ES-TABLISHMENT) was entered. The Establish substate is entered from a condition where the link has not been established. The Reestablish substate is entered when establishment has been lost and an attempt is in progress to reestablish the link. Finally, the Pending Release substate is used when, while attempting to establish the link, a primitive (event) requesting the release of the link comes into the protocol. These substates allow for a distinction to be made of not only "where" in the protocol we are but also "how" the software reached that place.

In summary, state data structures give the state identifier, environment, and history. These elements are used to determine what the appropriate responses to events will be. The next subsection will discuss the analysis of events.

Events and the event parser

The purpose of the event parser is to break down an event into all of the necessary components for the state machine to respond appropriately. A secondary purpose is to break out information from the event to be used, if necessary, in the response.

Parts of an event. An event must have an identifier. Within the ITU-T nomenclature, event identifiers are called primitives. A primitive can come in via any entry point to a protocol entity. In the OSI and ITU-T models, these possible entities are the layer protocol handlers and the planes (Control, User, Management, and Supervisory). The most generic event consists only of an identifier. Such a generic event will affect the entire software system—or the entity to which the primitive/event is addressed.

The next items that may be needed are association elements. These parameters (to a primitive) allow an event to be associated with a particular part of the system. Thus, a DL_DA_RQ primitive from layer 3 to layer 2 can be associated with a specific data link rather than being for the system as a whole. A primitive from layer 3 to layer 2 requires a CES, SAPI, and TEI. A timer event can apply to the TEI negotiation process rather than to the system as a whole.

The final area of information given by events is general data. General data may be in the form of parameters associated with the event (primitive), or it may be contained *within* the data and must be further parsed. Incoming data packets are the primary example of this.

Example of the parsing of an event. A PH_DA_IN primitive comes into the data link layer. There is one main item contained—that this is a

PH_DA_IN primitive. Other data may be associated directly with the primitive. This would be, for example, the physical link identification (if more than one BRI was supported) and the TDM channel (D-channel, B1-channel, or B2-channel). Last, some reference to the actual data must be conveyed from the physical layer to the data link layer. A useful method of presenting this information (see Chap. 11) is to give a pointer to the data and some type of reference to the buffer location being used.

We now know the primitive and the physical channel. We also know where the data are located and how to refer to them. Since the primitive is a PH_DA_IN, the next step is to parse the contents of the data. For Q.921 (as an example), this means that the address and control fields must be broken out of the data. The address field tells us three items of information. These are the TEI, SAPI, and C/R bit field (which is actually used in conjunction with the control field). The TEI and SAPI allow identification of the logical link for which the data are meant.

The TEI and SAPI must be used to obtain access to the state data structures associated with the logical link. The state of the logical link also implies what parts are currently defined. That is, a logical link in the TEI_UNASSIGNED state would have neither TEI nor SAPI assigned. A data packet coming in would be ignored in this state. TEI assignment needs to come from the management entity, either based on an MDL_AS_RQ based on a fixed TEI or as response to the ID_ASSIGNED management message received via a UI frame on the management link. So, for incoming messages, a TEI that has not been locally recognized and assigned is illegal. States 4 through 8 of Q.921 allow for frames to be accepted.

The address field of the incoming data is used for identification of the appropriate logical link. The control field (in conjunction with the C/R field of the address bytes) provides the information necessary for the response to be determined.

In summary, an event is made of a primitive identifier and, optionally, parameters that are associated with the event. If the event conveys data, the contents of the data must also be parsed in order to fully break down the components of the event. The act of analysis is to determine the proper response to the event, based on the current state of the protocol for that logical link (in the case of link-oriented events).

Responding to an event

Once an event has been parsed, it is possible to decide on the appropriate response to make. The response is associated with the full contents of the event and the current state of the protocol handler for the particular logical entity (if appropriate). The state table is designed to allow the correlation between precise event and state. This correlation determines the response.

A response may be null (no response), a change of state, one or more events initiated, or a combination of event initiation and change of state. A null response may not be completely without action. For example, a PH_DA_IN that is to be ignored still has to have the data disposed of in a manner that is appropriate to the system (probably some type of buffer recovery). A timer that expires may need to have internal data structures updated even if the timer, itself, deserves no response.

A change of state is a matter of changing the data structures associated with the addressed logical entity. If actions are also performed, location in the sequence where the state is changed may be important. For example, a recursive call to part of the protocol handler will use the state as it exists within the data structures. Should it be treated in accordance with the *new* state or the *original* state? Normally, actions should be associated with the original state—which implies that the state change should happen last in a series of responses. For some situations, this may not be the case. However, it is important in the design of the state handler to be aware of just when, and how, the state data structures should change.

Entry and exit points were discussed briefly in earlier chapters. An entry point is a place within a protocol handler where an incoming primitive is used as input. An exit point is a place where an outgoing primitive is issued as output. Any response that makes use of a primitive/event to access other layers or entities will make use of an exit point. It is often useful to route all primitives that are going to the same place through a common function. Thus, all outgoing DL_ primitives could be routed through a dl_out( ) function.

A response has three potential parts to it: internal state data structure changes, events initiated to the outside world, and change of the state identifier. There are also different types of events and responses. These fall into the categories of immediate events, delayed events, timers, and error conditions.

Immediate events.   An immediate event is something that can be acted upon immediately, without the need to wait for other subsequent events. An example of such an event might be a MDL_AS_RQ primitive. This primitive assigns a TEI to a logical link (with attendant CES and SAPI). The data structures can be changed upon receipt of the primitive. It is also possible that some other responses may be made (if the link was in ESTABLISH_AWAITING_TEI, for example), but such actions are not inherently part of the MDL_AS_RQ primtive responses.

Delayed events.   Let's suppose that the event to be processed is a DL_DA_RQ primitive. The primitive has a number of parameters associated with it. Among these are the CES, SAPI, and the buffer information associated with the data to be transmitted. The buffer information is sufficient for the purposes of a DL_DA_RQ primitive because the data will only need to be encapsulated within a frame and not parsed. (The concept of layering means that the lower layers should never need to know the contents of higher-layer data.) The CES and SAPI allow the protocol handler to determine the appropriate addressed logical link. Within the data structures associated with the logical link is the current state.

Let's assume that the current state allows transmission of multiframe acknowledged data. The current environment, however, indicates that k (layer 2 window size) frames have already been sent but not yet acknowledged. Thus, the DL_DA_RQ primitive is unable to be fully processed at this time. It must be stored (in order of receipt of data) in a data queue.

Later, an RR frame comes in to the protocol entity via the PH_DA_IN primitive from the LLD. This frame acknowledges one of the currently outstanding data packets that have been sent. Internal acknowledgment variables are updated and k frames are no longer outstanding—it is possible to send the next frame. So, the protocol first removes the acknowledged frame from the internal data queues (doing whatever is necessary to clean up the buffer space in use). Next, it searches for additional data to be sent. A new data item is removed from the front of the data queue (or shifted by use of pointers from the untransmitted portion of the queue to the waiting-for-acknowledgment portion of the queue). This data is encapsulated between a layer 2 header and the appropriate HDLC framing and sent off via a PH_DA_RQ primitive through the LLD into the network.

Eventually, this data frame will be acknowledged. The buffer space will be reclaimed, and the original DL_DA_RQ primitive will have been completely performed—no data will remain that was associated with the event. Many primitives can be done only in segments. One part can be done at the time of receipt of the primitive/event. This changes the state environment (the contents of the data queues, in this example). The change in the state environment affects how later primitives are handled. A cascading set of events causes the final desired results.

Timers. A timer is initiated by the protocol whenever it needs to place a limit on a situation. The situation may be to set limits on the time allowed for a response. The situation may be to force a periodic check. Using the ITU-T nomenclature, the primitives associated with timers will be _TM_IN and _TM_RS. The timer should be tied into the hardware to keep it reasonably close to the desired value. (If it is completely software supported, there is no way to prevent situations where the timer is not able to expire in time.) Often, there will be a software process activated by a hardware timer interrupt. This software process then sends appropriate events to the relevant software entities.

A timer response will have the same types of information associated with it as do most events. That is, it will have the event identification, and it may also have parameters that indicate the precise type of timeout and the logical entity with which the timer is associated. The state machine then interprets the event according to the current state.

If a timer is linked to the internal data structures (indicating that a timer is in progress), a timer response will, minimally, clear the data structures to indicate that the timer is no longer active. It will probably also cause a series of responses.

There are a series of operations associated directly with timers. Timers may be started. Timers may be stopped. An optional operation is to restart a timer. Restarting a timer is equivalent to stopping it and then starting it again. For efficiency reasons, it may be useful to combine these actions into a single restart operation. In some protocol entities, only one timer type is active at a time for a particular logical entity. In such a case, the timer event does not need to have the timer type associated with it—only within the internal data structures.

Error conditions.   When an event occurs, it may not be the "correct" event to happen within the state machine. We have already noted that it may be possible to receive a PH_DA_IN primitive within a state that does not allow reception of the data. In such a case, the data buffer must be reclaimed. It is also possible that the error should be logged in the system or accounting records updated.

In every error condition there are two requirements. One is that system data structures must be cleaned up so that the error does not affect the rest of the system. The other requirement is to take any responses (if any) required by the state machine. Data buffers must be either used or reclaimed. Timer structures need to be reinitialized.

In many instances, an error condition will trigger a series of responses. As an example, assume that a transmitted I frame has been lost in transmission (corrupted such that it did not meet valid frame requirements on the reception side). The receiving side will ignore the invalid physical layer frame. The physical layer of the transmitting side *may* automatically retransmit the frame if hardware notification of problems has occurred. If this happens and no other frames have gone out in the meantime (likely the case if the hardware has noticed the problem), the protocol scenario has not changed. The only effect to the system is that a frame has taken a longer period of time to be successfully transmitted.

Assume that the transmitting side has noticed a problem in transmission and has resubmitted the frame to the network. The receiving side (for whatever reason, successive errors that cancel the original error as an example) receives the original frame *and* the duplicate frame. The layer 2 protocol will notice a duplication in sequencing (if an acknowledged I frame). The protocol will then discard the duplicate frame and may start a resetting, or other error recovery, sequence.

On the other hand, assume that the transmitting side is unable to notice a problem. The receiving side will receive an I frame that is out of sequence; a sequence number has been skipped. Once again, the receiving side will initiate a resetting, or other error recovery, sequence.

Both of the preceding possibilities were concerned with acknowledged I frames. A (UI) frame does not have a sequence number. It will be the responsibility of the transmitting side (probably layer 3 or above) to recover from the lost frame. Timers are often used to prevent these types of problems. In the case of Frame Relay core functionality, data integrity is never guaranteed. Error recovery is always the responsibility of the higher layers.

Design of a State Machine

We've now covered, in detail, the definition and parts of a state machine. The next stage is to understand how to design such a machine. The first step is to fully understand the functions desired. What is the purpose? What are the boundaries to the system? With what other entities does it need to interact? What types of limitations are associated with the system? What are the resources available?

The next step is to define the events. What events are possible? What information needs to be associated with each event? How is the information conveyed? What are the contents of the data? Is it important? What aspects of the data are important?

The final step is to determine the responses to the events. These responses will also define the states needed for the system. What possible errors can occur? How should these error conditions be handled? What options are available? How is the appropriate option determined? In the following subsections, we will discuss these design aspects in greater detail

Purpose of the system

The first aspect of the design of a state machine system is to fully decide on the purpose of the system. In this book, and chapter, we are concentrating on the needs of data communications. However, the same principles apply to other systems—mechanical as well as electronic. A compiler will have various states as it is in the process of processing input within a single "pass" or over multiple passes. A word processor has various states possible. Still, protocol devices are the ones most directly associated with state machines.

The concept of the OSI layers helps in the process of defining state machines. Each protocol entity may be defined within its own boundaries rather than needing to incorporate every function needed for the entire system. Thus, the physical layer is designed to interact with the physical transmission medium, accept commands from (and send responses or indications to) the system, and pass data to and from the data link layer. This is a considerable amount but still much simpler than if the physical layer also had to deal with the contents of the data.

Let's use the physical layer as an example for the delineation of functions for a state machine. The physical layer has three entry and three exit points. One entry point and one exit point are concerned with the physical transmission medium. Another set is concerned with the management entity, and the final set interacts with the data link layer. This defines the boundaries of the protocol entity.

The primary purpose of the physical layer is to receive and transmit data using the physical transmission mechanisms. Thus, the first duty of the system is to get the physical transmission medium into the proper framing and synchronization to carry data. The next duty is to be able to translate data from the form carried over the physical medium into a form usable by other layers (and vice versa). The final duty is to be able to configure the use of the medium according to ongoing needs.

A LLD, working with the physical layer, will normally be designed to work with a single set of hardware that supports the physical interface that is involved. This is a limitation of the physical layer—it is hardware dependent. Since the physical layer is responsible for data transfer across the physical interface, it will require buffer space for data. This is a resource need. It will also need (for the BRI *S/T* interface reference point) timers, which is another type of system resource.

Events of the system

Once the boundaries of the system have been defined, the next step is to define the events. We will continue using the physical layer as an example. (It is useful to be working with an already defined system.) In the physical layer, we have events coming into the system from the physical transmission medium, the management layer, and the data link layer. Each has a set of events.

The data link layer needs to be able to do two things to the physical layer (via the LLD). It must be able to activate the physical interface (PH_AR primitive), and it must be able to request data to be transferred. An activate request (needed only for BRI *S/T* interfaces) is performed by the data link layer when the transmission line needs to be used. For European systems, this happens only when data are ready to be transmitted. (It is also possible for the peer entity to request activation.) An activation request (PH_AR) is a short event. It initiates a sequence of events that result in a synchronized transmission line capable of carrying data. Since it, in itself, is of short duration, it may be carried via a synchronous function call. If only one physical interface is present in the system, no additional parameters are necessary.

The other primitive from the data link layer is to request transmission of a data block. (For this to be achieved, activation of the line must have been accomplished.) If the transmission line is presently busy, the data block may need to be queued within the LLD. However, eventually, the data block will be transmitted unless activation is "permanently" (activation sequence has failed) lost. A PH_DA_RQ has a number of qualities associated with it. One is the priority of the data. For multipoint BRI, signalling data have priority over the D-channel. Another aspect is that of communicating information about the data. This will probably entail a length, a pointer, and some type of buffer reference.

A final attribute, or parameter, of the primitive is what is to be done with the buffer after it has been transmitted. If it is an acknowledged type of data, the data link layer must be responsible for the later deallocation of the data space after the data have been acknowledged. However, if it is unacknowledged, the data link layer has requested transmission and expects the data area to be reclaimed without further intervention from other layers. Thus, a PH_DA_RQ primitive has three categories of parameters. These are a priority (for D-channel traffic over multipoint situations), data reference items (dependent on the general system buffer design), and a designation of what is to be done with the data after transmission. If an LLD supports multiple TDM links, some form of identification of the appropriate TDM link to be used is also necessary.

The management entity does not have any direct request primitives that are universal for all LLDs. (The MPH_DEACT_RQ primitive is optional for TEs and depends on the needs of the network.) However, it may be expected that there will be a number of special primitives that are associated with the needs of the hardware. A timer response is likely to be a management primitive. One such event might be to initialize the LLD. Another might be to change the peripheral equipment associated with the LLD (such as a COder-DECoder, CODEC, device for voice signals). Management primitives are the mechanism available for customization of a system. The needed parameters (if any) will be associated with the general system.

Finally, events will occur on the physical medium. In ITU-T Recommendation I.430, these events are patterns known as INFO elements. A received pattern will alter the state of the LLD and indicate various stages of synchronization and framing necessary for data transmission use of the line. The process of receiving and transmitting patterns is the method used to achieve activation.

Responses of the system

The preceding subsection discussed the events coming into the system. What types of events need to be able to go out from the system?

(The next subsection will discuss the algorithms needed to link particular responses to particular events.) Once again, we are dealing with three entities: the data link layer, the management entity, and the physical transmission medium.

For the data link layer, we have two types of events. One is notification of the activation or deactivation of the physical transmission medium. The other is indication of data arriving over the link. When dealing with activation or deactivation, no other parameters are required. In some instances, it might be useful to convey information associated with *why* a deactivation occurred. In the case of I.430, this is not particularly important. In the case of data indications (PH_DA_IN primitive), there are two categories of data involved. One is identification. Which physical and TDM link did the data come in over? This was not important in the case of PH_DA_RQ primitives because the LLD can assume that the data link layer is addressing the data to the device and TDM link with which the LLD is associated (if an LLD handles multiple links, this *is* important for a PH_DA_RQ). The other item is the data information—how long, where located, how to reference?

The management entity requires fewer general primitives coming from the physical layer. (A request will have indication of success or failure associated with it.) Most primitives going to the management entity will be forms of indications. Such indications can be directly sent to the management entity, or they can be caused as a side effect of what is done with the data link layer.

It may be possible to have primitives going to the management entity in parallel to similar primitives going to other layer entities. These would be primitives such as informing the management entity of the activation or deactivation of a link. For such, once again, no other parameters are needed. It may also be useful to convey information concerning errors to the management entity. This is associated with situations that do not necessarily affect the data link layer or the rest of the system, but it is an indication of temporary (or permanent) problems with the system. For example, if line activation was temporarily lost (but could be recovered before affecting the rest of the system), a MPH_ERR_IN primitive might be useful for statistics. Such statistics might be useful to determine the need to examine the physical cable. A final primitive is indicated in I.430 that is really a subset of the activate and deactivate primitives. This is associated with power detection. I.430 differentiates situations where power has been lost from those where the power is available but synchronization has not been achieved. Any time that parameters might be needed to differentiate specific primitives, it is possible to split such into multiple primitives.

Finally, we have the inputs from the physical transmission medium. These will be patterns called INFO elements. These patterns can be detected by the physical hardware, noticed by the LLD, and used as part of the state machine to achieve desired results.

Algorithms of the state machine

A PH_AR primitive's desired result is the sending of a PH_AI primitive (and, possibly, MPH_AI primitive also). The construction of a state machine is the process of determining how each of the events are connected to the desired results. Let's examine each event and result to see how a general vision of the complete system can be synthesized.

PH_AR to PH_AI or PH_DI. A PH_AR is unnecessary unless the system is not deactivated. Thus, it is best (for system efficiency) that the data link layer not send a PH_AR unless the system is deactivated. This relies on the initial state and keeping track of the status (as supplied by primitives or other global data). Assume that the PH_AR arrives in a deactivated state (thus, we have our initial state for the state machine). The PH_AR tells the LLD to start to send a new pattern to request activation (this pattern is called INFO 1). Since the system has changed (we are now sending INFO 1 rather than the idle pattern, INFO 0), we are now in a new state—giving us two states. This new state varies from that of the deactivated state in that a different pattern is being transmitted *and* because a PH_AR is no longer useful.

The desired result of the PH_AR is a PH_AI primitive. However, it is possible to also obtain a PH_DI as a result of the request if it is not successful. How do we distinguish between these possible outcomes? A timer is going to be needed. This timer (called T3 in I.430) will provide a time limit on the activation attempt. If a timer is not part of the system, and activation is not achieved, there will be no response to the data link layer and no way to reset the device to an idle state. So, we now have two states and a timer.

The next part is concerned with the peer system. The response to our new pattern (INFO 1) will determine what happens next with the system. It is possible for the other side to send an idle signal (INFO 0), a synchronization signal (INFO 2), or a framed and synchronized signal (INFO 4). It is also possible (particularly at test sites) to receive other signals that are not in the anticipated group of signals. This is sometimes called INFO X.

We are now in the new state, sending INFO 1, and waiting for a response from the peer. Since INFO 0 is an idle signal (and we were previously deactivated, or idle), receiving an INFO 0 from the peer does not indicate a change to the system. Receiving an INFO 2, however, does indicate a change to the system—the system has been changed by what is being received from the peer. Also, since a response has been received, the INFO 1 request for activation is no longer needed, and we can start sending a synchronization signal (INFO 3) to the peer.

Both sides are now synchronized. We are now waiting for appropriate framing. If an INFO 4 comes from the peer, we have now achieved synchronization and framing and consider the medium to be activated. We can now send the PH_AI (and MPH_AI). This is a total of four states—idle, sending an INFO 1 and waiting for the peer, synchronized, and activated. In the process of determining how we will achieve activation, we have determined four necessary states, their inputs and outputs, and how the transitions are made from one to another.

This is a partial state table—one where a successful activation has taken place in the shortest period of time. A complete state table also needs to allow for unsuccessful events. We stated that, in the second state (I.430 state F4), three possible signals could be received from the peer. If an INFO 4 (synchronized and framed) signal comes in, an early transition to the activated state (I.430 state F7) can be made.

If an unknown INFO X signal arrives, however, actions will need to be different. We cannot send an INFO 3 because we do not know if the other side is synchronized. However, we also do not want to continue to send INFO 1 signals because it is possible that the other side is attempting synchronization. So, send INFO 0 once again and wait to see if a recognizable signal arrives from the peer. (Remember that the T3 timer is still active in case no recognizable signal ever arrives.) This is yet another new state—one in which we are receiving an unknown signal and transmitting an INFO 0.

What if the timer T3 times out? In this case, if we have not yet achieved activation (I.430 states F7 or F8), the attempt has failed. The system sends out a PH_DI (and MPH_DI) primitive, resumes sending INFO 0, and returns to the deactivated state (I.430 state F3). It is also possible, after changing to the synchronized state, to start receiving INFO 0 signals once again (synchronization was achieved but not framing). In this case, we also send the PH_DI primitive and return to the deactivated state.

These are five states, which encompass the possible situations on the way to activation. Transitions are possible from one state to another, including back to the deactivated state. One other state is possible. It is possible to temporarily lose framing after being activated. This is a sixth state, and these six states are sufficient for nonpower detection situations for a BRI LLD.

PH_DA_RQ. A PH_DA_RQ primitive may be synchronous from the point of view of the interface, but the actual mechanism will always be asynchronous. The PH_DA_RQ triggers a sequence of events that will initiate transmission. The transmission will be carried out autonomously without notification to the data link layer. The LLD will process this primitive while in the activated state. Since no change occurs to the general state of the system (data and INFO 3 signals are equivalent in purpose), this does not affect the design of the state machine—only the LLD and its use of the hardware.

INFO signals from the peer. A PH_AR can start a sequence of events to achieve activation. The peer, by use of particular INFO signals, can also activate or deactivate a line. The result is a situation where the "request to activate" state is passed over, and the remaining states are equivalent to a local request. This affects the state machine in that it requires acceptance of additional events to trigger transitions. It does not force new states because the old states are sufficient for either local or remote activation.

Use of a Predefined State Machine

In most cases, the implementor of a system will be using the state machine that has already been defined for use with particular protocols. In the case of ISDN, these are located as standards within the various ITU-T recommendations. The exact state machines implemented, however, are going to be based on the contents of the specifications of the particular network with which the equipment will be working. State machines can be defined in a number of different ways. The primary methods for doing this are in general language descriptions, SDLs, and state tables.

Language descriptions

The language description of a state machine can be looked at in the same manner as a "word problem" for mathematics. The purpose and the events are defined, and the results are mapped into the pattern of events. In general, this is the hardest form of state machine to implement, primarily because it is very difficult to define a state machine in regular language descriptions. Many "holes" can be left in the description that can be left to the implementor to decide on the appropriate option. This leads to interoperability problems.

Another major problem with language descriptions of state machines is that they cannot be directly coded into software or hardware. Normally, implementors will need to create their own types of structures that will reflect the actual needs of the software.

A language description will usually be based on primitives. That is, there will be sections that describe the use of a primitive. The primitive may be an incoming event or a response (which becomes an event at another layer). The event descriptions will detail just what the event means. It will probably also say in what states the event is applicable and what exception conditions exist. Error conditions, however, are often isolated from the event descriptions. There may also be other applicable exception conditions that are not listed in the same area as the event description. Thus, the job of the implementor is to round up all of the conditions, exceptions, and details associated with an event so that it can be implemented.

Responses to events will also be described in terms of what events can trigger them (and under what circumstances). For example, a MDL_ERR_IN in the ITU-T Recommendation Q.921 has a table. The table indicates what the various error codes mean and how they are used. They do *not* specifically say in what states this can happen. The implementor has the task of determining the precise situations where the error conditions can occur (unless the state tables are used).

Specification Description Languages

An SDL looks like a flow chart. The flow chart starts from a particular state and then branches into a *switch* or *case* situation where the various potential events within the state are listed. Responses are listed in sequence, and the final part of the graph will indicate the next state (it may be the same).

An SDL has an advantage, and disadvantage, within its form. This is that options can be listed. Thus, the flow of the chart may change based upon implementation options. It is an advantage because it allows the implementor to easily recognize the different possibilities that are allowed within the protocol. It is a disadvantage because it does not specifically state what the *preferred* option is for implementation.

The SDL can be very complex or very simple. Normally, it will not list the precise parameters necessary for particular events or responses. Responses are stated as a generic description—which may, or may not, be amplified in other portions of the SDLs.

Another advantage of SDLs is that they are especially useful to distinguish (within a page of SDLs) sections of code that can be implemented in common. As an example, within state *x*, two events, *A* and *B*, can occur. Event *A* requires a special sequence of actions. Event *B* requires a different sequence. However, each ends with the same sequence. In the SDLs, this commonality of function can be graphically expressed by having the code listed only once. This may be an indication of later *in-line* software, or it may be an indication that this area of code is a good candidate for separating into a function call that can be used within many diverse areas of the state machine.

SDLs express the protocol logic in the same form as possible software implementation. They also express implementation options. If they are complete, they may be the easiest form to use as a guide to implementation. If they are not complete, they can act as a basic structure for the protocol software. Since many events and options may apply to a specific state, SDLs are often split among various pages of text. This decreases the value of common code unless these sections are split into separate function SDLs (which exist as named separate protocol graphs).

State tables

A state table can also be described as a *state/event matrix table*. This form allows for completeness and is probably the easiest form to create from a language description. A state is listed (most often as a column). All the various events, along with any environment or parsed criteria, are listed as rows. The intersection between the state column and the event description produces the sequence of responses (and next state, if not different from the current state).

If a state table is being produced from a language description, it facilitates completeness in the analysis. Some event/state combinations may not be possible (or, at least, expected). These are indicated within the state table by means of a special code (often a | or / character). If an entire row of events is impossible, it means that the event is not necessary or useful. It is important that the implementor always evaluate "impossible" situations from the point of error recovery. Even though an event/state combination may be illegal, it may occur under specific circumstances because of the way the entire system has been designed. The implementor may decide to log such situations so that the software may be reduced—or the system analyzed for correctness. Still, the system must be able to recover. An impossible situation that contains data must free up that data—otherwise system resources may be depleted. An impossible situation that would normally affect the state is an indication of possible errors in implementation.

State tables have a virtue of completeness. That does not mean, however, that they will necessarily describe each action in detail. Important specific parameters for primitives will be listed. Standard parameters will usually not be described. The other attribute that is linked with completeness is that of conciseness. A complete description may not be concise—there may be many duplicated areas of description that will want to be merged within a software implementation. However, it should list *only* the events and parameters that are necessary to provide the event/state combinations that are needed for the protocol. So, if a parameter is not mentioned as needed (for example, the P/F flag within some Q.921 frames), it can be ignored within that section of the software.

Discrepancies

In the process of evolution of a protocol standard, the first description of a protocol will usually be in a language description. The next step will be to produce an SDL for the protocol. The final step will be to produce state tables. The reason for this is that these three description methods are in progressively greater detail. When an SDL is produced, the language description will be changed to reflect any new features or options described therein. The state table occurs when the protocol has stabilized—stops being modified. The state table will express the protocol, and options, as most often implemented.

During this process, discrepancies will appear. The SDLs may mention something that the language descriptions do not. A particular implementation may incorporate an option from the SDLs that is not listed in the state tables. Discrepancies disappear over time. In the meantime, the implementor needs to be aware that such can exist. It is useful, while implementing, to compare all three (or however many exist) descriptions. The language description will usually take precedence over the other forms—but it is also possible that options have been eliminated over time.

Methods of Implementation

A state machine can be implemented at various levels and with different amounts of abstraction. The levels correspond to the OSI model in their distance from the hardware. A physical layer device will usually be controlled by a combination of hardware autonomous actions, hardware access, and software translation of commands and data. This type of software is sometimes called an LLD. Design aspects of LLDs are discussed in Chap. 11.

In theory, any algorithm that can be implemented in software can be implemented in firmware (software that has been encoded into a Read-Only Memory, or ROM, configuration) or hardware. In practice, this will be determined by the need for speed, complexity of the system, and stability of the algorithm.

The level of abstraction involves the amount of generality incorporated into a state machine. If a software implementation is capable of running any protocol with only changes in the data used within the system, it can be considered fully abstract. If the software implementation directly follows the needs of the specific protocol (using any type of state description as design criteria), the system can be considered nonabstract.

In general, the more abstract a system is, the less efficient it is for a specific protocol. However, if the system is fully understood by the implementor, changes to the protocol are easily made. The automaton and object-oriented methods are both degrees of abstraction—with the automaton model giving the highest degree of abstraction. There is thus a linear correspondence of abstraction to level. Hardware implementation is the least abstract, and automatons are the most abstract. In theory, an automaton could be written to implement all types of protocols. In our discussions later in this chapter, we will see that automatons are normally not abstracted to this degree.

The language of implementation for a state machine depends primarily on the needs of access by the language to various resources. Most high-level languages are sufficiently flexible to allow implementation (COBOL, FORTRAN, PL/I, ALGOL, PASCAL, ADA, and so forth). Examples will use the C language because this language was developed with hardware access specifically as a design criterion. The actual language used is immaterial—the important aspect is that the language have access (via compilation or assembly) to the hardware machine language instructions necessary to do the tasks needed by the protocol.

We will discuss some aspects of designing a state machine directly into semiconductor devices. Next, we will discuss the use of protocol automatons, table-driven software, and aspects of applying object-oriented software approaches to state machines.

Hardware State Machines

A protocol state machine is fastest when implemented in hardware. The device can be fully dedicated to the protocol and customized to the needs thereof. However, it is also the most difficult to modify. This means that it is normally only useful as a method of implementation when speed is critical, the protocol is very simple, or the protocol is highly unlikely to change.

Speed-critical situations

An example of a situation where the speed is critical is that of video processing. Video signals require large bandwidths over prolonged periods of time. The data necessary to make use of the signal is usually completely embedded within the signal. In other words, any optional parts to the protocol are set up only at the beginning or end of the signal. The signal, itself, will indicate the nature of the data. These data include indications of voice, audio, picture components (red, blue, green, hue, and intensity), location or synchronization of picture elements, and so forth. This type of processing also has a very simple structure—with one primary event. The device examines the data stream and uses particular fields or field values to drive other hardware devices (electron beam for a monitor, digital-to-analog converter for sound).

A video processing device can be considered a very simple state machine. Interactive video adds an extra level of complexity. In such a system, data will have to go in both directions. Thus, the number of possible events will increase. There will be the incoming data, the outgoing data, and also control commands. All this must be done without interruption of the real-time aspects of the data. Multiplicity of events adds to the complexity of the device. So, in many situations, what will be provided will be *chip sets*. These chip sets are groups of specialized devices, each independent in function and command usage, integrated into a single device or group of devices. In this manner, the complexity of the system can be limited without setting a similar limit on the functionality.

Simple state machines

An example of a very simple protocol would be that for control of an appliance. The possible events are strictly limited by the designer, and (currently) it is unlikely to have to interact with any other devices. You press a key on a microwave oven and the display changes, indicating additional required information. This is a change of state based on an event. Once this change of state has occurred, the events that are allowed have changed. An automatic cooking cycle may not allow a time to be set—only a weight and type of food. On the other hand, starting a timed cooking cycle may not allow weight to be entered as a factor.

An example of a situation where the protocol is fairly stable and speed is important is that of ITU-T Recommendation I.430. Note that we mentioned protocol stability *and* speed requirements. Even if the protocol is very stable, it is unlikely to be implemented in hardware unless it is either an isolated state machine situation or has speed requirements. One exception to this kind of a rule would be a mass-produced device such as a digital watch or calculator.

Cost factors

Use of a digital watch or calculator as an example calls out one additional decision factor about how to implement a state machine. This is cost. A specialized piece of hardware will cost a certain amount to design, fabricate, and distribute. Many of these costs can be distributed across the lifetime of the product.

Assume that a device requires $300,000 to design and prepare to manufacture. In unit costs of 1000, each device costs $5 to manufacture. If only 1000 of the devices are produced, the cost to the distributor would need to be $305 each for the designing company to break even. Add in packaging, distribution, marketing, sales, shipping, delivery, and other miscellaneous costs *plus* reasonable profit percentages and you may have to charge $1200 per device to make sufficient profits to justify the design and manufacturing. Now add this to the total cost of the product using the device and compare it to the cost of a generalized microprocessor and some ROM and Random Access Memory (RAM). It will not make sense to design the hardware specially *unless* generalized hardware does not meet the criteria of the system (such as speed).

However, if a device is used within millions of products, the unit cost becomes quite reasonable. The same situation as given before, for 5 million units, requires a per unit cost of $5.06 to break even. When manufacturing enters into this type of quantities, the $5 manufacturing cost will also probably be significantly lower. So, maybe we have a $1/unit item—when the quantities are in the millions. Quantity, design criteria, and final cost all factor into the decision to put a protocol state machine into hardware.

Stability also enters into the cost analysis used to determine whether hardware should be used to implement a state machine. Assume that a unit price has been determined based on a minimum expected sales. Before the device has paid back on the investment, the protocol has changed. Inventory, marketing, packaging, and a good portion of the design costs are now impossible to recover. This is one reason that newly produced devices usually have a high unit cost—to avoid unanticipated changes that may limit the marketability and to reclaim development costs early in the product cycle.

Event analysis

No matter what form an implementation takes, the requirements for a state machine remain the same. It must be able to interpret the events, store the state information, and respond appropriately to events. The case of hardware implementation has the same requirements.

In preceding discussions, we mentioned some of the analysis for events that a physical layer device may make concerning events that occur on a physical interface. It will also be necessary to be able to take note of other events that are occurring in the system. A hardware device will usually do this by means of registers. A register is a logical area of data to which the software and hardware both have access. Methods of access are discussed in Chap. 11.

Register usage. Registers fall into categories of the ability to read and write the contents. A register (or fields thereof) may be read-only. This means that it is an indication register (to use the ITU-T nomenclature). It will be the responsibility of some other application (such an "application" could be a human noticing that a message has appeared on a control panel) to read the information and react appropriately. A read-only register, or field, is meant for information about the system.

A read-write register allows for bidirectional communication. Normally there will be some type of protocol associated with the setting of data. For example, a single bit field may be used as a semaphore. If the field is 0, the software can read from, or write into, the register without fear that the results will be unstable due to changes by the other side of the interface. If the field is 1, contents are available to be used by the hardware without fear of changes.

Every register is actually a read/write register for some entity. Otherwise, how could the data be changed—or used? The terminology usually refers to the software access to the registers. A read-only register is written by the hardware and only read by the software. A read-write register can be read or written by either the hardware or the software.

A write-only register is a special situation. It is still really a read-write register, but access to the data is not enabled by the hardware on a read cycle. A read of the data by the software is undefined—but it may be presumed that the hardware is reading the contents to make use of the information.

Event recognition.   An incoming event either occurs on a physical interface, with interpretation of the event done solely by examination of the format or a command is entered via registers. The sequence will be that various registers are programmed with all of the parameters to the command, and then a command register (or bit field of a register) will be set to request execution of the command. The command register (or bit field) will be cleared, or a separate command acknowledge bit set, to indicate that the operation is being (or has been) performed.

State information

There will be two categories of state data kept by the hardware. One will be solely for internal use. This is the equivalent of internal data structures used by a software state machine. The other is available for access by other entities through the use of registers.

Let's use the example of the hardware implementation of the ITU-T recommendation once again. An INFO signal comes in over the physical interface. This causes a change of state and, perhaps, a change in the response over the physical interface. It would be possible for the hardware to indicate a PH_DI and an MPH_DI in succeeding indication registers. This would be read by some other piece of software and sent to the appropriate software entities. It is more likely that only the new state would be given in a read-only register. The LLD would then use this new information to determine that PH_DI and MPH_DI primitives should be sent to the data link layer and management entity.

If a hardware device handles all aspects of the protocol internally, an LLD is not necessary as a separate software entity. Separation of the functions that are most needed within the hardware device from other, less critical, functions allows the hardware to be much simpler (and less expensive and less prone to change). Combinations of LLDs and hardware make a competitive, cost-effective, solution.

Responses to events

This is the area in which a hardware implementation can truly prove its effectiveness. In a software situation, an event will come in, will be parsed and compared against the state information, will have various responses indicated, and will send out the responses. A software situation can require quite a few microprocessor cycles to effect such a scenario. A hardware implementation may require only a small fraction of the same time to effect the desired result. The process is the same—the speed is drastically different because the hardware has been designed to treat the event as a special command.

If a software language had an instruction that said, "Receive INFO 2 according to current state" and there was a machine instruction that supported this command, close to the same results could be achieved in software as in hardware. This is a description of *microcode*. Microcode allows new pseudo-machine instructions to be recognized and acted upon. This is an intermediate level of efficiency and speed. Microcode can reduce a series of software checks and instructions into a single command. It can*not* actually create new hardware to effect the command. This is the difference between a true hardware implementation and a microcoded version—the hardware implementation is designed to perform the actions, and the microcoded version uses the general capabilities of the system.

Timer integration

Timers need to be initiated by hardware support to keep the accuracy needed. If a state machine is implemented in hardware, it is very useful to integrate the timing needs into the device. This serves two purposes. It allows the internal state machine to process according to internal, rather than external, events. Second, the hardware device can control the timer as part of the response sequence. Timers can be started, stopped, or restarted as part of other response sequences.

One disadvantage of timer integration is that a separate mechanism is needed for each potential timed event (or a single timer can be used if various events are mutually exclusive). This is reasonable if only one event is to take place (such as the T3 or T4 timers for I.430 BRI activation). If a hardware device implemented Q.921 (LAPD), separate timers would be needed for each possible logical link. At this point, the advantages of an external timer mechanism may start to be more attractive on a system basis.

An external timer may be generic and multipurpose. The timer generates an interrupt and causes a software task to update data structures and send off appropriate events. The timers on such a software data queue may be for any type of event for any number of logical entities. This is another aspect of general function versus specific use.

Limitations

Hardware implementation is limited only by the cost factors associated with the development. Anything that can be implemented in software *can* be implemented in hardware. Is it worth it? Many devices are available for physical layer partial implementation (they require the use of LLDs). This is useful for speed, and the complexity of the device is limited by isolating the time-critical functions from the hardware/software interface use by the LLD. Some devices exist for subsets of the data link layer. These may be useful for systems that require only the limited subset of functions available. Video processing chips exist and hardware *bonding* (using multiple bearer channels as a single data stream) devices are produced. As standards become stable and speeds increase, more hardware devices will be produced to implement state machines.

Maintenance

Maintenance of a hardware implementation falls into three categories. There are those that may be "corrected" by use of hardware patches (*whitewiring* or *bluewiring*). There are those that may be corrected by use of software methods (called *workarounds*). Finally, there are errors that can only be noted as errata and which need to be fixed in the design and manufacture process (called hardware revisions). (It is hoped that the first two categories of errors will also be fixed in any such revisions.)

Maintenance is the process of keeping something working in its proper fashion. In a software situation, this means that the software is changed and new copies of the code are sent as updates. In a hardware implementation, the only way for diversions from the correct functionality can be fixed are for new devices to be produced. If all of the existing errors can be corrected by hardware or software changes, the device is usable. If important deviations from the optimal performance are unable to be corrected in any fashion, the design is a failure. Thus, maintenance of hardware devices must be done primarily *before* the device is released.

Advantages and disadvantages

A hardware implementation of a state machine has no flexibility past the original design. It is also hard to maintain—within the system (corrections can be made to supporting hardware or software). It has the highest efficiency possible for a system. It is thus ideally suited for systems that are fully stable and have strict efficiency requirements. It may also be useful for stable systems that are useful in great quantities (such as calculators or watches).

Automatons

An automaton is defined as a *self-acting machine*. In other words, once you tell the machine to start, it continues until it has completed its programmed sequence of actions. So, how does this differ from other software methods of implementation? In certain ways, it does not. However, an automaton for a state machine means that it does the *same* sequence of actions for any event. That doesn't mean that the responses are necessarily the same for any two events (or even the same event in two different states). It means that the actions have been split into a series of identical procedures. These procedures may invoke customized responses, but the sequence stays the same. It will also usually make use of a stack data structure. For this reason, it is also sometimes called a *push-down automaton*.

The basic flow of an automaton is steady. The first step is to equalize the entry into the *motor*. The motor acts as a self-contained software processor. When the motor returns, the event has been processed. So, all events must be mapped into a set of parameters that are fed into the automaton. These parameters usually contain the context (equivalent to the state information for a link or system) and the event.

The motor has now been invoked. If this is the first call to the motor, it will initialize a stack structure (or reinitialize it if the motor was called for a previous event and the data structure can be reused). If it isn't the first call, it may check the stack integrity to make sure that the stack contents are correct.

The next step is to start a loop that will parse events and "push" further actions onto the stack. The first time through the loop, the incoming event will be the item to be processed. The event is parsed in connection with its context. Analysis will first take place to break out all useful elements of the event. The event and context are then compared to a set of criteria. This criteria will usually be located in a data structure—one set of data per unique combination. Part of the data will indicate what should be done if the parameters and context match. This may be given in the form of a function pointer or an index into a series of action tables.

The events, parameters, and context have been matched against a data structure. Some action is indicated. Next, the action is performed. During the process of performing the action, further events will be pushed onto the automaton's stack. Eventually, all direct actions associated with the event have been accomplished and the automaton returns to the loop.

The loop checks the stack to see if anything is left on the stack to perform. If so, it does an analysis of the "top-most" item on the data stack and matches the analysis against the event, which then performs the associated action, which may then push additional items onto the stack. Eventually, each action will either create a primitive that gets passed to another entity or changes the context. When all items from the stack have been parsed and acted upon, the motor has completed.

Thus, an automaton has three components—a generic engine, a set of data structures that determine behavior, and specialized functions that are appropriate to the responses needed. If the specialized functions are eliminated, it is a fully generic automaton. Change the data structures and the purpose of the system has changed. An automaton without specialized functions requires *all* data to be located in the tables.

For example, say that a given event in the context causes a T200 timer to be started. An automaton could map the event and context to a special function that says "start T200 timer." It could also call a *start timer* function with parameter data that indicates that the T200 timer is to be used. It could also call a *start event* function with parameter data that indicates it is a timer and that the specific timer is the T200 timer. It could be further broken down to a *process event* call with parameters that indicate it is a *timer,* it is to be *started,* and it concerns the T200 timer. Each change makes the system more general but adds to the amount of data passed *and* adds to the complexity of the function that is invoked (it must determine what actions are appropriate based on the increased amount of data).

We will now discuss the various parts of an automaton in detail. The first item is the motor. The next items concern the data structures. The final item of the design covered will be function interaction with the rest of the system.

Central software motor

The central software motor is a device that analyzes, compares, reacts, and continues. Figure 10.2 shows a simple example of a motor and parser. The initial entry into the system "primes" the automaton by passing the triggering event to the automaton. This is then analyzed (or parsed) into a form that is common to all of the possible events. The data thus obtained is compared to the possibilities, as given in the event data structure array, and this results in an action. The action may include putting new events onto the stack to be parsed. The example gives a very simple scenario with a limited number of actions.

One of the problems with an automaton is that it must convert all events into a common format. A PH_DA_IN primitive may have a C/R flag, a command field, a P/F flag, a receive sequence number, a send sequence number, and data reference information. It will also have internal state variables associated with the link—what state is it in, what are the internal sequence numbers, what timers are active, and so forth. A PH_AI primitive may need none of these parameters. However, to use a common method of reacting, the data structures

```
entry_point(struct event *event_ptr)
{
    struct context *context_ptr;
    struct event_info event_info;

    parse_event(event_ptr, &event_info);

    get_context(event_info, context_ptr);

    motor(context_ptr, &event_info);
}

motor(struct context *context_ptr, struct event_info *event_info_ptr)
{
    extern struct analysis_table Table[];

    struct stack_info stack;
    int i;

    push(event_info_ptr, &stack); /* Initialize stack data structure */

    while(stack.event_info_ptr)
    {
        for(i = 0; i < MAXTABLE; i++)
        {
            if(Table[i].primitive == -1) /* End of table */
                break;
            else if match(Table[i], stack.event_info_ptr)
            {
                Table[i].function(&stack, context_ptr);
                pop(&stack); /* Finished with this event */
                break;
            }
        }
        if(i >= MAXTABLE) /* Event not found in table */
        {
            log_error_in_system(stack.event_info_ptr);
            clean_up_event(stack.event_info_ptr);
            break; /* Do not attempt to continue automaton */
        }
    }
}
```

Figure 10.2 A simple automaton software skeleton.

would need to have all of these factors passed (or null references as parameters) to be able to use the automaton. This means that the work and data structures are all rounded up to the worst possible scenario.

One method of decreasing this disadvantage is to split each event into a *primary* and a *secondary* condition. Thus, the automaton might parse out only a subset of the necessary information needed to properly do the comparison. Let's say that the event is broken into the primitive, address, and context. The PH_AR primitive is then compared against the PH_DA_IN primitive. The automaton calls separate functions for each primitive. However, if the PH_DA_IN primitive was received, additional parsing will be necessary and a new table comparison made. The trade-off is the depth of parsing (with attendant stack use in the software base) versus eliminating unnecessary parsing at the beginning.

Event data structures

As can be seen in Fig. 10.2, a primary data structure for an automaton will be something that allows comparison of event factors, combined with a method to produce appropriate reactions to the events. The structure seen in the example is such as would be used for a linear search of the tables. If two additional fields were added (unsigned short go_if_false and unsigned short go_if_true), the linear structure could be converted into a structure that could be searched in a binary-tree fashion.

Each data structure will be organized around specific operations that are to be performed with the data. A list of data structures is also sometimes called a data table. A linear search algorithm is very simple—but, with large amounts of data in a list of structures, it may be very inefficient. A tree-search algorithm is slightly more complex, but it can significantly reduce the search time through a table. A binary search pattern means, however, that matches may occur more than once—with only the final match being a *complete* match. Figure 10.3 gives an example of how a linear table can be converted into a binary-tree searchable table.

The example in Fig. 10.3 is a simple one. The real savings exist in complex states where many complex event sets must be considered. Note also that the ordering is important in both table examples. Events should be kept in the table in the order of frequency of occurrence.

Stack usage

An automaton requires some method of storing events to be analyzed and processed. A stack is a very efficient method of storing information that needs to be accessed later. A stack is also called a *push-down, pop-up,* or *Last In, First Out* (LIFO) structure. This is similar to the situation that exists with a stack of dishes. This also means that the implementor needs to code actions in the reverse order to which they need to be evaluated. If it is necessary to start a timer, send a primitive, and change the state, these events must be pushed onto the stack in the opposite order.

A queue preserves the order of events. It is also slightly more complicated since it has a *head* and a *tail.* If order needs to be preserved, a queue is a better alternative. If additional events need to be inserted into the middle of the data structure list, an ordered linked list may be desired. The data structure should be mapped to the functionality needed. The proper data structure saves coding and preserves information.

LINEAR table structure:

struct analysis

```
{
        int primitive;
        int P_F; /* Poll/Final status */
        int T200_retry_less_than_N200;
        int (*function) (struct stack_info *, struct context *);
        int next_state;
};
```

/* AWAIT_ESTABLISHMENT TABLE */

struct analysis Table[] =

```
{
{DL_U_DA_RQ, FALSE, FALSE, action_100, AWAIT_EST},
{UI_IN_QUEUE, FALSE, FALSE, action_100, AWAIT_EST},
{MDL_R_RQ, FALSE, FALSE, action_101, TEI_UNASSIGNED},
{PH_DI, FALSE, FALSE, action_101, TEI_ASSIGNED},
{SABME_C, FALSE, FALSE, action_102, AWAIT_EST},
{DISC_C, FALSE, FALSE, action_103, AWAIT_EST},
{UA_R, FALSE, FALSE, action_104, AWAIT_EST},
{UA_R, TRUE, FALSE, action_105, MULTI_FRAME_ESTABLISHED},
{DM_R, FALSE, FALSE, no_action, AWAIT_EST},
{DM_R, TRUE, FALSE, action_101, TEI_ASSIGNED},
{UI_C, FALSE, FALSE, action_106, TEI_ASSIGNED},
{T200_EXPIRED, FALSE, FALSE, action_107, AWAIT_EST},
{T200_EXPIRED, FALSE, TRUE, action_108, TEI_ASSIGNED},
{-1, FALSE, FALSE, no_action, AWAIT_EST}
};
```

Binary Table structure:

struct analysis

```
{
        int event_type;
        int event;
        signed short go_if_true; /* if index == 0, execute function */
        signed short go_if_false;
        int (*function) (struct stack_info *, struct context *);
        int next_state;
};
```

/* AWAIT_ESTABLISHMENT TABLE */

struct analysis Table[] =

```
{
{PRIMITIVE, DL_U_DA_RQ, 0, 1, action_100, AWAIT_EST},
{PRIMITIVE, UI_IN_QUEUE, 0, 1, action_100, AWAIT_EST},
{PRIMTIVE, MDL_R_RQ, FALSE, FALSE, action_101, TEI_UNASSIGNED},
{PRIMTIVE, PH_DI, FALSE, FALSE, action_101, TEI_ASSIGNED},
{PRIMITIVE, SABME_C, FALSE, FALSE, action_102, AWAIT_EST},
{PRIMTIVE, DISC_C, FALSE, FALSE, action_103, AWAIT_EST},
{PRIMITIVE, UA_R, 1, 3, no_action, AWAIT_EST},
{P_F_FLAG, FALSE, 0, 1, action_104, AWAIT_EST},

{PRIMITIVE, UA_R, 1, 3, no_action, AWAIT_EST},
{P_F_FLAG, FALSE, 0, 1, action_104, AWAIT_EST},
{NO_CHECK, TRUE, 0, 0, action_105, MULTI_FRAME_ESTABLISHED},
{PRIMITIVE, DM_R, 1, 3, no_action, AWAIT_EST},
{P_F_FLAG, FALSE, 0, 1, no_action, AWAIT_EST},
{NO_CHECK, TRUE, 0, 0, action_101, TEI_ASSIGNED},
{PRIMITIVE, UI_C, 0, 1, action_106, TEI_ASSIGNED},
{PRIMITIVE, T200_EXPIRED, 1, 3, action_107, AWAIT_EST},
{RETRY_LESS_N200, TRUE, 0, 1, action_107, AWAIT_EST},
{NO_CHECK, FALSE, 0, 1, action_108, TEI_ASSIGNED},
{-1, FALSE, FALSE, no_action, AWAIT_EST}
};
```

Figure 10.3 Linear-to-binary data conversion example.

Responses activated by the system

After analysis and comparisons, the final steps for the automaton to achieve are to effect the responses that are needed for the particular event within the state. By use of the stack mechanism, a single event will be expanded into a series of actions. Each action may be recursive—adding additional events to the stack. It may be a direct change to the context, or it may initiate communication with the other layers of the system.

Let us say, as an example, that a DL_DA_IN primitive has arrived at the automaton. The DL_DA_IN will be parsed to the point that it can be given over to the motor. The motor compares against the data structures and produces a desired action. This action changes the internal state variables (context). It also creates a N_DATA_IN primitive to pass the information up to higher layers. This N_DATA_IN event is pushed onto the stack to be eventually parsed and acted upon. It would also be possible to act upon the event immediately, but this would create a more specific automaton.

Maintenance issues

Maintenance of an automaton is easy and difficult. It is easy to change the tables, and the final functions that are to be invoked to respond to an event. It is also easy to add completely new events to the system with attendant new functions (if any are needed). However, the debugging of the system may be difficult. The structure of the automaton will be general and will *not* be directly linked to the protocol needs. Someone who is very familiar with the code structure may be able to know precisely where to add debug information and what data can be ignored.

Thus, there are two components to maintaining an automaton. The first is to understand the full mechanisms used within the automaton. The second is to understand the needs of the particular protocol for which the automaton is being used. The second requires the first. An implementor fully familiar with ISDN and the protocol needs will be helpless without understanding the structure and use of the automaton. This situation is similar to that which is encountered in object-oriented systems (as we will see later in the chapter). The required knowledge base is greater to maintain an automaton-based general system than for a system that is implemented specifically for a particular protocol (such as an ISDN layer protocol).

Flexibility of use

We have indicated that the maintenance aspects are greater for a general system such as an automaton. This is balanced, to a certain degree, by the flexibility of the system. An automaton can be changed to support a completely different protocol. There can be one set of data structures and a file for specific action functions for LAPD and a separate set for Q.931. The same automaton may be able to be used for ITU-T Recommendation X.25 with both layer 2 and layer 3. Thus, an automaton is reusable to a considerable degree.

Efficiency

Flexibility and efficiency are often found at opposite ends of design. This correlates to the difference between specific and general design. An automaton has extra overhead at every step of the process. It requires a fixed number of parameters to be passed and compared in the usage of the data structures. The passage, and comparison, of unnecessary parameters adds to the number of microprocessor cycles needed to process every primitive except for the most complicated situation. (The most complicated situation is the worst case for which the structures must be designed.)

Even for the most complicated event, the automaton has additional levels of overhead as compared to a state machine that has been specifically designed for the protocol. This is inherent in a general system because there are procedures necessary for the maintenance of the system in addition to the protocol needs.

Advantages and disadvantages

The primary advantage to an automaton is its flexibility. It can be modified for any protocol. Changes are also easy, but debugging and general enhancements require extensive knowledge of the workings of the general system. Efficiency is the greatest disadvantage of the system. It is ideally suited for systems where there is no need for real-time responses and where recursive use is an inherent part of the protocol (such as for a compiler). It is poorly suited for systems that require real-time responses (such as low-layer protocols). The flexibility may cause it to be used for higher-protocol layers where efficiency is not required as much.

Table-Data State Machine Drivers

The previous type of state machine was driven by the data contained in its data structures. So, what is a table-data state machine driver? Basically, it is a direct mapping of the state table of a protocol into a higher-level language. It can be implemented from either SDL form or from the state table matrix format.

The first step for such an implementation is the same as for every type of state machine—determine the precise event. Once this has happened, the event can be given to a function that handles events for that state or a function that determines responses according to event. This is part of the reversibility of a state matrix and will be covered shortly in a subsection.

The next step is to determine the responses. This is done by implementing the intersection of an event/state matrix situation into a higher-layer language. Each intersection has a set of software that corresponds to the actions listed. In one respect—that common code can be (and should be) kept in common between sets of actions—correspondence to the SDL is more accurate as a description. This can be done via common function calls or common in-line language statements. For example, if there are two events that vary only by one possible parameter (the P/F flag setting) and the actions are mostly in common, the software can be used in common with only an If statement used to determine which of the different branches of code is to be used. Fig. 10.4 gives a short example from ITU-T Recommendation Q.921.

Reversibility of the state matrix

A two-dimensional matrix has a *row* and a *column*. It could also be said that each intersection is the result of a function incorporating two factors. In a state matrix, these two factors are the event and the state. The combination creates a unique situation that allows for designation of a precise set of actions. The driving factor can be either one of the two combinatory factors—state or event. A state-driven implementation will perform all the possible actions for a given state based on the event that has been received. That is, it is equivalent to a column taken out of the matrix table. An event-driven implementation performs all the possible actions for a given event based on the current state—equivalent to a row taken out of the matrix table.

Let's examine the aspects involved with the decision as to whether the table should be row implemented (by event) or column implemented (by state). A major factor is that of complexity. The state is a single variable that can take any one of a fixed set of values. It may also have a substate involved. This means that two possible values will interact for certain circumstances (in the ITU-T Recommendation Q.921 state tables, these substates have their own separate columns).

Events are highly variable. The primitive will vary. The parameters will vary. The number, and use, of the parameters will vary. Event parameters are determined from internal conditions in addition to those which are contained as part of the event coming into the system. In short, an event is the complex part of a state machine.

If we implement according to the event, we will need a function per event set (or it could be broken down into separate steps similar to what was done within the automaton). In ITU-T Recommendation Q.921,

```
/* Excerpt from a MULTI_FRAME_ESTABLISHED state handler */

switch(event_ptr->primitive)
{
      case UA_R: /* UA response frame */
            if(event_ptr->P_F_flag)
                  send_mdl_primitive(MDL_ERR_IN, ERROR_C);
            else
                  send_mdl_primitive(MDL_ERR_IN, ERROR_D);
            break;
      case DM_R: /* DM response frame */
            if(event_ptr->P_F_flag)
                  send_mdl_primitive(MDL_ERR_IN, ERROR_B);
            else
            {
                  context_ptr->state = AWAIT_ESTABLISH;
                  context_ptr->substate = REESTABLISH;
                  send_mdl_primitive(MDL_ERR_IN, ERROR_E);
                  context_ptr->retry_count = 0; /* reset N200 counter */
                  send_lld_primitive(PH_DA_RQ, SABME_C, 1); /* P = 1 */
                  if(!context_ptr->remote_busy)
                  {
                        send_mdl_primitive(MDL_TM_IN, STOP_T203);
                        send_mdl_primitive(MDL_TM_IN, START_T200);
                  }
                  else
                        send_mdl_primitive(MDL_TM_IN, RESTART_T200);
                  context_ptr->local_busy = FALSE;
                  context_ptr->local_reject = FALSE;
                  context_ptr->remote_busy = FALSE;
                  break;
      case UI_C: /* Unacknowledged Information command frame */
            send_dl_primitive(DL_U_DA_IN);
            break;

      /* And so forth for other events */
}
```

Figure 10.4   Example of state table translation to software.

there would be about 80 rows to be implemented as an event-driven implementation. There are 8 possible states for Q.921 (with an additional 19 possible substate combinations). All of the states could thus be implemented in a total of eight functions—each handling its own state (and associated substates, if any). Generally, implementation by state will create fewer needed modules than implementation by event.

A second factor (beyond complexity) is redundancy. If a given event, or event set, causes the same set of responses in every state, the software implementing that set of actions will be duplicated in each state if the system is state driven. If the system is event driven, only one set of codes is needed for all states. A state matrix that has only a few differences for a small number of event sets may be very efficiently implemented by an event-driven system. Eighty calls may be needed, but that will be the extent of the complexity. Redundancy can happen between states (within rows) or between events (within columns).

Complexity alone would argue that all table-driven state machines be state driven. The factor of redundancy modifies this decision. If a state matrix is highly state redundant (duplication of actions for many states), an event-driven system will be the most efficient.

Our discussion so far has inferred that the system can be implemented only in one direction or the other. This is not true. It is possible to implement a system that is *primarily* from one direction but has exceptions that are implemented in the other direction. For example, upon entry to the protocol system, the event is parsed (and may be partially analyzed). If the event is one of a small set of state-redundant events, it can be immediately acted upon—without invocation of the state functions. This combination is often the most efficient way to code a table-driven state system. The word *exception,* however, is the flag for danger in this approach. Exceptions reduce clarity and increase maintenance problems. Exceptions are permissible if they are few in number. If they are extensive, it is an indication that the architecture may have been poorly chosen.

Entry points

One of the primary areas for any type of state machine is the concentration of data into a single handler. It is possible to have multiple handlers—each of which handles a particular subset of possible events. (For example, one handler might work only with DL_ primitives.) This is a possibility for implementation, but it effectively splits the state table into subsections. If this is desired, it should not make much difference in maintenance issues and should only affect the efficiency and code size a small amount.

An entry point has two tasks. These tasks are the identification of the appropriate context (for the data link layer, this would involve isolation of the appropriate logical link) and the breaking of the event into the important components that will later be analyzed and used to determine the response set.

In the design of a state machine, the event components need to be analyzed. This, in turn, allows construction of the appropriate data structures and methods of breaking down the events. A pointer to the context (or state) information is probably an excellent idea. This reduces software stack usage and allows for consolidation of related information. If the context is extended to add this temporary information, only one pointer is required for later actions access to the data.

There are no particular requirements for the entry points for a table-data state machine over those of other forms of state machines. However, the form of the data will change. For a hardware implementation, the final form is that of register access. In an automaton, the various event components must be split out so that data structure comparisons can be made. In a table-data situation, the only requirement is that the components be available to the state handlers.

Maintenance issues

Maintenance for table-data state machines is a very strong point. There is a direct correlation between the specification as given (once it has been translated into SDLs or state tables) and the software. The main state handler is basically a translation of the data into software functions and code. The data flow is particularly easy to follow. The only requirements are a knowledge of the high-level language and the protocol.

Since the data flow can be followed easily, debugging and changes are also easy to create. A potential problem is dealing with options—different specifications within the same module, for example. This can obscure the mainstream software for a specific version. In an automaton (and hardware implementation, though in a different way), any options or variations are hidden in the data structures.

Support functions

The actual state machine handlers are almost identical to the specification. In order to get to this point, however, there will be various functions and data structures that are specific to the system. Some of these are involved with system resources and management primitives. These are discussed in Chaps. 11 and 12.

Other functions are involved with the parsing of the events, location of the appropriate data areas, and access to the exit points. Parsing of events is the first step to the analysis of the events. This accomplishes the task of breaking the event into its components. Some analysis may take place at the same time—determining whether the event is legal in form. If it is data, is it too long or too short? Does it have a known illegal address (as opposed to a potentially legal but unknown address)? Is the command a legal one? Is the primitive recognized? These are aspects of analysis that are not necessarily part of the state machine. It is possible to refer to this as preliminary analysis.

The location of the appropriate data areas involves use of the association information. For LAPD physical layer primitives, this would be the TEI and SAPI. For Q.931 incoming packets, the CRV is the main associative identifier. For X.25 layer 3, the LCI is the main point. No matter how the appropriate data unit is identified, the event must be associated with the appropriate data structure.

In some instances, there will be no existing data structure in use. An incoming SETUP message arriving at Q.931 will contain a CRV that has not been used before. Thus, there is often a "special case" where initialization of a new data structure must be done. The general algorithm is to check to see if a data structure already exists and, if not, to see if the event is one which legitimately may require a new data structure. If neither is true, the event must be discarded in accordance with the needs of the protocol, the needs of the entire system, or both.

Exit points are functions which allow for all primitives that are to be sent to a particular entity. A LAPD implementation, for example, will have a send_dl_primitive( ) function to send DL_ primitives to the network layer. The use of a common function for all primitives to a given entity allows the software to make use of code that is common to any primitive being sent to the entity. A send_ph_data_in( ) function might be able to do the entire layer 2 header processing for the primitive. So, upon entry, an event is broken into its diverse components. At an exit point, the event is encapsulated into the structure that incorporates the components.

Options in data structures

A table-data state machine is a very useful form for similar, but not identical, protocols. Take the Q.921-oriented protocols of LAPD, LAPB, and LAPF and V.120 layer 2 as an example. LAPB was actually created before Q.921 but many aspects of it were used as the basis for ITU-T Recommendation Q.921. An additional structure member can be used to state the specific variation being used for a logical link. Then, at divergent sections of the state tables, the protocol type can be used as an additional event condition to determine responses.

Other options (for any type of implementation) include the modulo sequencing, DTE/DCE part of a protocol, and negotiable parameters (such as window size, packet/frame size, and so forth). Options give the possibility of implementing one default version and multiple subversions within a single module. Options may also decrease readability and increase maintenance.

Modularity and flexibility

In a table-data state machine, the event/state responses are centralized (either by state or by event). The data flow is according to the specifications. The system is not easily modifiable to a different type of protocol (although, as mentioned above, it is very easy to adapt it to similar protocols). That is, a LAP module is not usable as a Q.931 module. (For an automaton, the change is relatively easy.)

Advantages and disadvantages

Maintenance is the strongest point for a table-data state machine. The data flow is easy to follow, and this allows easy debugging and modifications. It is moderately flexible and fairly efficient. A table-data state machine is best suited to protocols that are well defined according to SDLs or state tables.

Comparison of Approaches

In the preceding sections, we primarily discussed three factors pertaining to the approaches for state machine design. These factors were maintenance, flexibility, and efficiency. Maintenance refers to the ease of following the data flow, debugging, and correcting problems that are encountered. Flexibility is associated with the degree that the device can be reused for other protocols or the addition of new states or events to an existing module. Efficiency is concerned with the amount of overhead (time spent doing work not directly associated with the protocol needs).

As can be seen in Table 10.2, a hardware implementation is the most efficient type of state machine design. It is also the least flexible and hardest to maintain. An automaton is the easiest to change—giving it the best advantage in terms of flexibility. It has the worst efficiency and a medium level of maintenance requirements. It is *possible* to debug and correct an automaton, but it requires extensive knowledge of the automaton mechanisms as well as knowledge of the protocol for which the automaton is in use. Finally, the table-data state machine driver is the easiest to maintain because it is directly based upon the specification information. It has average efficiency and flexibility. It is probably the best mechanism, overall, for implementation of state machines—unless the criteria indicate that efficiency or flexibility are of higher priority.

That is the crux of the matter. What are the criteria for the system that is to be implemented? If speed is critical, it may *have* to be implemented as a hardware device. If flexibility is very important, an automaton may be desired (or object-oriented approaches used on the table-data method). The architectural decisions should be based on the criteria.

TABLE 10.2   Comparison of State Machine Methodologies

Method	Maintenance	Flexibility	Efficiency	Comments
Hardware	3	3	1	Best suited for stable protocols, with speed requirements
Automaton	2	1	3	Best suited for situations that require great flexibility
Table-data	1	2	2	Best suited for maintenance needs

Object-Oriented State Machines

Object-oriented programming is not a focus of this book. There are many good books available about this subject, and we will not try to duplicate their efforts. This section primarily concerns the applicability of object-oriented techniques to state machine software devices.

Object-oriented programming is a methodology of abstraction. As such, it adds the potential of increasing the flexibility and reusability of a module of code. It is not, in itself, a type of state machine design. It may be used with an automaton, and it may be used with a table-data state machine driver. It adds one additional level of flexibility and (for virtually every object-oriented compiler) decreases the efficiency. Maintenance issues are variable. Without tools, the level of abstraction provided by object-oriented programming gives the same type of hidden structure that an automaton inherently produces. With tools, the system can be maintained and changed with minimum effort.

In some ways, protocols are perfect examples for applying object-oriented programming. We have events coming into and going out of the system. An event could thus be a perfect class. There will be specific types of events that are well suited to be instances of classes. There will be instances of creation of events as well as their destruction. Finally, each instance of an event will be associated with a particular set of operations that may be public or private.

Summary of Chapter

A state machine is the heart of a protocol module. The combination of the event and state provides a unique set of responses that create the sequence necessary to provide the final (desired or undesired) result. Every state machine has a number of design aspects in common. They each need to be able to accept and send events to other entities. They need to be able to break out the essential components from an event. They need to be able to store information concerning the current state (and other information) about the logical entity for which the event is intended. Finally, they need to have a method of determining the appropriate set of responses to a given combination of event and state.

There are three primary methods of implementation. One is hardware oriented. This method is best used when speed is critical or when the advantages of mass production can be used. The other two are software methods. One, the automaton, is data-structure driven. The other, the table-data state machine driver, is a direct translation of the event/state information provided in the specification. Other variants exist but combine aspects of the main methods.

The next chapter will concentrate on the needs of the general system. In particular, aspects related to real-time programming and problems will be addressed.

Chapter

11

Real-Time Programming and Problems

A real-time system is one that requires the ability to react to events in accordance with chronological time. In other words, if an event is to occur in 30 s, it should occur in 30 s in accord with the time kept outside of the system. Such a system must be able to service interrupts. (Only an interrupt can be assured of being able to be serviced at specific time periods.) The mainstream software, which is most often used, will also have to be efficient.

Problems exist in real-time systems that do not exist in other situations where elapsed time does not affect behavior. The first problem is that of speed. The system must be able to respond in a sufficiently efficient fashion to meet the criteria of the system. The second is that interrupts must be able to be serviced at any time. A variation of this is to have "preemptible" tasks in the system that force abdication of the currently running task to allow real-time-sensitive tasks to run at greater priority. This is a variant of direct interrupt handling since the operating system will need interrupt access in order to preempt the running task.

The next problem is that the process of locating an error in the system may change the behavior. It is possible to add some software that allows debugging and have the problem that is being tracked disappear. It is also possible to have the debugging software move the problem area such that much time is spent tracking a problem that exists only because of the debugging software. Real-time debugging may be considered a simple version of a Heisenberg situation—the observation of a problem can make the problem change or disappear.

Real-time programming is, thus, a set of techniques that allow system access to the needs for real-time systems. Real-time systems will need to access hardware. This adds the requirement that the high-level language used allow access to the hardware devices.

An LLD is a special form of software that creates an interface between a hardware device and the software system. It may work in conjunction with a state machine built into the device to provide a full protocol. It is also possible that the LLD will provide the full state machine—with the hardware device only responding to commands and giving indications in its register sets. Generally, this will not be sufficient for real-time requirements. Very few hardware devices implement the full protocol within the device.

Hardware devices (or semiconductor chips, if you prefer) are very important in the world of real-time design and programming. Specialized functions are provided that would be difficult, and sometimes impossible, to implement in software. The CODEC is necessary to translate voice signals from digital to analog (or vice versa). Video processing chips are mandatory to decode video (usually multiplexed with audio) signals at real-time speeds.

This chapter will discuss the various aspects of real-time programming and problems. The first area to be covered will be that of system resources. All software systems need resources, but some resource requirements used by real-time systems are particular to their needs. Resources include buffer management, timers, and data space utilization.

Since a real-time system must be efficient, there need to be methods to determine the efficiency of different parts. This is needed to determine bottlenecks and to provide direct feedback about how changes affect performance in a functioning system. This is sometimes referred to as *profiling*. The generation of a profile on software provides information to be used by the system designers.

Every complex system will initially have problems. As stated above, finding problems in a real-time system can cause problems beyond those that may already exist in a system. It is important to make use of nonintrusive, efficient techniques and to also have methods of determining the cause of problems that have been noticed.

A simple system (such as that for an appliance control system) may not need an operating system. More complex systems do have need of such. Task scheduling and the management of system resources are part of the efficiency, and portability, of a real-time system.

A real-time system may directly interact with the outside world (which is used for the I/O providing commands and results). It may also be split into a system that handles the real-time-intensive parts of the protocol and other less real-time-intensive applications on other microprocessor platforms. If a separate processor is supported, the design may make use of shared memory that is accessible over common data buses, or it may make use of common port protocols (discussed in Chap. 9).

This chapter will finish with a discussion about the design of LLDs and the functioning of interrupt processing routines. This is partially a hardware/software interface issue. It is also about integrating the hardware as part of an efficient, overall, protocol system.

System Resources

A resource is anything that is finite in supply. Resources can include items such as microprocessor cycles or general power. For most protocol systems, the general electrical power is not an issue. Microprocessor cycle restrictions are just another way of stating that efficiency is needed in a real-time system. If a microprocessor has "idle" cycles left (usually these are spent in the operating system), it means that the system is not processor-bound and that the bottlenecks of the system reside in another part of the environment (perhaps the disk system or the physical medium). Note that just because a system has bottlenecks does not mean that it is poorly designed. Every system has bottlenecks, and the goal of system improvement is to keep migrating those bottlenecks while continuously adding to the overall performance of the system.

Another classification of resources is memory. The memory used within a system will usually include a combination of ROM, RAM, and possible Electronically Erasable Programmable Read-Only Memory, or EEPROM, (for static, changeable, settings that need to be retained during nonpower situations). The RAM may also be in a *shared memory* configuration. Such a configuration allows access to the data area from more than one processor system. The shared memory may be arbitrated by hardware (preferable), or it may require the use of semaphore areas to keep the data stable as it is accessed.

Memory use is an important attribute of a system. Access to ROM will normally be faster than access to RAM (due to needs of refresh cycles for the memory). However, all of the memory can be RAM if it is necessary for general system requirements. Reduction in the amount of memory needed will correlate to savings in the cost for the system, size of the product (if the board size, itself, is of importance), and complexity of the control circuitry. In general, the system should have a general guideline of how much memory (and in what form) will be available. The goal of the designer is to then use only about 90 percent of this memory—making allowance for some necessary modifications and unforeseen requirements. Even if memory is not a constraint for the system, many compilers can produce smaller, faster, machine language instructions if the memory use is curtailed.

Another area of resources for the system is the availability of hardware devices. In a general system, such resources might include SCSI (Small Computer System Interface) ports, serial ports, parallel ports, disk drives, screen I/O, keyboard and mouse input, and so forth. When devices are shared, allocation of the resource is important to the general system performance.

In an ISDN system, the most obvious restricted hardware is the interface. A BRI has two 64-kbps bearer channels and a 16-kbps multiplexed D-channel that may potentially also be used for data. PRI has more bearer channels available, but the amount is still limited. One of the methods of making more versatile use of the bearer channels is called *bonding*. Bonding allows groups of bearer channels to be used as an aggregate grouping. Thus, the two B-channels of BRI may be bonded together to provide the logical equivalent of a single 128-kbps channel. Bonding may be done at the hardware level or by use of software protocols (such as X.25's MultiLink Procedure or the MultiLink Point-to-Point Protocol currently being developed).

Other hardware devices are also in contention. However, the one area most likely to be a limitation of the system is the interrupt processing. As we shall see later in this chapter, it is possible to obtain information from the hardware via interrupts or from software polling of the registers. Each has its place in an architecture. However, the amount of time spent in interrupt processing is a limit to the amount of time available for the rest of the system. In recognition of this dilemma, many hardware manufacturers have developed products that act in parallel to the microprocessor—they may be said to have their own processor. This devoted processor is not normally a general one (it is allocated to the specific needs of the device). However, it reduces the amount of time that the main processor needs to interact with the device, thereby freeing up the main processor for other tasks.

Some resources are variable. They exist as part of the general system. For example, a system may have 64 kbytes of data space available in RAM. The allocation of this memory is up to the designers of the system. However, it is still a finite amount of space, and whatever is allocated to one part of the system is unavailable to other parts of it.

This variability allows *tuning* of the system. Let's assume that a particular data queue is 10 structures in size. The maximum number of structures actually used is seven. In this case, the queue should be reduced to eight—freeing up the space of the two unneeded structures.

Buffers

For a system that is involved with software transmission of data, buffer allocation and use is the second-most important aspect of system design (the state machine is still the most important). Buffers refer to the area of memory in which the data are located. For some systems that rely only on hardware processing of data, the buffer use may be of less importance. For example, a system that supports only voice services will use buffers only for signalling support. Signalling occurs at sporadic periods and does not take a lot of the general resources of the system. Thus, buffer management is less vital. This would also be true if the data were processed as video or other audio.

Assume that the ISDN system being developed *is* a software data transaction system. There are a number of buffer-related aspects of the system that will affect overall performance. The first is allocation of the buffers. The next is how information about the buffer is passed from one entity to another. This is directly related to the way that the buffer is used. Finally, the buffer must be deallocated so that the space is available for other data use.

Buffer allocation. A buffer is allocated from what is called the *buffer pool*. Such a pool is an area of memory from which particular sections can be assigned for use by software until the area is specifically returned for general use. It is important that the data area be managed such that a single buffer is never in use by more than one entity (or logical part thereof) at a time. This is called buffer management and includes allocation, referencing, and deallocation.

Any type of data structure can be used as the buffer pool. Two categories are considered to be *fixed-size* buffer allocation and *variable-size* buffer allocation. In a fixed-size buffer pool all the sections available for allocation are the same size within memory. This causes potential unused segments of each buffer but simplifies management techniques. The variable-size buffer pool will make optimal use of the memory but also has a higher overhead for general buffer maintenance.

Let's examine possible alternatives for, and advantages of, fixed-size buffer pool management. The first design aspect is that of the receive direction. When a buffer is allocated for receiving data, it must be of the largest possible size that may be expected. This is because the actual size will not be known until after the data have been received. Thus, for an ISDN system, the data buffer must be at least N201 bytes long. This N201 is the limit of legal frames received at the data link layer. Depending on the use of the data buffers by the hardware, there may be need of an additional 2 to 4 bytes for storage of the CRC and delimiting flags. This amounts to 264 bytes for the D-channel. The maximum length of data for the B-channel depends on the type of circuit and agreed upon parameters. (A packet-switched circuit will normally enforce a smaller maximum limit than that of a circuit-switched connection.)

Reception requires allocation of a maximum-sized buffer space. What about transmission? Transmission requires an extremely variable amount of data. It may be as small as 3 bytes for a supervisory frame in Q.921 or 2 bytes for X.25 layer 2. It may also be the full maximum size. Assuming a uniform mixture of packet and frame sizes, the average size needed would be 132 bytes. If all of the buffers allocated for transmission were 260 bytes (assuming that the CRC and flags do not need buffer space), an average of 128 bytes would be wasted in every buffer (actually ranging from 257 bytes to 0 bytes). The data space can be more profitably used for other things.

However, the mix of buffer use will not be uniform. Signalling messages will usually be less than 10 bytes (plus 4 bytes for the Q.921 header). Data blocks will usually be of the maximum size—with only ending blocks taking up less than the maximum possible size. So, if we split the allocation of fixed-size buffers into two sizes, we will waste much less space. One buffer pool is used for layer 2 supervisory frames and most signalling packets. The other, larger, buffer pool is used for receiving data and for transmission of data packets and occasional longer signalling messages.

If buffer sizes are fixed, the best method of allocation is to have a fixed-sized two-dimensional array of character blocks. Buffer[0] indicates the first possible buffer, buffer[1] indicates the second buffer, and so forth. This is the actual memory space reserved for the buffer use. There must also be some type of buffer management list that indicates which buffers are in use. The algorithm for allocation is very simple. The buffer management list is checked to see what buffer is still available for use.

In a linear search method, the first available buffer is given to the requesting entity for use. Unless the system is very large, with large quantities of buffers available, this simple type of system will work best. A simple system is usually more efficient.

It has been shown that the reception of data requires a fixed-size buffer of maximum size. We have also seen that waste can be minimized for transmission by allocation of two different fixed-size buffer pools. However, we have not shown that fixed-size buffers are best for transmission. This depends mostly upon the degree of variability. If, as stated above, most large buffers are fully utilized, variable-size buffers are difficult to justify. If, however, the system use is such that the size is truly variable at all times, a system decision must be made as to which resource is the most valuable—processor cycles or memory space.

Allocation of variable-sized buffers takes place out of a "heap." This heap is a large array of all memory available for this particular purpose. For example, an area of 32 kbytes might be allocated out of RAM for buffer allocation. Each request has a particular size associated with it. Each is allocated, as available, from the data heap. Thus, the first allocation might be from address 30000 hex through 30040 hex, the next allocation might be from 30040 hex through 30060 hex, and so forth. Since each buffer is allocated to be the size requested, there is no wasted space within the buffer (except on the receive size, which still must allow for the maximum possible size). However, a variable data heap still has the possibility of wasted space. This occurs in a process called *fragmentation*.

Fragmentation occurs in a running system when buffers are deallocated (more concerning this topic in a bit). It is possible that, out of the 32 kbytes of data area, 5 kbytes are still available. However, this space is not contiguous—it is available only in small blocks between buffer areas that are still in use. If a large buffer request arrives, and no sufficiently large contiguous area is available for allocation, the request must be refused. (There are possibilities of moving data buffers around during run-time to allow more contiguous space, but these methods are not practical in a real-time multilayer protocol system.)

A variable-length buffer pool also needs a buffer management structure. It must be considerably more complicated than that of a fixed-size structure. A fixed-size pool requires only a list of the buffers possible with indication of their availability. The buffer management for a variable-sized list requires the beginning address and size of each data block allocated. (Other structure members may be very useful for deallocation and reclamation as we shall soon see.) Figure 11.1

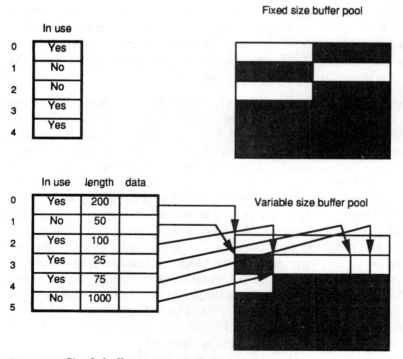

Figure 11.1 Simple buffer management structures.

shows an example of fixed-size buffer allocation structures and those of a variable-size buffer allocation scheme.

Buffer information.   Buffers will be allocated in some manner. However, once allocated, they are the "property" of the allocating entity until passed off to some other entity. In some circumstances, one entity may store information about the data buffer before passing it on (the data link layer does this with any I frames before passing on the data as part of a PH_DA_RQ primitive).

There are two items of information needed for an efficient system. These are the buffer reference identifier and the pointer to the data area. It would be possible to not have to have both—the reference identifier may be recreated from the address pointer. This is not ideal, however, because it would require the buffer management routines to be considerably more complicated. It also causes problems when an entity changes the data pointer (which is likely to be very useful as headers are added and taken away).

There is a third piece of information needed in an active buffer reference. This is the length. When the buffer is allocated, a length will be indicated (unless the system has only one size of fixed-length buffer available). However, at this point, the data buffer is not in use. Each entity that makes use of the data buffer may (optionally) add or delete information from the data buffer block. Thus, a complete buffer reference will include a reference identifier, a data pointer, and a length. These items of information will be passed along whenever the data block is sent from one entity to another.

Buffer usage.   Buffers are allocated, and they are passed from one entity to another. However, to be useful, they must contain actual data. The copying of data from one block to another is one of the most processor-intensive operations likely to happen in a system. It is possible to use special machine instructions to do "block copies" or to copy data in larger sections (such as 4-byte words rather than single bytes). It is best if the copying is minimized as much as possible.

One efficient way of doing this is to allow room surrounding data for additional use. For example, data being passed from the higher layers to the network layer (and then to the data link layer) may require space at the beginning of the block for header information. It would be possible to copy data from one block to another at each data insertion. The data buffer from the higher layer to the network layer would contain only the high-level data. The network layer allocates a new block, puts in the appropriate header information, and then copies the data into the packet. This is one method.

A much more efficient method would be to allow extra room before the higher-layer data. In general, a higher-layer application will know what the supporting lower layers require (and what functions, such as retransmission, they will do for the application). If extra space is allowed, the network layer can back the data pointer up the required number of bytes and then insert the header information. Only one buffer allocation is needed (at the application level), and there are no redundant data copies necessary. This can be done at all lower levels, but it does require knowledge of the maximum possible header (and trailer, for some protocols) space needed. It is important that each layer know where the "real" beginning of the buffer exists and that the true beginning of the data (and true total length) be passed down to the physical layer when the data are ready for transmission.

In the direction of reception, no extra space is needed—but it is very useful for the data pointer to be able to be shifted. Thus, to remove the layer 2 header information before passing incoming data to layer 3, the data link layer needs only to parse the header and increment the data pointer to the beginning of the contents of the information packet.

These are reasons why the entire use of the system should be kept in mind while designing portions. If we were not aware of the possible efficiencies to be obtained by making use of data pointer changes, we might have decided to make the data address fixed.

Sometimes a buffer is needed only for a temporary storage of data. A local stack structure may be allocated for this. If the data needs to be passed "up" (to a calling function), the data will need to be allocated at the highest point of the function tree and the address passed down to routines that will make use of the data area. Global data may also be useful—making certain that this data will not be used by any other software at the same time.

Buffer deallocation.   Buffer deallocation is very simple in the case of fixed-size buffers. The buffer management array just marks the buffer as no longer in use. In a situation where efficiency is very important, there is only one way to improve on fixed-size buffer pool use. That method is to use a fixed buffer. It is only practical, however, if only one buffer is in use at a time or when explicit acknowledgment is given when the buffer is no longer needed.

Variable-size buffer pools are more difficult to deallocate. In the allocation description, we simplified the discussion to only the aspects needed for allocation. For this, only a note of the data area and length are needed. For recovery of the data space, it is necessary to know the precise areas that are *not* in use.

In the example shown, the data areas that have been allocated and freed continue to occupy a space in the buffer management list. The structure size and address mark areas that are available for allocation. What happens when an adjacent area is deallocated? It would be possible to combine the lengths of the two adjacent buffer management descriptors into one. This reflects the actual free space available. However, the second of the two descriptors will now be noted as having a data length of zero. This means that this descriptor will no longer be useful and cannot be reclaimed until the descriptor following it has been deallocated.

So, let's take the buffer management list to the next step. This is to make it a linked list of allocated and free buffer areas. The linked list starts with a single structure. This structure lists as available all of the data heap. When a portion is allocated, a second structure is set up and linked to the first. The first is marked as allocated for the size requested (and has an identifier which was passed to the requesting entity). The second is still free and has available the rest of the data heap. This process continues as more data buffer requests are processed.

When a buffer is deallocated, the identifier and the buffer are marked as unused. Figure 11.2 shows a more complete variable-size buffer pool management structure using linked list buffer management. When two adjacent buffer descriptors are both unused, they are merged together (by adding the size of the second to the first and then freeing the second buffer descriptor). This allows an accurate representation of the state of the data heap and also keeps the list to the same number of allocated and free buffer sections in the data heap.

It may be noted that the data heap is still fragmented. It would be possible to gather together free space *if* the data segments could be

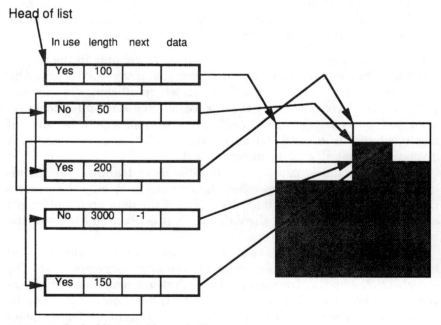

Figure 11.2 Linked list variable size buffer management.

moved. This is called *garbage collection*. However, in a real-time system, the control of the buffers is passed to the requesting entity at the time of allocation. This precludes the possibility of moving data segments by the buffer manager.

Timers

Timers are an important part of any protocol system. They are even more important in a real-time system. So, efficient and accurate timer support is highly needed in a real-time protocol system.

Timers have operations associated with them. They also have their own data structures. Finally, they must be connected into the hardware system. A real-time timer must have a reasonably accurate external clocking mechanism. This is usually provided by means of a timer interrupt mechanism that is defined according to the system clock lead.

Hardware interconnection. A semiconductor chip requires a clock lead to support the timing to synchronize the separate logic operations. This clock can be used to trigger specific events as well as for synchronization. Thus, many processor chips directly support additional timer interrupts. Those that do not still allow additional timer devices to be added onto the clock lead. Sometimes more than one timer interrupt will be available. This has potential use in a software system by dedicating one interrupt to a special, time-critical, area.

The programming of the timer interrupt falls into two sections. The first is the programming of the interrupt. This entails setting up the type of interrupt (automatically restarting or a one-time interrupt) and setting up a divisor register, which allows the clock cycle to be used as a base for the timer. Thus, the setting of the register will depend on the frequency of the clock on the system. The second involves setting up the interrupt processor for the interrupt.

Timer interrupt. The timer interrupt will normally invoke a routine to do whatever processing is needed. A special interrupt will do the specific operations needed for the timer event. An example of such a special interrupt might be for use at the physical layer—the T3 timer. A special interrupt has the advantage of being more precise. Since it is invoked only for the specific event, the timer hardware can be programmed for the precise needed interval, and there is little software overhead to make use of the interrupt.

General interrupts are used more often. The hardware may support several timer interrupts, or it may support only one. However, it will *not* support a separate interrupt for every possible timer situation of the system. It is therefore necessary to make the single interrupt able to be used for many different purposes within the system.

A general timer interrupt routine will normally maintain a system software clock. This clock is incremented each time the interrupt occurs. The value is used to determine when timer events need to be sent to other entities within the system. If no timer event needs to be produced, the interrupt routine only needs to increment the clock and check the pending timer event list.

The general timer interrupt can also be useful in systems where the hardware devices need to be polled. In addition to checking on timer events, devices can be polled at intervals. Thus, one interrupt can be made to service many of the periodic needs of the system. There are two categories of data structures needed. These are counters needed for polled events and a timer structure associated with the system timer events.

Data structures. The timer structure needs to be designed with two things in mind—minimization of the time needed to be spent within the interrupt handler and the ability to process multiple interrupts that are independent of each other. The fact that each event has a relationship to the other (some occur earlier and some occur later) and yet needs to be independent argues for a linked list of events. A linked list allows for ordering—but also provides for removing or inserting items without significantly changing the other items on the list.

One method of ordering is that of incremental times. Assume that five timers need to exist. One is for 500 ms, one is for 10 s, two are for 30 s, and one is for 1 min. The timers are listed in ascending order, but they may have been started in any order. An ascending incremental timer list would have each timer value affected by the preceding timers. Thus, the first timer would be set to expire in 500 ms. The next would be set to expire in 9500 ms after the first timer (500 + 9500 = 10000 ms = 10 s). The following timer would expire in 20 s after the second, and so forth.

Each timer is affected by the time of those timers preceding it, so the timer interrupt routine needs to only be concerned with the first item on the list. When 500 ms have elapsed, the first item is ready to be sent to the appropriate entity. The next item is still set to expire in 9500 ms (the correct remaining amount of time). This is very efficient in two ways—it allows the timer interrupt to do minimal work, and it keeps each timer independent of the reactions of the rest.

Since the list is based on preceding values, the insertion or deletion of an event will affect the following timers. Let's continue the above example and insert a new event that is to expire in 20 s. We go through the list, adding up time intervals as we go. We find that this new event needs to be inserted between the second item on the list (10 s total) and the third. We put it in at that point and set the incremental value to 10 s (10 + 10 = 20 s total). We *also* need to change the incremental value of the item in front of which it is inserted. It used to be 20 s (10 + 20 = 30 s). Now it needs to be 10 s (10 + 10 + 10 = 30 s). We have added a new event to the list by adding in a linked structure and changing the timer interval of one existing event.

Deletion operates in a similar fashion except that instead of decrementing the timer interval of the item that used to follow the deleted event we increment it. Let's take the original list of five events. We delete the second item in the list. We need to increment the *new* second item by the interval value of the old second item. This gives us a new interval time of 29500 ms (9500 ms + 20 s = 29500 ms). Of course, if an event is (or used to be) the last item of the list, there is nothing to be changed.

Using the timer event. We have mentioned several structure members that need to be part of the timer structure. One, explicitly mentioned, is the timer interval. The other is a link from one item to the next (and possibly one that points backward if a double-linked list is desired—not truly needed so far). The rest is concerned with the *content* of the event.

A timer event will have a destination. It may also have data associated with it that is to be used by the destination process. Remember, however, that we are still within the timer interrupt service routine. We want to keep the time spent within the interrupt routine to a minimum (while we are there, other interrupts, and scheduled system operations, are unable to take place). Thus, it is not a good time to evaluate the precise actions needed for this particular event.

Take the expired event off the list and ship it off to be processed when it is convenient. There is a possibility that the event might be canceled in that interval between shipping it off and having it processed. Remember, however, that it has *already* expired—no matter where the final processing takes place. So, we have structure members that are used for the maintenance of the timer event list, and we have information that is to be sent off, and used, by some other noninterrupt routine.

The event arrives at the intermediary processing task. It now examines the event. Some timers will be associated with the system. These will effect changes to data structures and calls to system routines. Other timers will be associated with a particular entity. So, a system start-up delay timer is processed directly by the task. Another might be sent off as an MDL_TM_RS primitive to the Q.921 software which will, in its turn, do the final processing of the timer event.

Operations. Operations on timer events have already been briefly mentioned. They consist of starting, stopping, and (possibly) restarting a timer. In the discussion on data structures, we talked about potential methods of organization that allow insertion or deletion. Insertion of a new event is equivalent to starting a timer. Deletion of an old event is equivalent to stopping a timer. Restarting a timer is the same as moving the timer event within the list. It is probably most efficient, even if a separate restart operation is supported, to change the timer event list in the same manner as deleting an old event and starting a new one. The data structures changed will be the same, and keeping the operations distinct will improve simplicity, speed, and general maintenance concerns.

Data space

We talked about memory use in the beginning part of this section. In this subsection, we will discuss parts of data space and how they can affect the overall performance of the system.

The first segments of memory that will be important are the initialized and uninitialized data space segments that are located in the global area of the system. Some data, such as constants, will be used as storage for use within specific functions and are thus available for access by the function when the compiler associates the address with the appropriate machine language instructions. Uninitialized data is normally global data within the system—it may be used by any entity that needs access to the memory space.

The initialized area is filled with constants. These are probably mostly character strings. If each string is used only once within the system, the only method of reducing the data space used is by keeping the strings short. Brevity should be balanced against clarity. Sometimes, it is possible to use coded versions of strings that can be expanded later in a situation where real-time demands are not being made.

Sometimes a particular string will be used multiple times. A string such as ERROR: item number nn might be part of a system error-reporting mechanism. If this message is used 100 times (with varying nn values for the number) within the system, it will occupy 2000 bytes (plus the variable amount for the number). If, however, the string is used only once as part of an error-handling function (which does other actions in addition to the message string), you can save 1980 bytes of memory that can be used for other purposes.

If data structures are needed, they can be either local or global. Global data will take space at all times, but it may simplify access and, thus, reduce the amount of object code space needed for the software. If the memory is available, and the information is to be used for more than one entity, global data can help in system efficiency. Global data can be used in a modularized, or layer-independent, fashion. This is controlled by determining which entity has control of the data and allowing only that module to "extern" the data.

Global data that can be accessed by any instance of a module or the system as a whole does have its dangers that need to be considered during the design of the system. The main concern is that of reentrant code. If it is possible for *two* Q.921 modules to be running and they are using the same global data area, it is possible to run into conflicts and inconsistent data structures when one instance modifies the data as another instance uses it. This potential problem can be reduced by making each task *non-preemptible* (see the discussion on operating system use). If each task is nonpreemptible, data remains consistent within the task use. Thus, we reduce the possible conflict areas to that of interrupt software.

If the data are not global, they will be local. Local data use normally means that it resides on the stack available for the task. This stack size is determined by the operating system, but it is often (for "private" system operating systems) capable of being modified by the system designer. Space that is not used is wasteful; lack of adequate stack space can be disastrous and very difficult to determine. It is usually best to start out with a generous stack size and reduce it to a more efficient size after the software is debugged and being prepared for release.

Global or local data space use requires memory resources (although local data space can usually be utilized by more than one task as one finishes and another begins). The next step for reduction of memory use is to design the structures according to need. If a value in a data item can be only in the range of 0 to 54, it is wasteful to allocate data space for 4 bytes. If a linked list is always searched in one direction, it may be unnecessary to create it as a doubly linked list. If fixed-size arrays of structures are to be used, keep the number to just a bit larger than needed.

System Analysis

The analysis of a system—any system—depends on its configuration and available tools. A system that runs in an environment where there are other systems in place (such as a word processor on a multitasking system) is one type of configuration. The firmware that runs on a TA is another type of configuration. A system that runs on a coprocessor (plug-in) board that is connected via data and I/O bus interfaces is yet another configuration. Every type of configuration has its own constraints. In many instances, the testing and development configuration will be different from the final operational configuration.

Tools also vary tremendously. If the system is meant to run with other processes, there may be tools already built into the more general system for the analysis of new processes. There may be supplemental tools. In-Circuit Emulators (ICEs) are used as microprocessor replacements during development on devices that are meant to be autonomous in the finished version. An ICE allows direct access to all of the memory, registers, and processor capabilities of the microprocessor. This allows full examination and manipulation of the software system. It may also allow access of peripheral devices (such as a semiconductor chip that provides the BRI S interface protocol). Since the memory can be examined directly, it is possible to use local memory for temporary debugging help.

External debuggers can also be used on many systems that will eventually be autonomous. They usually rely on special interrupts that are available on the microprocessor. The primary purpose of these interrupts is to allow the debugging task, which has been compiled into the system to be developed, to control the microprocessor briefly. This task then allows the system to be examined. It acts as a separate system but is combined with the system to be developed. It is primarily useful in real-time systems for "snapshot" information. The use of this type of debugger will likely prevent the system from behaving in a "normal" fashion but, by letting the system behave normally up to the problem area, the debugger allows full access to the current state.

In the discussions to follow, a lot of time will be spent on the "low-level" types of debugging. These are methods that can be used with minimal tools. It is possible that, in the process of developing a system, you will have sophisticated tools available that will allow you to achieve the same goals in an easier fashion. In that case, the same types of actions will be performed, but they may be hidden by the interfaces of the tools. An examination of the actual methods used at the lower layers allows evaluation of tools and a better understanding of how to use the tools to do the task.

Profiling

A profile is a representation of a system that hides many of the details. It gives data about certain aspects of the system behavior and, by *not* giving other details, brings them out as data to be studied. In general, the profiling of a system is oriented toward specific system states. These may include how much the stack is used, the time spent in various parts of the system, the number of times a function is invoked, or any other specific piece of data that is useful for evaluation.

Profiling can be done by having the compiler insert special instructions automatically into the object code when the source is being compiled. It can also be done explicitly by the software programmer—inserting profile-oriented instructions into the source code. In either case, the extra instructions will affect the behavior, and efficiency, of the system. Thus, it is important to do profiling in such a manner that it can be "turned off" for the final product. It is also useful to create several timed benchmarks for the system when it is performing without using the profiler so that the time used, with the profiler, can be compared. This comparison allows for a correction of absolute values of data obtained. Sometimes this is unimportant because only the relative values are needed. It should be kept in mind at all times while profiling.

Profile data collection

How is profiling actually done? There are two portions to profiling. One is the maintenance of a database (local to the system or communicated across a special communication link, often a serial port) of the data collected. The other part consists of the instructions that collect the data. A final part is some type of tool that correlates the data collected into a form that highlights the information needed.

Function statistics. A common need for profiling is the determination of how long various functions take to complete and how often they are called. This is done by collecting data at two points. At the entry point, a call is made to the database indicating that the function has been called and including a time stamp of the time at which it was called. The second collection is made at each exit point (before a "return" statement or the end of the function). This second call to the database marks the completion of the function, and a time stamp of the time that the exit is made.

The difference between the time stamps of the two data entries gives the period of time that it took for the function to operate. It is not precise. If inserted manually into the source code, there will be stack maintenance operations performed both before and after the function. If it is inserted automatically by the compiler, or other tool, there will also be a certain amount of code that is not included. This is not particularly important to the profiling because it is constant overhead. If absolute times are needed, this overhead can be calculated and added to the differential time (some tools may add this back in automatically).

If interrupts are possible in the system, you may find that the time spent within the function varies a lot. This is because the time spent in the interrupt handler resides "between" the entry and exit points. This can be compensated for by triggering a counter in each interrupt routine. This counter accumulates the time spent in interrupt routines. When the second database collection call is made, at the exit point, this time can be decremented from the time-stamp value (or note of the time can be made in a separate data value to preserve the accuracy of the true time) and the counter cleared for future situations.

These two simple database collections also give the possibility of additional useful information. One is the frequency of the calls to the function. Rather than comparing the time stamp of the entry point to the exit point, a comparison is made between the entry points to subsequent calls. This allows determination of frequency. Counting the number of entry points gives a running statistic of how many calls in total have been made. So, by using these two calls (with the addition, perhaps, of an interrupt counter for time correction) we are able to collect information of frequency, duration, and absolute counts for function use.

Data space usage.   The determination of function usage is very useful for profiling a system. Other aspects of the system are also important. One such important item is variable data usage.

This is done by marking levels. When a data space item (such as a system stack or an intertask message queue) is added to a variable-length data structure, data collection is performed. This data includes the identification of the data structure and the current usage. For example, on stack usage, the data might give the current task identifier and the amount of data used on the stack. The collection of data on subsequent uses of the stack by this task will give a "high-water" mark that indicates the most data actually needed by the task. It will also give the ability to calculate an average data space used.

Another example would be on an intertask message queue. Useful statistics might be how many items are in the queue, what are the most items in the queue, and how long is each item in the queue? Answering all of these questions requires two types of data collection calls. One is done each time an item is added to the queue. The data collection call gives the queue identifier and the current length. The other data collection call is associated with a message. A message identifier (probably composed of a message queue identifier and a message counter) is logged into the database along with a time stamp of when the message was queued. When the message is taken out of the queue, a second call is made to log the information. Similar to the use of calls concerning function use, this call gives the message identifier, again along with the ending time stamp. This gives the time spent in the queue. In this case, no correction for interrupts is likely to be desired because the interrupts will be part of the operating system scheduling mechanisms and should be included as part of the intertask message delay.

Resource and event counting.   Sometimes it is not important how often something happens or how long it takes. It is only important that an event actually happens. This is a type of resource or event counting. Most profiling information is useful only in the period during which the system is being prepared for release (although some statistics may be worth the decrease in efficiency for ongoing performance measurements). Counting information is often retained as part of the general statistical accounting for a system.

When an event occurs, the important piece of information is the identifier of the event. This may not even be retained specifically in the data. Rather, a counter of the number of events is kept. Thus, there may be a "number of PH_DA_IN primitives" counter. There may also be a "number of N_CONN_IN primitives" counter. These counters vary from profiling data because the important aspect is to keep track of one type of data—number of occurrences.

Errors also fall into this category. It may be important to keep track of how many errors or how many of a particular type of error occur. A variation of this is how many occur per unit of time. If the system is running for 24 h and 100 errors have occurred, did the errors occur over the entire 24 h or just within the last 30 min? The former may indicate a "normal" level of errors within the system and network. The latter may be a sign of impending failure or a change in general condition of the system. Thus, periodic statistics may be important.

There are several ways of approaching periodic statistics. One is to keep a set of counters. One counter is for errors in a certain period of time. Another is for a subsequent period of time. If each counter is for a 15-min interval, four counters allow for an hour's worth of statistics. A better strategy for this would be to have five counters. The first four counters can then be used for the previous hour, and the fifth can be used for the current 15-min period. Then, when the 15-min interval has elapsed, the first timer can be used for the next interval, and four counters always contain a snapshot of an hour's worth of statistics.

Another method is to keep a time stamp with each error indication. This requires more data space but allows a full analysis of changes in frequencies. Perhaps errors occur at half-hour intervals. This may be valuable data for tracking down the root causes of the errors. Use of time stamps with errors also allows easier correlation between different items of data. If error A always occurs 2 min after event X, there is a good chance that error A is a side effect of the way that event X is being handled (locally or by the peer system).

Function trees

When we discussed the use of starting and ending event markers within functions, we left out one important aspect. Most functions will call other functions. In fact, virtually every system will consist of an entry point that calls functions which in turn call other functions and so forth. This cascading sequence of functions may be called a function tree.

This fact also affects the analysis of the data accumulated via the starting and ending data collection calls. Within the entry point, the statistics will actually reflect the entire time spent within the task— the accumulation of time spent in the entry handler and that of every

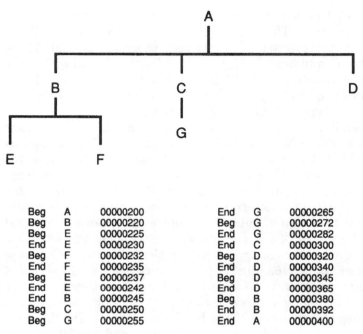

Beg	A	00000200	End	G	00000265
Beg	B	00000220	Beg	G	00000272
Beg	E	00000225	End	G	00000282
End	E	00000230	End	C	00000300
Beg	F	00000232	Beg	D	00000320
End	F	00000235	End	D	00000340
Beg	E	00000237	Beg	D	00000345
End	E	00000242	End	D	00000365
End	B	00000245	Beg	B	00000380
Beg	C	00000250	End	B	00000392
Beg	G	00000255	End	A	00000400

Figure 11.3 Example function tree.

other function that is used while the task is active. This may be useful information, and it may be the only information that is necessary for your particular analysis of the system. However, it does *not* allow you to determine the performance of each individual function (except for the ones that do not call any other functions).

In order to correct for this situation, the tool that processes the data needs to make corrections. Let's use an example of a short function tree (as seen in Fig. 11.3). In this tree, routine A is the entry point (or root of the tree). It calls functions B, C, and D. Function B calls functions E and F. Function C calls function G. Functions D, E, F, and G do not call any further functions. The second part of Fig. 11.3 gives an example time-stamp sequence of the calling of these functions.

The true time needed for the code residing within function C can only be calculated by subtracting the time for function G. The true time needed for the code residing within function A can only be calculated by subtracting the time for all functions within the tree. A's total can be determined by subtracting the times for functions B, C, and D. The time spent within the other functions should not be decremented since they are already included in the time for their calling functions.

Thus, by varying the data collection tool's algorithm, the total elapsed time for a function or the time spent only within the software residing within the function can be determined. The data do not vary—the analysis does.

On another note, the data can be used in a completely different fashion. It may be used, in itself, to create a function tree. Many tools exist to create function trees from the source code modules. This, however, allows the implementor to create a function tree of functions that are actually used in the real run-time situation. It is not unusual to find that certain "paths" (potential function tree branches) are not used at all for normal run-time situations. They may be used only for error conditions that have not occurred during that particular analysis. It is also possible that they aren't used because they aren't really needed.

Use of the data

In the preceding discussion, we talked about what types of data may be accumulated and what types of statistics may be derived from the data. How do we make use of the data and the statistics that are thus obtained?

Frequency. The frequency of invocation of a function is a direct reflection of the reliance the software makes upon it. If this information is used in conjunction with a function tree, you may find that a function is used many times, but it is only called from a few (or even one) functions. Each function (or subroutine, in some languages' terminologies) invocation requires time. There is time to push the parameters onto the stack and to prepare the stack for the function's data space needs. There is also time (and object code) needed to pop off the function's stack use and restore registers to the expected values of the calling function. This is overhead that is repeated every time a function is invoked.

This type of situation is often a clear indication that the function should be brought *in-line*. In other words, the software that exists within the called function should be brought directly into the calling routine—without the intermediary state of function invocation. This would increase code space slightly (if it must be incorporated into more than one function) but will result in a frequent operation being performed more efficiently.

Even if a function is called from many locations, the frequency is an indicator of the need for it being efficient. A function that is only called once every 5 min may be inefficient without hurting the overall system performance. A function that is called hundreds of times every 5 s will have a considerable impact on the system. Perhaps the function should be rewritten in assembly code. Perhaps the data structures used by the function should be examined more closely. Perhaps it is already as well written and as efficient as is currently possible. The function times are used to help determine that information.

Function times.   The frequency is an indication of reliance on a function (as opposed to the importance, which is an indication of whether success or failure makes a difference). The amount spent in a function does not, in itself, give an indication of efficiency. However, it is necessary to have a method of determining whether changes have decreased, or increased, the time needed within the function. It is a feedback mechanism.

Function times are also important to determine system task allocation. If most of the time is spent within one entity, several possibilities exist. One is that this entity is the least efficient. Another possibility is that much of the work is happening within the module. It is possible that reorganization of the functions that make up the entity can improve overall system performance. This involves delayed actions. For example, if many statistics are being collected by an entity, it may improve general performance if the data is written to global memory and some other task acts upon it when the system is less busy.

Event occurrence.   Events may be normal, or they may be indications of error situations. Errors may be expected to occur at intervals. For example, there will occasionally be times when the physical layer loses its synchronization. If this happens frequently, it may be a sign that there are physical line, or device, problems. The frequency of the error (and, sometimes, the periodicity) determines its importance (as long as it is recoverable). In some situations, an error cannot be recovered from, and such cases require debugging rather than profiling.

Normal event occurrences are important only for statistical purposes. They can be used as indications of traffic load or the efficiency of the given protocol. However, they do not normally affect the actual performance of the system, and the logging of such should be minimized in the final system to keep performance as good as possible.

Summary of profiling

Profiling is a method of determining statistics about events that can be useful for analysis of a system. Normally, it is not good to leave profiling code in a system after it has been optimized because the profiling software will decrease the efficiency of the system. It can be used to determine appropriate variable data sizes and to note sections that should be improved. The next section will discuss the process of debugging. Many of the specific techniques of profiling can be used in debugging also—but the purpose is different.

Debugging

Profiling is associated with determining just what is happening within the system and how much time is spent doing various tasks. The process of debugging is associated with the task of discovering just *why* the system is not performing in the manner that is expected and desired. The term *debugging* was first used in a situation where periodic problems in the system were found. The problem ended up being a dead insect that was disrupting some of the circuits. The term *debugging* was henceforth used to describe the process of finding, and eliminating, problems within a system.

There are a number of techniques that can be used in debugging. Most of them are oriented toward getting data about the current state of the system. This data, in turn, can be used to determine just what deviances from desired behavior have taken place and, eventually, the modifications needed to change the behavior into what is desired. There are thus two aspects of detailed debugging. These are determination of the needed data to locate the problem and appropriate location of data collection to retrieve the data at the appropriate time.

Understanding the system

The first step necessary to do debugging is to understand the system that is being debugged. It is important to understand the boundaries and the interfaces as well as the data flow. Boundaries are involved with just how much of the total system is needed to understand it. Interfaces allow a method of dealing with entities outside of those boundaries (and for entities *within* the boundaries but separated by operating system interactions or other internal primitive protocol boundaries). The data flow allows the debugger to place debugging code into places to "intercept" the data flow and obtain information about what is going on that can be compared to what is desired. An understanding of the data flow is also needed to know what should be expected.

Boundaries. A boundary to a system depends on the stability. If you are working with a non-real-time system that is being used on an existing stable hardware platform, and being used under a previously used operating system, and dealing with known interfaces and peripherals, there is a much smaller system that must be understood. In contrast to this, the designer of a real-time system that is to operate on new hardware, making use of a new operating system and a new interface into the rest of the system, has a much larger system to understand.

Normally, the development of a large system will be distributed. One person is in charge of Q.931. Another person is in charge of LAPD. Yet another person is involved with the hardware architecture. One approach would be to set the boundaries in alignment with the responsibilities. This can work on stable, existing, platforms. It does not work well with newly developed systems. This is because the interfaces have not yet been proven. It is impossible for one person to be sure of exactly where the problem is located.

If the hardware person is working with the physical interface and the peer level of the interaction is "proven," it is relatively safe to assume the problem lies on the local side. (This isn't always the case, although it is likely that the local side will be the side that has to change to adapt to whatever variances exist on the peer side.) If, however, the problem lies at the physical layer to data link layer side and both sides are new, it is difficult to determine where to make changes. Pointing fingers is not productive—the task is to find where the problem is located and to fix it. This requires that the physical layer person understand the data link layer sufficiently to know what reasonable commands may be expected and the reasons for their use. It also requires the data link layer person to understand the expected flow of information from, and to, the physical layer. The "how" is unimportant to each person, but the "what" is very important.

Interfaces. The above description emphasizes the importance in debugging (and in system design) of the interfaces between parts. One of the most difficult coordination requirements is that of two or more people assigned to the same task without explicit interfaces between segments. This is why such coordination can easily add up to 1 plus 1 equals 1.5. The amount of time spent in coordination can easily take up a quarter of each person's time. The situation gets even worse if more people need to coordinate without explicit interfaces (1 + 1 + 1 = 2?).

Explicit interfaces are mandatory for efficient design, implementation, and debugging. The OSI layers help to provide a default layering or sectioning of tasks. Each layer has its own responsibilities and primitives. The expected behavior, and results, from each command are known. However, the actions across the interfaces are not in isolation. The sequence of events that are to happen is also important. This is why many standards and specifications include example data flow scenarios. It allows an extra understanding of just *what* is supposed to occur across the interface (but hides the how, which is relatively important within this entity).

Data flow. Someone who is debugging in a new system, thus, has a need to know the data flow—when are events expected into one part of the system and what events do they trigger? If you have tracked the events across the interfaces, you are able to give the person in charge of another section of the system the information needed to track what is going on within that section. Of course, the problem may (and likely will, since you have noticed the problem) exist within your own section of software. In this case, you have tracked the problem to an event (or set of events) that does *not* cause the expected responses.

Let's use an example of some data flow situations within ISDN. An N_CONN_RQ primitive comes from the higher layers to the network layer. The network layer issues a DL_EST_RQ primitive to start making use of the data link layer. The data link layer needs the physical layer to be activated and *also* needs a TEI. It thus sends an MDL_AS_IN primitive to the management layer to request a TEI. The management layer sends a UI frame over a broadcast channel to request a TEI using a DL_U_DA_RQ primitive. The data link layer still needs physical layer activation, so it sends a PH_AR primitive to the physical layer. It also either queues the UI frame (waiting for the PH_AI primitive) or sends along the PH_DA_RQ primitive to the LLD to be transmitted after activation (the standards are not clear as to the preferred sequence of events).

At any rate, as a result of the N_CONN_RQ primitive from the higher layers, we are now in a situation where we are waiting for a PH_AI primitive to arrive. Let's say that the primitive doesn't arrive. The network layer is faced with the situation where a primitive has not arrived and could not be responded to. The data link layer knows that a PH_AI is needed to handle the request. It is impossible to know from this brief description whether the problem lies at the physical interface, physical layer, or data link layer. However, at this point, the network layer knows that the problem does not directly exist within its module. This could not be known in isolation from an awareness of what the data flow should be.

It is also still possible that the network layer is involved in the problem. It is possible that something about the DL_EST_RQ was wrong—perhaps the wrong link identifier was given, causing the data link layer to attempt establishment on a nonexistent physical link (which, in turn, confused the LLD). However, the point of debugging now lies in the hows of the data link layer and the physical layer.

Techniques

Most techniques end up being a decision as to where to collect data and what data is useful. The easiest (but sometimes still very frustrating) problem to locate (and fix it is hoped) is where the system stops. A command is entered (or the system is started), and the system stops. This is a static problem because the system must, by definition, stop at a particular place (this may not be at a particular place in software—it may be after a particular time or a specific sequence of events). The debugger tracks such a problem by figuring out just where it stops. The area is split into smaller and smaller areas.

Logging. The main independent tool of the debugger is that of logging (other, specialized, tools also allow data collection). Most of the discussions in this section will use logging as the method to obtain data—because logging is the lowest-level method that can be used. Data collection involves taking data from one area and making it accessible. This may be done in many different ways. In one system, the only accessible information for a hardware device being developed was that of some programmable Light Emitting Diodes (LEDs). A set of LEDs was available, and they could be turned ON or OFF by accessing particular locations in memory. So, in this situation, logging involved turning ON or OFF certain LEDs to give patterns. Since four LEDs were available, sixteen separate patterns were possible.

Often, logging is involved with memory areas. A special area of memory is set aside for use in logging data. These data can then be accessed via an ICE or, in shared memory cases, by a separate processor. The advantage of this is that the form of the data can be as needed. American Standard Code for Information Interchange, or ASCII, (or IA5) character streams can be used for easy readability. Numeric data can be translated into character format for ease of use. The amount of data can also be changed.

Assume that only a limited amount of logging memory is available. It is very useful to compress the data so that more information is available. A separate tool (or just the experience of the person examining the data) can be used to uncompress the data into a useful form. If much memory is available, or the data is being sent over a serial port to another device, the information can be more verbose.

Still, logging (as is true for profiling) takes away time from the system that may be needed for other activities. It is not uncommon to insert logging code into a real-time system and have a problem change in its symptoms. This may occur because of changes in timing, or it may occur because the additional instructions have changed the alignment of the machine code instructions and the use of the data stack.

Narrowing the search. At first, the problem area may exist in any part of the system. Logging is inserted into the system at the point before start-up of the operating system, at the beginning of each possible task of the system, and at the beginning of each possible interrupt (assuming, for the moment, that the operating system is fully functioning). If the first log does not come out, the debugger knows the problem exists in the start-up procedures before the operating system is invoked.

If an entry point (or interrupt entry) is the last thing seen in the log, a next step needs to be taken. This is marking the exit point of the function. If the entry is logged and the exit is not, the problem has been defined to be within that entity (probably). If the entry and exit points are both logged, a different set of potential problem areas exist. One is the operating system—the next task is never scheduled for some reason (either due to the implementor's use of the operating system or a problem within it). Another is that of data stack usage. It is possible that the boundaries of the usable stack area have been exceeded. This did not cause problems while the task was being executed, but it overwrote necessary data for further execution.

Once the potential problem area has been narrowed, the next step is to further narrow the possibilities. Placing log messages within the main entry point starts the process of eliminating function branches. Log messages continue down the function tree until the precise problem area is located.

Moving problem areas. We mentioned briefly that it is possible that insertion of log messages may change the behavior of the system. This needs to be addressed before further debugging can take place. Is it a problem of the logging taking too much time, or is it a data usage problem? If it is the former, the debugger needs to reduce the time taken (this scenario occurs frequently when logging occurs in interrupt routines). Reduction of data is the first step. The next step is to use alternative logging schemes. Memory can be used as a "logical LED." That is, a logging activity can exist as the setting of a bit in memory (probably the fastest method possible).

If decreasing the time taken does not help, it is probably a memory problem. These are the hardest problems to track in a system and will be addressed in more detail later in this section.

Choice of data. Logging may just mention that a section of code has been reached. At some point, however, you will have found *where* the problem takes place, and you will need to know *why* it exists. Sometimes the error will be obvious. There may be a mismatch of parameters in a function call. A pointer may be used without initialization. A counter limit is uninitialized or past the boundaries. Often, the situation will require knowledge of the contents of the data being used. At this point, logging of data in addition to location is needed.

Backtracking. Backtracking is the next operation. The problem area locates where, and what, data are inappropriate. The next problem is to determine *why* the data are in this inappropriate state. Rather than put logging statements into sections of code based on location, it will now be important to track the data. Logging statements (with the contents of the surveyed data) are located in places where the data are changed. If it turns out that the data changes "inexplicably"—that is, in one logging section it is fine, but in the next it is now wrong—it is an indication that data structures are being overwritten (assuming that the logging has been done correctly). Most of the time (it is hoped), there will be a section of the system that is using the data incorrectly and causing the problem. It can then be changed and the results verified for correctness.

External debuggers

An external debugger can allow the software to be left unmodified (no explicit logging statements inserted into the system). This can greatly speed up debugging, although it does not actually add functionality to the debugging process (anything that can be accessed by the external debugger can be accessed by the internal logging statements). External debuggers provide several types of tools. These include the setting of breakpoints and the ability to examine and change the contents of processor registers, stack variables, and global memory.

Breakpoints.   A breakpoint is a place where the system is told to stop running. In non-real-time systems, it is usually possible to then continue running the system until the next breakpoint. In real-time systems, particularly in those with hardware interrupts enabled, this is unlikely to be the case. The delay taken by use of the debugger to examine the process is likely to cause interrupt problems and change the entire debugging scenario to the point of being unusable. However, the breakpoint (equivalent to inserting special "infinite loops" into locations internally) does allow access of information.

Most external debuggers use the memory maps and symbol tables of the software running on the system. This gives the debugger symbolic information, so the user may state "set breakpoint at function send_dl_primitive." It may also be used for memory access. Sometimes an external debugger allows multiple triggers for breakpoints. This would be an instruction such as "set breakpoint on send_dl_primitive when stack variable *primitive* is equal to DL_EST_IN." Of course, the actual language used with the debugger will depend on its user interface.

Sometimes external debuggers will also allow breakpoints to be set on multiples of events. Assume that you are checking out what is happening when a DL_EST_IN primitive arrives. You set a breakpoint at "receive_dl_primitive when primitive is equal to DL_EST_IN." When you arrive at the breakpoint, examination of the system indicates that there are no problems at this point. Still, you have evidence that a DL_EST_IN primitive is triggering a problem in the system. The breakpoint set in the real-time system caused the further behavior to be unreliable. So, you start the system again, and this time you indicate to "set breakpoint at second occurrence of receive_dl_primitive when primitive is equal to DL_EST_IN." This time you have an error situation (and you start the backtracking).

Memory manipulations.   An external debugger can be extremely useful in manipulating memory. This allows a "what if" situation to be processed. Data are examined and the contents determined. If the values are not those that are expected, or desired, the contents can be changed and the code allowed to progress. This technique is useful in a real-time system only within an entity-processing routine. For example, if a DL_DA_RQ primitive arrives, this can be used for processing the DL_DA_RQ primitive—but once this primitive has been processed, the rest of the system behavior must be considered unreliable.

Register accesses are useful in checking for illegal memory space usage. For example, the stack pointer may point to an area of memory that resides in code space. This will cause severe problems. Backtracking can be used to determine when the register's contents were erroneously set.

Once again, nothing can be done within an external debugger that cannot be done by explicit instructions within the software (although register accesses may be easier via the external debugger). The external debugger gives the advantage of speed since the software does not need to be recompiled to gather additional information.

Hardware interactions

Most microprocessor systems start up, in a "power-up" situation, by setting the instruction address pointer to a fixed location (often this is address zero of the memory space). To the systems developer, it means that this is the beginning of the software portion of the system. The instruction pointed to by the first address (usually a "jump" instruction to a initialization function) is the first place where a log statement may be entered (actually, it is much better to insert it within the instructions that make up the start-up routine).

Other hardware interactions include interrupt and hardware/software routines. As mentioned before, logging instructions should be limited in such situations since this is an area that is likely to affect system performance drastically. Many external debuggers are limited in the amount of access that they have to any peripheral devices that are used by a microprocessor. This is due to the use of memory mapping or I/O bus address use that the generic external debugger is unlikely to be able to access (a specific external debugger may have access to routines linked into the system that give peripheral access).

One useful technique while debugging hardware/software situations is to do register dumps of the peripheral device. A dump should be placed in such a location (such as at the end) that the effect on the rest of the interrupt processing is minimized. It is still likely that it will adversely affect the behavior of the system, so calculation of specific trigger conditions is very useful to get the information only at the critical location.

One of the basic requirements while working with hardware/software interactions is to make sure that register access is possible. This can be done by writing to a hardware register and then reading back the register (make sure that the addressed register is read/write).

Another is to make sure that interrupts are working. This can be done with a simple log message at the beginning of the interrupt routine. If the interrupt should have occurred and the log never appears, make sure the device is configured correctly. Check that the interrupt mask is set correctly, that the interrupt controller has interrupts enabled, and that the event really should have occurred. The next step of debugging such a situation is to verify the configuration registers (this will be different for each device and needs to be in accordance with the hardware connections used)—it may be necessary to do this with an electrical engineer or other person who understands the connectivity and how to test it.

Hardware logic can only be verified. Software logic can be changed. That is why many "workarounds" are used for hardware problems—it's the only way to do it most of the time. A register is set, a command is issued, and nothing happens. Was it a problem with the hardware access, the order of the commands, perhaps a need to wait a short period of time between commands, or is the user's manual for the device wrong? It's hard to determine, and although it becomes much easier with experience, debugging at the hardware level will always be one of the most time-consuming parts of a new system.

Processes and intertask debugging

What does the process scheduling have to do with debugging? Primarily, it is involved with the data flow. The data flow, while debugging, will go from the entry point of one task to the exit point (with possible side trips via interrupts). At this point, another task should eventually be scheduled, based on the events issued as responses, by the task. If the scheduling is not known to the debugger, it may be possible to set breakpoints at the wrong place.

A process, or task, running under an operating system has a number of attributes. One is whether it is preemptible. A preemptible task may be stopped and another task allowed to run if it has a higher priority. Another possibility is for the task to be allowed to run for a certain period of time (or less, if it is stopped waiting for a resource). This is called time slicing. Real-time systems, and data communications protocol systems in particular, usually can be set to allow each task to run to completion. This is because all of the events interact. Why delay the processing of a PH_DA_IN primitive when all that it would do would be to delay the arrival of a DL_DA_IN at the network layer? If there is no inherent priority assigned to a specific task, it is more efficient to allow the task to run to completion.

Completion of a task normally means waiting until another event has occurred. A primitive arrives, and it is processed. The task then calls the operating system to request another primitive. Under most operating systems, even if a new event is waiting at the moment the task requests another event, some other task will be given the chance to operate. (Of course, if no other task has anything to do yet, the same task may be rescheduled.)

Scheduling is the organization of the tasks that may be invoked by an operating system. There are two main criteria used to determine what task runs next—its priority and its readiness. Readiness is an indication that the task has the resources necessary to be executed (it has an event waiting, in the case of a protocol system). Priority has a couple of components. One is the overall importance assigned to the task. If two tasks have a high priority and there are five other tasks set to have a lower priority, the five tasks will run *only* if neither of the high-priority tasks is ready to execute. If the high-priority tasks both have a lot to do, the system may never run the low-priority tasks. Under such a system, low-priority tasks may be called *idle tasks* because they are executed only when the system is otherwise idle (no high-priority tasks ready).

There are lots of ways to distribute the scheduling, and operating system books go into the various scheduling schemes. For a real-time protocol system, it is often best to run the *simplest* scheduling algorithm because it involves the least overhead for the operating system. This is a non-preemptible *round-robin* equal-priority scheduling. Think of a ring of people holding hands. One person is asked if he or she is ready to do something. If not, the next person in line is asked, and the next, going around the entire circle one person at a time. A person who is able to do something is allowed to complete that task before going to the next person to see if he or she is ready. People, like software tasks, may end up interrupted while in the midst of doing something—but, also like software tasks, they are allowed to continue after the interrupt has completed.

Run-time problems

Much of debugging is the same within a real-time system as in a system that is not real-time critical. However, there are also issues of timing and handling interrupts that affect real-time systems that are not otherwise important.

System speed. Let's assume that characters are coming into a physical layer entity at the rate of one character every 125 µs (this is approximately equivalent to a 64-kbps transmission line). If there is no queuing built into the hardware, this means that the character must be read every 125 µs. This is not a long period of time and will be difficult to achieve for many systems. What are the effects of not being able to service the hardware quickly enough? The first is that an *overrun* condition is likely to occur. Most hardware data processing chips are able to check to see whether a new character has arrived before the previous character has been "read." The second is that characters will be lost—a character string that reads "the quick brown fox" might arrive as "thequck bron fx."

During testing, it should be noted that characters from test messages have been lost. (This is one aspect of testing that should always be incorporated into a test plan—use known data.) However, placing log messages into the data handler will probably only make the problem worse. This is why it is very important to take advantage of any error conditions that the hardware is able to inform the software about. If an error has been logged that says "overrun," the implementor knows that characters are being read too slowly. Without this error message, the debugger will need to eliminate lost character possibilities at every point.

This error (and others) is of inherent service to the developer and debugger. It tells what and why. The next step is to correct the situation. It means that each character must be read more quickly. Changes in data structures may be needed. It also reflects the importance of using the right hardware for the right needs. A hardware device without a queue means that handling time is equivalent to the data rate. This may be all right for a serial port interface running at 300 bps (26.7 ms per character). It may be totally impossible to achieve successful transmission or reception at a higher speed.

Assume that a queue is available in the hardware device. The queue holds 32 characters at a time. This means that the queue can be serviced once every 4 ms (32 times 125 µs per character). (Note that, once the queue is full, there are still only 125 µs before it is necessary to start reading from the queue.) This is more in line with the software possibilities for service time. The queue is read into internal data structures, and another 4 ms (minus the amount of time needed to read the queue and do other processing) is left before it must be serviced again.

It is also possible, in very real-time-critical areas, that the section may need to be rewritten directly in assembly language. This is usually a last step. Most compilers are sufficiently good at optimizing the final machine instructions that it is rare to improve performance in this manner more than 10 percent or so. If this is not sufficient, the hardware design must be changed—so, it is important to understand the limitations of the hardware *and* software while implementing.

Limited logging.   Most problems inherent to a real-time system involve interrupts and timing problems such as we have just discussed. Because of this, logging must be limited within the real-time-critical portions of the software, or it will cause more problems and be of little help in tracking the existing problems.

Logging should be restricted to very small amounts of data at a time. It may be that many attempts will be necessary to accumulate the full amount of data needed. Use of reserved segments of memory may be useful in debugging real-time-critical sections. This is similar in effect to that of the LEDs mentioned earlier.

Data conflicts

Data conflicts are the most difficult part of debugging a system. The data obtained during debugging may be, in themselves, suspect. There are two main categories of data conflicts within a system. One is overwriting stack segments. The other is overwriting global data. The first causes unpredictable (but sometimes reproducible) effects within the local function. It is also possible, if the stack is disturbed badly enough, that the return stack elements will be destroyed. This is actually a blessing because it will mean that the problem will have been reduced to the simple case of the software stopping. The danger exists that, in destroying the software process integrity, it may be difficult to retrieve the data being kept in the debugging effort. In order to avoid this, it is necessary to stop the system before the function is entered—and then track the instructions within the function.

If the stack overwriting does not destroy the process integrity, the function will do unpredictable things based on the erroneous contents of local variables. If the data can be observed to be incorrect, the place where they first become incorrect can be located by tracking the contents of the data.

In the case of global data overwriting, debugging takes two steps. The first step is the hardest—locating the area being overwritten. Once this is done, however, the first thing to do is to examine the memory map of the system. If the preceding area of memory is an array, there is a good chance that the array definition is being exceeded (if the array has five members, six are being written). If it is not (or the array boundaries are checked and proven correct), it is important to go back through the memory map to discover the first area that appears to be correct. The likely problem area will be the first global data structure following the last, still correct, data.

Summary of methods

Debugging is the process of locating a problem and fixing it. Real-time systems vary from other types of systems only in the aspect of time-critical regions and interrupt processing. In each problem, the tasks involved consist of first determining *what* is wrong and, then, *why* it is wrong. This is true for logic, timing, and data overwriting problems—although the precise methodology varies slightly for each one.

The process of debugging mainly takes patience and perseverance. An external debugger can decrease the amount of time needed to obtain more information. This can be very helpful. The same information can be obtained in other ways.

Operating System Use

An operating system is able to provide many different functions to a software system. These include the scheduling of tasks, provision of system resources, and commonality of system interfaces. We have already discussed the various aspects of simple scheduling in the debugging section. System resource management is an aspect of providing an entire system with common access to general resources. One of the most important of these is the provision of mechanisms for one task to communicate with another. Other resources include semaphore handling to provide synchronization during certain event processing. The commonality of system interfaces provides easier portation from one hardware platform to another.

Intertask message queues

A message queue is a resource that is associated with a task. Other tasks may enter messages into the queue, but only one task is normally expected to be able to remove them. In actuality, this is not strictly enforced by many operating systems. Rather, each task and message queue has an identifier. This identifier is used for operations on the message queue. The coordination, and access permissions, are enforced by following the appropriate system restrictions as developed by the system designer. Sometimes access permissions are associated with resources. This is set up at initialization and maintained by the operating system. This is only a facilitation of access enforcement, however, because it still ends up being a matter for the implementor to determine access.

Operations.   The operations on a message queue primarily are that of sending, receiving, and sometimes inserting the message as a priority message. Any task may send a message. The exact format will depend on the operating system. Normally, the operation will include a message queue identifier and a pointer to a data structure. This is an asynchronous operation that makes the message available on the queue for a request operation.

The receiving operation also has a message queue identifier and a pointer to an area where the retrieved message should be placed. A final parameter that often exists is that of a status flag that specifies what to do if a message is not currently available. Two options are to wait forever until a message arrives and to return immediately if a message is not available. The second option can be useful to query the operating system for events that are available for use (sometimes an operating system will have a separate primitive for querying resource status). The first option is what is usually used in a multitasking system. If something is now available (and the scheduling mechanism permits immediate access), the task retrieves the message and begins processing. If the message is not present, the task is made nonready, and other tasks in the system are allowed to run (if ready).

Inserting a message at the front of the message queue allows varying priorities for messages. The same type of thing can be done in software without operating system assistance. Two queues are used—one is high priority and the other is lower priority. A receive operation is performed on the high-priority queue with the option set to not wait for a message. If the high-priority message exists, it is received and processed. If not, the task proceeds to wait for a regular-priority message.

This is slightly different from a single-queue system since, if the message queue is empty, the single-queue system will allow waiting for the next message independent of priority. The double-queue system gives first priority to the priority queue as long as high-priority messages exist but will wait for regular priority messages otherwise; therefore a new high-priority message will have to wait until after a regular priority message has been processed. This may be a "fairer" algorithm since it increases the chances of processing regular-priority messages, but it may be undesirable if the high-priority messages are time critical.

Implementations. Message queue systems can be implemented in a variety of ways. An efficient method is to maintain each message queue as a separate data structure. Another method is to make the message queue part of a task's data stack area. In either case, the actual data may be maintained in the structure, or a buffer may be allocated for the message and only a data pointer and reference identifier kept in the message queue. This second alternative keeps the message queue data space small with additional data space allocated only when needed.

If the queue is strictly implemented (no insertion into any place other than the tail of the queue), a single array can be maintained per message queue. Head and tail indexes are kept for insertion and removal operations. If insertion operations are allowed, a linked list mechanism may be desired.

Semaphores

Semaphores are sometimes called simple resources. One resource is allocated, and competing tasks must wait for another task to release the resource before they are allowed to continue. This can be used for coordination of global resources that may potentially be accessed by multiple entities at the same time (preemptible tasks or between tasks and interrupt procedures).

The main difference between a semaphore operation and the autonomous use of a shared variable is that of being guaranteed noninterruptible access. Consider the act of doing this in software. The variable would need to be read and the value compared against an "accessible" value. If it was accessible, it could then be set so that any other entity reading the value would find that it was *not* an accessible value. If it was done as an interruptible operation, it would be possible that *after* the value was checked (but before setting it), the value might be checked (and found accessible) by some other entity—allowing access by multiple entities and defeating the purpose. The checking and setting must be guaranteed to be noninterruptible.

Let us assume that an interrupt procedure accumulates information (not contained within a message queue). It does not want any other entity to gain access until the data are complete. It requests a resource semaphore. Once the data are complete, it releases the semaphore. Any other entity may now use the data when it requests the semaphore and obtains it. This type of situation really requires two semaphores—one for the interrupt routine to know that it is safe to put new data into the block and one for the tasks to know that it is safe to use the data. One "semaphore," however, may be part of the data block.

Data structures

The data structures used within an operating system will vary from system to system. There will be a structure containing information about all potential tasks in the system. This will either be initialized when the operating system starts up or will be started as an empty queue that requires specific task-oriented operations to create new structure instances on the task queue. There will also be resource structures for message queues, semaphores, and multiple-item resources.

Each task must also have a data space that can contain its stack when it is running. Some operating systems may allocate a separate stack space for each possible task. Some may reuse the same stack for all nonpreemptible tasks. Either is workable, but the second method will use less total data space. A separate stack area may be maintained for use by interrupt routines.

System interfaces

Operating system interfaces fall into the categories of static and dynamic. Static operations are performed when the system is compiled or activated. This may include things such as queue sizes, task stack sizes, the maximum resources of the system, and so forth. They may also include data structures that are compiled into the system that specify tasks and existing message queues.

Dynamic operations perform within the maximum amounts allocated within the system. If, for example, the operating system is defined to support a maximum of 10 tasks, it would be possible to start up 10 separate tasks, but the attempt to start an eleventh task would cause a system error. Sometimes an operating system will maintain a message queue that is inseparable from a task. A single operation will create both within the system. Sometimes these two resources will be created separately.

Operations within an operating system will fall into general categories. These might include memory management, task scheduling, semaphore and other event management, timer management, and intratask operations.

Memory management. Memory management within an operating system is useful if the system will always be used with the same operating system. The mechanics of such are sufficiently different between operating systems, and the use sufficiently dispersed within a system, that use of the memory management routines will effectively lock the system into use of the same operating system. This may be fine, but it should be taken into account during the design of the system. Operations will include allocation and deallocation of memory. They may also include the possibility of setting aside additional memory for specific purposes such as a new task stack.

Task scheduling. There will always be a need to initialize a task. This associates a software entry point with a task. It will also include the task resources, although these may be implicitly defined as part of the system resources. Sometimes the initialization of tasks will be performed from data structures without separate operations needed. If dynamic task creation is allowed, there will also be the possibility to stop the task. This may be considered to be a "permanent suspension" of a task such that the task resources are not reclaimed. A true stopping of a task will enable the resources to be used for a different task.

Semaphore and other event management. These operations consist of noninterruptible requests for one, or more, resources managed by the operating system. You must be able to both request and release resources (semaphores or multiple resource items). This can usually be duplicated by other methods if the operating system does not directly support semaphores—for example, by using a special message queue.

Timer management. Timers fall into the same situation as does buffer management. These operations are likely to be widely dispersed throughout the system (though less than buffer operations) and can lock the system into a single operating system. Once again, this is fine if that is a design criterion for the system.

Timer operations may include the ability to set a system clock. This system clock is not mandatory for all systems—or for protocol systems. It may be useful to coordinate a separate coprocessor system with some type of network clock. Whether the clock may be set or not, some type of system clock is necessary if time-sliced task operation is to be used.

The types of timer operations mentioned in the protocol sections in this book require the ability to trigger specific events at a given interval. Thus, the operating system operation must be able to set up an event (semaphore setting or message transmission) to be enacted after the time interval. It must also be possible to cancel the timer.

Intratask operations. These operations are executed *within* an executing task. We have already seen how a request for a semaphore, multiple event item, or message queue item can cause the temporary suspension of a task until the resource is available. Other operations may include putting the task to sleep for a given time interval or changing its priority to get more system cycles or give them to other tasks. It may also be possible to make the task nonpreemptible during critical regions where shared data are being used (and then preemptible once again after such use has been completed).

Stand-Alone, Integrated, and Coprocessor Systems

A stand-alone system is a fully autonomous unit. It may be used as a TA, in the case of a protocol system. An example of an integrated system might be a workstation that incorporates various peripheral interfaces into its overall design. A coprocessor system is like a *plug-in card*. Parts of the system are designed to operate on a separate hardware board incorporating a microprocessor and various required peripheral devices. For high-speed systems, a coprocessor system allows the data bus to be shared between devices and, thus, increases the potential throughput to that of the general system.

Stand-alone systems

In an ISDN system, a stand-alone system will usually contain the entire chained layers set of protocols. The higher layers, including the application, may be part of the system, or they may be designed to access the lower-layer protocols as part of an R interface reference point procedure. The next major section will discuss these higher-layer entry points in more detail.

A stand-alone system has advantages in portability. Since it has no common data bus and is independent of the higher layers and the applications, it may be used with multiple systems. This is similar to that of a modem being used with different computer systems and data communications applications. It does require a standard interface into the system, but there are many different interfaces that may be supported (as discussed in Chap. 9). It will also require *front-panel* access. This is a method to set the device up for the specific use of the application.

Front-panel initialization in an ISDN system (or TA, in this case) requires a method to set up the network identification parameters such as the SPID or LCI ranges (in the case of an X.25/X.31 application). It may also need to be able to be programmed with other network-related information.

Integrated systems

An integrated system controls the protocols, higher layers, and applications within the same system. This means that the microprocessor being used must be able to handle all the demands of the system. The limitations of such a system are that it mainly relies on the power and speed of the microprocessor. A voice-only ISDN system can probably be easily handled in such a situation. Hardware-intensive protocols (such as video or high-quality audio) can also be handled. Other software-intensive data protocols (such as X.25) can be handled as part of a single-task application situation. Multitasking workstations that are to handle software-intensive data protocols will need a powerful microprocessor.

There are system advantages to an integrated system. The application may be able to be compiled directly with the chained layers. This is useful in data-intensive applications such as video. It also means that some potentially duplicated hardware can be eliminated (such as special RAM and microprocessors). It does reduce versatility unless the protocol system is designed directly into a portable operating system (such as UNIX)

Coprocessor systems

A coprocessor system splits the tasks over multiple processors. Usually this means that the application, and other higher layers, will reside as part of a general computer system. The protocol layers may be split at any layer, but it may make the most sense to keep the chained layers together on the coprocessor system. This allows some integration of the lower-layer protocol data needs and also maintains the ITU-T primitive interfaces without a need to modify them for interprocessor communication requirements.

The splitting of the system into two parts allows the host processor to concentrate its power on the application needs, including maintaining file system integrity for data transmissions and receptions. The ability to split the disk I/O from the process of maintaining the data line protocol is particularly useful for safety. There are also advantages of maintaining application interfaces to different I/O boards. For example, a system might have coprocessor boards that support N-ISDN, Frame Relay, Ethernet, and so forth. It would be possible to have all these data transport protocols on one coprocessor board—or on separate boards.

Higher-Layer Entry Points

Within any system, there must be a way to communicate information between the lower layers and the final user (or an independent upper-layer gateway application if the data is not for this particular node and is operating as a gateway). There are several possibilities for this. One is a routing application that operates as a gateway to forward data between two networks. Another is as a TA which communicates across an R interface reference point.

In the case of a coprocessor board, there are various possibilities. These include communication across the data or I/O buses of the system. One common method for special data-intensive applications is to use a *shared memory* (sometimes called dual-port RAM). Another possibility is to use an I/O port on the main processor. For example, in a DOS system, communication can take place via a COM port. Shared memory systems need to be able to communicate the information that an event is available for processing. This can be done via polling or interrupts.

Gateways

The purpose of a gateway is to provide a routing from one network to another. It is probable that there will be no human user directly involved. Data will arrive via (for example) ISDN and be sent to the higher layers as N_DATA_IN primitives. The connection information is used with the data contents to then call another protocol stack to transmit the data using a different protocol.

In a stand-alone system, there would be a series of tasks and interrupts associated with each protocol stack. A high-layer coordination application acts as a router for the data and as a translator for primitives. The address and subaddress fields could be used from an N_CONN_IN primitive to generate a new connect request to the other network. It is also possible to have the application examine the data for routing information. This will depend on the protocols involved.

In a coprocessor system (or TA), the primitives must go across the processor boundary to the host application. From this point, the behavior is similar to that of a stand-alone system. The main difference in gateways is that there is no human intervention in the active system (though administration may be needed).

Terminal adaptors

We discussed the case of TAs in Chap. 9 at length. In this case, there is a translation of the ITU-T primitives into a set of commands that is understood on the other side of the R interface reference point. It may be a synchronous or asynchronous interface, and certain events may not necessarily be part of the primitive translations. However, there will be control information as well as data transfer.

Let's assume that we have an asynchronous interface crossing the R interface reference point. Primitives come up to the coordinating entity, which recognizes that a serial port is being used for this particular connection. It passes the primitive to the TA task of the system. The role of the TA is to perform the necessary translations of the primitives. After processing the primitive, a LLD is called to send the information to the TE.

The higher-layer functions are actually performed by whatever application is part of the TE. The coordinating entity and TA task act as routers and translators. For some situations, the TA may be able to perform autonomously. A TA set in "auto answer," for example, might automatically send a N_CONN_RS primitive in response to a N_CONN_IN incoming data call primitive. Some indication of the active link is still necessary and may be performed by changes in interface leads or in character strings sent across the data transmission leads.

Shared memory

A shared memory interface between coprocessor and host processor systems requires a protocol. Semaphores cannot be used directly since there is little way to coordinate the two processors. Thus, a polling interface may be needed with added constraints to provide data integrity. Let us presume that there are two "mailboxes." One mailbox is used from the host processor to the coprocessor. The other is used from the coprocessor to the host system. Each is like a mailbox slot through which "letters" may be passed in one direction only.

Interprocessor mailboxes.   An N_CONN_IN primitive arrives. The coprocessor-to-host mailbox is checked to see if something is in it. (It does this by checking an acknowledgment bit, or field, within the mailbox.) If the mailbox is available, the primitive members (including identifier) are copied into the mailbox area, and the acknowledgment area is marked to indicate that there is a message present (similar to raising the flag on a standard mailbox). As long as the flag is raised, no more primitives may be placed in the area.

The host processor then examines the mailbox. It may do this when prompted by an interrupt from the coprocessor, or it may do it as part of a polling situation. At any rate, once it has noticed that the mailbox has information in it, it reads out the information and processes it. After the information has been copied, or in some other way no longer needed, the acknowledgment area is reset (flag lowered) to indicate that the mailbox may be used once again.

This protocol is safe because setting the acknowledgment area is a noninterruptible operation. It is important, however, to make sure that the mailbox is used in the proper sequence. To use the mailbox, the acknowledgment area is checked, primitive information is inserted, and then the acknowledgment area is set to indicate information is available. On the receiving side, the acknowledgment area is checked first, then the data are read (or otherwise used), and then the acknowledgment area is cleared for more primitives.

Operation in the other direction is very similar. The sending side may write data (but only when it is indicated that any prior information has been processed). The receiving side may read the data.

Data buffers.   We discussed the overhead of data copying earlier in this chapter. It would be possible to have each mailbox have a section large enough for the biggest data buffer. This is a simple method and works directly within the mailbox scheme. However, data would need to be copied from the local processor to the mailbox and then copied out before the next primitive could be processed. This copy operation is a high-overhead situation. The question is, is this needed?

If sufficient memory is available in the shared memory area, the answer is No. We talked about buffer pools, but we did not say specifically where the buffer pools were located. What if the buffer pool was located in the shared memory? (Note that only large buffers would need to be so located since small buffers are used only for buffers going out on the data transmission line.) This would allow a single copy of data. The application can write directly into the shared memory buffer and pass the buffer reference to the coprocessor system. The same can happen in the other direction.

We now have a point of possible data contention. The buffer manager must be located on one processor system even if the data being controlled are located in shared memory. Otherwise there will be possibilities of contention and algorithms to solve this would add overhead. (Overhead, especially in a real-time system, must always be reduced as much as possible.)

If the buffers are managed locally on the coprocessor side, the host-to-coprocessor direction is easy. Before the host-to-coprocessor mailbox is marked as available, a data pointer which points to a buffer inside of the shared memory is placed within the mailbox. This data pointer can then be used (perhaps with modifications to the address dependent on the memory map) directly by the host application. On the receiving coprocessor side, the buffer reference (and data pointer) is passed along as part of the primitive entering the protocol stack. Notice that not all primitives will require a buffer. The use of a data-length member in the mailbox can indicate whether the buffer has been used (with a length of 0 indicating that it has not been used).

The coprocessor-to-host data use is a bit more complicated because of needs to deallocate the buffer. Once the acknowledgment area has been reset by the host processor, the data buffer (if length is nonzero) can be deallocated by the coprocessor. Although it is not absolutely necessary, this indicates that it may also be useful for the coprocessor to examine the acknowledgment area of the coprocessor-to-host mailbox even when additional primitives are not pending to be forwarded. This would allow faster reclaiming of a data buffer for potential use by the rest of the system.

I/O bus usage

There are a number of buses available on a general computing system. These include the data bus, and I/O bus, and probably a control bus. The data bus is used for shared memory access and in other situations where data are being transferred between system components. The I/O and control buses allow access of specific registers on various system boards.

A data bus will usually be the fastest means of transferring data in a computer system. This is because it will allow parallel data transfer. (Instead of sending 8 bits of data in a sequential fashion, it may send all 8 bits at the same time. Data buses may be any width—so the 8-bit-wide example is just that—an example.) The I/O bus will normally be a serial bus that allows data to be carried only in a sequential fashion. Nevertheless, it can still provide data transfer, and the speed may be great enough for many applications.

One advantage of using the I/O bus is that there may be standard protocols available on the system to make use of it for accessing peripheral boards. (The parallel bus would possibly make use of the mailbox interface as described above.) A particular register is a read-character register on the coprocessor board. If the host processor writes the register, the datum is available on the coprocessor.

Bonding

Chapter 4 discussed the process of ITU-T Recommendation X.25 MLP. This is one form of software bonding. Bonding occurs when two or more bearer channels are used together to provide an increased aggregate bandwidth. It is also possible to perform hardware bonding. Hardware bonding synchronizes the multiple bearer channels into a single channel at the LLD point. In both cases, there will be multiple channels set up through the network.

It is possible to perform call setups independently on all bearer channels that are to be used. This is actually the simplest method because it requires minimal knowledge by the network of the use of the channels. It is also possible to use the additional bearer services if such are supported by the network.

Let's consider the situation where multiple channels are set up independently. In this case, there will be multiple CRVs. Each channel will be set up, and torn down, in an independent fashion. However, since they will be used together to form a single apparent channel, the N_DATA_IN and N_DATA_RQ primitives should be submitted over only one of the connection identifiers (unless the bonding is performed outside of the protocol stack). It will then be the responsibility of the software or hardware bonding mechanisms to distribute (for transmission) the data equally over the different physical, or TDM, channels. In the case of reception, all data will arrive at the higher layers as part of a single connection identifier N_DATA_IN primitive.

If the hardware performs the bonding, the data stream going through the software layers will appear as if the data had arrived on a single channel. If layer 2 does the bonding, a single channel will be seen at layer 3. If layer 3 does the bonding, the higher layers will see only one channel. Of course, if it is done on a host processor, the primitive data flow will remain the same as it would be for multiple independent channels.

Low-Level Driver Design

LLDs are a software type that is unique to real-time programming. The LLD acts as the interface between the peripheral hardware (as opposed to the processor) and the software. For proper layer independence, all commands for a particular type of hardware should enter through that LLD's entry point. This layer independence allows easier changes in the hardware part of the system (such as changing the system to use an S rather than a U interface).

The LLD has to have full access to the register interface of the hardware. It will also be programmed in accordance with the electrical characteristics of the system. This programming will mainly be done at initialization time, although some changes may take place during run-time. The LLD has a full complement of entry and exit points, as does any other layer. The LLD also acts as the software portion of the protocol implemented. The events from the physical medium that cause responses in the system must be noticed and acted upon. This can be done in an interrupt-driven or polled fashion.

Hardware register access

An LLD must have access to the functions of the hardware device that it controls. This means full access to the registers of the device. This is done in two primary ways. The first is through memory mapping; the second is via a control bus.

Memory mapping of hardware devices. Memory mapping allocates various regions of memory to different devices. The memory will be allocated into regions. One region will be for the ROM, another for the RAM, and possibly a third for shared memory (if any). After the memory regions, memory segments will be allocated to peripheral devices. Some peripherals may be part of the integrated circuit. Many others will be separate devices that are interconnected over the data bus. Figure 11.4 shows an example of the memory mapping on a possible device.

Memory mapping provides the range of addresses for the device. Within this range, the hardware manufacturer will map different registers to different addresses. For certain hardware configurations that do not allow byte accesses over the data bus, these addresses

0x00000000	ROM
0x10000000	RAM
0xC0000000	Peripheral 1
0xD0000000	Peripheral 2
0xE0000000	Peripheral 3
0xF0000000	Shared memory

Figure 11.4 Example memory map.

may need to be multiplied by 2 to give a 16-bit address per register. Thus, if the "offsets" from the device memory base are 0, 3, 4, and 5 bytes, this double-byte configuration would require the address offsets of 0, 6, 8, and 10 to be used.

Register access via memory mapping is used identically to that of any other memory access. It may be used as an address pointer for reading or writing. It may be necessary to declare any variable that is used during a read operation as "volatile" (or the equivalent in languages other than C) to ensure that the compiler does not optimize out an operation that apparently has no direct use (but triggers a hardware function). For example, the line:

```
local_var = *(unsigned char *) 0x60000020;
```

would normally cause a read operation from some register at an offset of 20 off from a base address of hexadecimal 60000000. However, if the "local_var" is not later used, some compilers will delete the entire operation during optimization (if the result is not used, it is assumed the operation is useless). The keyword *volatile* (in C) tells the compiler that the variable causes side effects.

Control bus access to hardware.   In a control bus situation, access to a register requires two sequential operations. The first sends a command across the control bus to indicate that a specific register needs to be accessed. The second sends (or receives) the data from the previously indicated register. Control bus access depends on the commands available for the particular bus configuration.

Control bus access has some advantages to the system design. It allows a set protocol for the different hardware devices. Since multiple accesses may be needed (dependent on what registers are given direct access and which ones have an indirect operation), it is normally less efficient than memory mapping.

Macro use for hardware access.   During software implementation, the final hardware access method may not be known (or it may be changed later because of added, or deleted, components). It is also possible that the same LLD may be needed on multiple hardware components. For this reason, it may be useful to use a macro for the actual hardware definition of the access method.

A macro, as can be seen in Fig. 11.5, allows substitution during compilation. The macro can be used in all of the software, and by changing the macro definition before compilation, the compiled software will automatically change to fit the new usage. Thus, different data bus configurations or memory-mapped versus control bus access can be hidden from the software, and changes are centralized.

```
#define        DEV_BASE     0xD0000000

#if defined(MEM_MAP)
#if defined(DOUBLE)

#define Read_device(reg, value)      value = *(unsigned char *) (DEV_BASE + (reg * 2))
#define Write_device(reg, value)     *(unsigned char *) (DEV_BASE + (reg * 2) = value
#else

#define Read_device(reg, value)      value = *(unsigned char *) (DEV_BASE + reg)
#define Write_device(reg, value)     *(unsigned char *) (DEV_BASE + reg) = value

#else

#define Read_device(reg, value)      *(unsigned char *) 0xC0000010 = reg; \
                                     value = *(unsigned char *) 0xC0000020
#define Write_device(reg, value)     *(unsigned char *) 0xC0000010 = reg; \
                                     *(unsigned char *) 0xC0000020 = value

#endif
```

EXAMPLE of USE:

Read_device(INT_STATUS, isr);

Write_device(INT_MASK, 0xE4);

Figure 11.5   Example macro use for device access.

Hardware requirements

A hardware device will be accompanied by a *data sheet* and probably some type of user's manual. These documents will discuss the interface leads, timing requirements, and general signals that are used. They will detail the functions available with the device and the registers that offer access to commands and internal options available.

Many semiconductor chips are integrated packages of separate circuits. One circuit may do audio processing, one may provide multiplexing/demultiplexing, one may do HDLC processing, and yet another may provide for the hardware support of the physical layer protocol. Although the same device incorporates all of these functions, the use of each will be handled in an autonomous manner, and each is likely to have its own separate set of registers. The data sheet will probably present a block diagram that indicates how the different circuits are linked together to provide access to the external interface leads which allow control and access to the microprocessor, buses, and physical line interfaces.

The next part of the documentation will describe the purpose, and perhaps the electrical characteristics, of the signal leads that are connected to other discrete components in the hardware system. These are usually classified as input, output, or I/O leads to categorize the direction that signals come into, or out of, the device.

Two registers (or sets of registers) act as complements to each other. These are the *interrupt mask* register (or registers) and the *interrupt status* register. These may be cascaded per operational block. For example, if the audio processing component is not being used, any interrupts from that section of the device, as a whole, can be turned off. However, if the section *is* to be used, there may be additional mask and status registers to allow specific actions to cause interrupts, or not.

Interrupts exist for error conditions (overrun, underrun, aborts), expected events (last byte transmitted, last byte received, closing flag transmitted), and status situations (buffer available, data available). Interrupt status registers are normally implemented such that interrupts may be turned off (via the mask registers), and the status can still be read via a polled method.

The registers in each functional block fall into the categories of status, command, and configuration registers. A status register provides an *indication* type of information from the hardware to whatever software is using the device. The above interrupt type conditions are a subset of such possible indications. Command registers allow requests for the hardware to perform specific tasks. These tasks may have eventual indications (responses) that indicate the performance of the task, or they may be without explicit response.

Configuration registers allow the software to set up the device to operate in a certain manner. Perhaps the type of flag generated between frames can be changed. It is possible the device will allow different control buses to be used or the use of push/pull lines rather than open-drain ones. A register bit might be used to put the device into a "loopback" mode for testing. Configuration registers are primarily used for initialization, but certain commands (such as loopback) might be used during run-time for special situations.

Initialization

The process of initialization begins at power-on time. At this point, the instruction pointer will be set to a predetermined location (often address 0), or the instruction pointer will be set to the contents located at a given location (such as address 0). This is sometimes called the *start-up* location. It is also possible that initial control will be given to a ROM program that will then deal with some of the hardware aspects and transfer control to the software. No matter which way the system control is set up, eventually some file (usually in assembler) will get control.

After the processor initialization has finished (within the start-up routine), the software initialization needs to be performed. Calling a software routine to do the software initialization from the end of the start-up routine is a good way to localize these important functions. Software initialization includes LLD software which will also set up the appropriate hardware register configuration. Finally, interrupts will be enabled, and control will be passed to the operating system.

Start-up routine.   The initial routine performed by the system (whether it is in RAM or ROM) is usually a routine coded in assembly language. This is because only the instruction pointer is usually initialized by the microprocessor (of course, this is not a requirement, and it is possible for the microprocessor to have all of its registers initialized to default locations). At any rate, it is very important that the initial instructions of the system know exactly what is available. This normally means that it can use only processor registers and has no access to any type of stack area.

The very first step is to call whatever machine language instruction that disables microprocessor interrupts. The next step in the start-up routine is to initialize segment, index, or other registers that the instructions will need. The following step performs a software mask of all interrupts such that, when the hardware interrupts are reenabled, interrupt processing can be controlled. Even though most processors will power up with all interrupts turned off, it is vital that no asynchronous commands enter the system until after initialization occurs.

At this point, it is safe to perform general initialization. Various things may be desired. If interrupts are to be used, the interrupt vector table needs to be loaded with the entry points. Interrupts also often have priorities that may be set along with the way that the hardware has been configured to trigger the interrupt (edge-triggered, level-triggered, etc.). Sometimes the microprocessor will have different modes (supervisory versus user mode, for example), and it is important to set the mode into the appropriate state.

The next step is language, and compiler, dependent. Most compilers for a particular microprocessor will expect the compiled software to have certain registers available for standardized use. This standardized use implies that certain values be set in the registers. This may include a stack pointer for task, and stack, use. There may also be a data pointer which is expected to give the address of the start of the global memory. (The compiled instructions normally have addresses calculated based on certain indirect addresses—this allows relocation of code.) It may also be useful to initialize all of the global uninitialized data area to a set value (all zeros, for example). Finally, the processor is set up into the state expected by the compiled code, and the initial software initialization routine (or, sometimes, operating system entry routine) can be called.

Software initialization.   Interrupts are still software masked (although interrupts may have been reenabled at the hardware interrupt controller level before calling the software initialization routine). This is necessary because the necessary data and register initialization has not yet taken place.

Software initialization for tasks is concerned with setting up data structures that will be used by the tasks once the operating system is told to begin process scheduling. It will also be concerned with LLD initialization. This will include data structures and also the setting of appropriate configuration register values.

Once the data structures and configuration registers have been initialized, interrupts can be unmasked. It is also appropriate to begin operating system task scheduling.

Run-time changes

Initialization sets up the system into a stable, beginning, state that is in agreement with the hardware configuration. As soon as interrupts are enabled, and task scheduling has begun, data structures will begin to change. Most of the configuration register contents will remain the same during the entire running of the system (since neither the hardware nor the general LLD software expectations will change). Some items may change, however, because they are not determined until later in the running of the system.

One type of run-time change is the result of call setup and teardown. It is not desirable to have a CODEC active before a voice call is present. During the call setup of a voice connection, information will be included about the use of the call. Part of this (as per the bearer services) will indicate whether it is voice, unrestricted data, or some other service. Another part will indicate the appropriate bearer channel. This information must be passed along to the LLD in order for it to set up the hardware in the correct manner.

Another type of run-time change is for maintenance and testing. A device may be set into loopback mode to allow internal and interface tests without interference with the network. Different flag patterns may be tried and collision situations artificially introduced to check collision algorithms and recovery.

A final category of run-time change is associated with when initialization takes place. The hardware device that controls the interfaces must be initialized at start-up to provide communication with the peer entities. A hardware device that controls B-channel activity may be totally unneeded until the signalling channel indicates that it is to be used. The TDM channels may need to be demultiplexed and routed in a different manner. All of this is unnecessary (and, perhaps, detrimental to the system) to do at the beginning.

Entry points

The entry points are similar in number to those of any other protocol layer entity. There are three. One leads "downward" to the hardware, and the other two lead to the management entity and the data link layer. In the case of UART (or other R interface reference point) LLDs, the higher-layer entity may be a TA entity. The primitives from the higher-layer protocol entity are, for ISDN systems, defined in the ITU-T recommendations. Most of the indications are also thus defined, as are a couple of the commands and responses. However, the primitive flow from the hardware is completely hardware specific. Many of the management commands are also hardware specific.

As mentioned briefly earlier, the indications and responses entering into the LLD from the hardware consist of changes in status registers. These must be responded to, according to the appropriate protocol state specifications, and will initiate further commands to the hardware or indications to upper layers. They will be defined according to the manufacturer's specifications.

Management commands are also mostly specific to the hardware and its purpose. There may be an initialization command. There may be a multiplexing, or loopback, command. All such commands are related to the functionality of the device and the parameters determined according to the needs of the hardware.

Because a system may use different hardware based on bearer services or other criteria, a level of abstraction may be useful between the higher layer and the LLD. In this way, a particular channel has an LLD rather than having the data link layer permanently attached to a specific device. Using this abstraction, a reference to "Device_driver[1].ph_cmd_handler( )" would indicate the data link layer to physical layer command handler for B-channel number 1. Since the PH_ primitives are standard for HDLC devices used within ISDN, the actual device is inherently transparent to the data link layer.

Management-layer commands to the LLD can also be abstracted in this fashion. It is much less useful, however, because the management layer will normally need to know which device it is addressing. This can be identified in an abstraction such as "Device_driver[1].hw_identifier," but it is of dubious use unless the device is run-time switchable. In such a run-time situation, the knowledge of just *which* hardware device is in use must be kept someplace. Keeping it in an abstracting data structure may be a cleaner alternative.

Exit points

There are also three exit points—leading to the physical hardware, the management layer, and the higher-layer entity. The use of the physical layer interface has been discussed briefly above and also in Chap. 10. The management entity and higher-layer entity exit points have also been discussed.

Interrupts versus polling

Timer interrupts were discussed above. There are also other possible specialized interrupts within a system—such as an interrupt from a host processor to a coprocessor system. Timers require interrupts because of their direct links to the real time of the system. Other interrupts may be polled or interrupt driven depending on the time constraints involved in servicing the hardware.

In a real-time system, there will be a time limit on how long it can take before the hardware event is serviced. This limit may be of any length—from microseconds to minutes. If the limit is in the microseconds range, interrupts will almost always be required. For other, longer, periods, it may be possible to use polling methods.

Definition of an interrupt.   Interrupts have been discussed briefly. An interrupt is a hardware-initiated break in the software sequence. Sometimes the term *software interrupt* is used within a system. A software interrupt is one where some entity, often the operating system, invokes an interrupt routine based on a software-defined event. However, in order to truly interrupt, the software task that initiates the software interrupt must be hardware interrupted.

For example, in an operating system that is able to "time slice" the scheduling of processes, a hardware timer interrupt will cause operating system functions to be performed. At this time, the operating system may examine system resources and determine that a software interrupt is to be invoked (or it may determine that the currently running task has completed its allocated time). The hardware interrupt is mandatory to provide truly asynchronous interrupt capabilities—whether it is a direct hardware function or an indirect software interrupt.

A hardware interrupt occurs when something "triggers" it. The device causing the interrupt will create a signal on a circuit lead that is attached to an interrupt controller (which may be integrated into the microprocessor). This signal causes the interrupt controller to calculate the appropriate offset based on just which interrupt has occurred. In turn, the offset is used to decide on a location in memory which will contain the address of a function to be invoked when the interrupt has occurred. The further workings of an interrupt will be discussed shortly.

Definition of polling.   Polling is similar to going to the mailbox every 15 min when you are expecting a letter. The scenario is to check something (a register or a set of registers in the case of hardware polling) every so often. If an event does happen, the correct procedures are then followed.

It would be possible to do polling "every once in a while." That is, it is possible to poll on an irregular basis. A polling routine might be placed at the end of an unrelated task or within the operating system when one task has finished and the system is deciding what task to run next. If the allowable time interval before servicing is sufficiently long, this may be feasible. It will depend on the worst-case scenario— what is the *maximum* amount of time that can expire before the polling routine is called? If it is less than the required servicing time, this method is adequate.

In most cases, the worst-case scenario is longer than allowable for hardware servicing. Polling can, in such a case, be "piggybacked" onto a hardware interrupt that is not directly related but is dependable to happen in less than the required service time. The timer interrupt is ideal for this.

If polling is done outside of a hardware interrupt, it is often necessary to disable hardware interrupts while the polling is taking place. This is determined by the required service time (interrupts will increase the delay before servicing) and the amount of shared data and resources (if any) that may be accessible to both hardware devices.

Pros and cons of the two methods. An interrupt will almost always service the hardware more promptly than polling (the exception to this is when a hardware event occurs very soon before the polling takes place). So, why would an implementor, or system designer, ever decide to use polling?

Polling has a set amount of overhead every time the polling routine is called. If a polling routine is called every 10 s, this does not amount to a very large percentage of the system time. If it is called every 10 or 20 ms, the total percentage of overhead becomes sizable. An interrupt routine also has overhead. Actually, the overhead for an interrupt routine is considerably greater than that of a polling routine (for reasons that we will soon discuss).

Let's assume, as an example, that the overhead for an interrupt service routine is 10 times as much as that for a polling routine. If the polling routine is called once every $\frac{1}{2}$ s (with a required service time of 0.6 s) and the event being checked for happens, on average, once every 2 s, the polling routine will have less overhead for the system. If, however, the event occurs, on average, once every 6 s the interrupt routine will have less overhead.

Therefore, two factors must be taken into account. These are average periodicity of the event and the *interrupt latency* time (hardware and software overhead needed to invoke an interrupt) compared to the polling overhead. It is also necessary that the polling take place within the required service interval—if it cannot, it is mandatory to use an interrupt routine.

There is one final aspect to be considered when deciding on using interrupts or polling. This is the matter of system resources. If the interrupt controller can control only four interrupts (plus the timer interrupt) and there are five hardware devices, there are only a couple possible options. One hardware device can be polled (the device should be chosen based on the maximum required service time from the range of possible devices). Another possibility is to "cascade" the interrupt circuit leads. In this manner, an interrupt can mean *either* one device or another. When the interrupt occurs, both devices will need to be polled to determine which one needs servicing (with the possibility of both requiring service). This is a hybrid type of polling.

We used a factor of 10, in the example above, for comparing the overhead of interrupts and polling. This is an arbitrary number. The actual ratio can only be determined by analysis on the hardware being used. Interrupts will always have higher overhead. On some systems, the interrupt overhead may not be much worse than polling. (Plus, remember that scheduled polling occurs within a hardware interrupt routine, so there will be a hardware interrupt overhead already incurred.) The randomness of most hardware events normally favors use of interrupts if possible. It is necessary to determine the actual overhead to make a design decision.

Compiler caveats

The volatile data type was mentioned earlier as something that is useful for avoiding compiler surprises. LLD compilation is something with which many compilers have trouble. This is probably because hardware access and interaction is a fairly specialized form of program and, thus, is not one of the primary test cases for new versions of compilers.

How does the implementor track down compiler surprises? The primary method is via disassembly of the object module. This may be done with a disassembler or, for many compilers, by giving an optional command flag to the compiler to request an assembly listing to be produced rather than the final object form. This second method is preferable because comment lines are usually inserted that contain the original language statement, or statements, followed by the assembly code produced. A disassembler is likely to lose any of the original high-language information unless it has access to the symbol and map table of the compilation.

What types of problems may be located? Well, the volatile problem was noticed by seeing that a high-language statement was ignored. Sometimes the compiler will produce a warning statement (and sometimes it won't) if it has discarded a statement that it feels is unnecessary. Such warning messages should be tracked down. In most instances, they are due to situations where the logic does not allow the following statements to be executed.

For example, take the statement "if(NUMBER_BRI ! = 1)." NUM-BER_BRI is a *defined constant* which can be set according to the desired substitution string in a header file or locally. In the software, the logic may be desired to be able to handle a single BRI or multiple BRIs. However, if it is equal to 1, the above example degenerates into "if(1 ! = 1)." Except in old versions of languages where numbers can be reassigned, this conditional statement will never be true. Thus, the compiler can safely not compile the test and any instructions that are conditional upon the test. This does not mean that the instruction is not useful—just not for this specific compilation.

Other problems that may be the result of compilation include register use. Register use within compiled functions can speed up the execution and reduce the stack size needed. The number of registers available for local use depends on the processor and compiler. Normally, if a function declares a number of variables to be stored in registers, the compiler will assign registers to as many of the declared variables as possible and then place the rest into the local stack area. However, the attributes of the variable may be lost in some situations. A "register volatile local_var" might lose the volatile status and, thus, cause statements to disappear.

Sometimes there are compiler problems that are independent of the use of the function. With luck, these will be such that they can be located by use of the debugging techniques described in this chapter. If a section of software has been found to be the problem, and there appears to be nothing wrong with the statement or statements, a disassembly may be the easiest way to find the problem. It may still be a logic error in the programming, but the variance in form will be useful in looking over the statements carefully.

Many of the problems above (and others that have not been specifically mentioned) come from the use of optimization. Optimization is useful. It decreases code size and the use of data (depending on the options chosen for the optimization). It also, inherently, changes the software logic from what was originally written. In most cases, these changes will be harmless and may reflect a better organization of the software. However, it will be *a change*. This is very dangerous in LLDs or other software that deals with hardware. It is likely to be useful, therefore, to always disable optimization on the LLD until the system has been tested and debugged. After the system is stable, it may be compiled with optimization reenabled. Then, if any problems (that did not exist previously) occur, it is relatively easy to locate the problem.

Interrupt Processing

There is a series of stages in interrupt processing. The first stage is the enabling of the interrupt. This is done via the processor, interrupt controller, and hardware control registers. The next stage is the hardware event triggering the interrupt. Then comes the interrupt controller handling the specific hardware interrupt and vectoring it to enact the associated interrupt processing routine. At this point, it is necessary to set up the system to be ready for the interrupt. Finally the interrupt is processed. The stages from the interrupt trigger until the interrupt is processed make up the interrupt overhead. The stages from the interrupt trigger until the interrupt processing routine is in control of the processor comprise the interrupt latency that is an inherent part of a hardware system.

This section discusses only the software-accessible interrupts of the system. Other interrupts exist that are internal to the hardware devices. These are processed according to the internal circuitry of the devices based on timing patterns used for their functionality.

Interrupt enabling

The enabling of interrupts follows a strict hierarchy. The microprocessor has overall control—allowing, or disallowing, interrupts depending on its current state. There is normally a machine language instruction that lets software change this state. The interrupt controller (which may be integrated into the rest of the processor) monitors a series of possible interrupts. For some interrupts, it will be possible to "mask" the effect of the interrupt—causing it to be ignored. Others cannot be masked and are, thus, always monitored (when interrupts are allowed by the processor). The lowest part of the hierarchy consists of the hardware devices (or, sometimes, special operational sections of the processor such as timers or integrated peripherals). At this point, the interrupts are enabled to be set.

When interrupts are initially set up, the processor must be set to disallow interrupts until the initialization has been completed. However, during the normal processing of interrupt events, the hierarchy is approached from the low layers up. If the hardware device interrupt is enabled, the hardware event triggers an interrupt signal. In turn, if this specific interrupt is currently set to be processed (via the mask register), the interrupt controller provides the system handling needed for the interrupt. The processor has only an ON-OFF capability (unless integrated with the interrupt controller), and it does not actively interact with the interrupt processing.

Interrupt triggering

A hardware event occurs. The hardware device sets a status bit, or field, in the appropriate status register. It then checks (if appropriate) the interrupt mask register, or registers, to determine whether an interrupt is to be generated for this event. If so, it sends an interrupt signal over the output control lead designated for this function.

This output control lead will act as an input for the interrupt controller. The signal will have a specific format. It is often designated as *active low,* which means that a low-valued signal (usually measured in terms of voltage) indicates the presence of an interrupt. There will thus be a transition when an interrupt is placed onto the control lead. It will change from a high value to a low one (or vice versa, if active high). This transition period is called the leading edge of the signal. Once the transition has been made, the interrupt signal will remain at the new level until it has been acknowledged. This steady signal is known as a *level* signal.

Most interrupt controllers can be set to trigger as edge or level triggered. The decision for this depends on the hardware design. If the signal is generated only after all internal registers have been set, the interrupt can cause an edge trigger. If, for instance, the signal is generated when the event occurs but before internal processing has taken place, making the signal level triggered will allow an extra amount of time for internal changes before interrupt processing may occur.

Interrupt handling

Interrupt handling is done by the interrupt controller. It will constantly be examining the current state of all nonmasked incoming interrupt control leads. Interrupts will always have priorities. One possible interrupt will have a higher, or lower, priority than another. Sometimes the interrupt can have its priority set during initialization. Also, it is possible to have groups of interrupts at the same priority.

Picking an interrupt. An interrupt signal arrives at the interrupt controller. It will be noticed only if that interrupt control lead is nonmasked. If it is noticed, a series of tests will take place (in hardware circuitry, normally). The first test is whether another interrupt is currently being processed. If not, this interrupt can be handled. If there is another interrupt active, the interrupt controller checks to see if that interrupt is preemptible. If not, the new interrupt remains in the current active state. If the existing interrupt *is* preemptible, the relative priorities are checked. If the new interrupt has a higher priority than the old one, the interrupt controller begins the process of handling the new interrupt (causing a *stacking* of interrupt processing). If it is a lower priority, the state remains the same. Figure 11.6 shows a flow diagram of this logic. Remember that the interrupt controller is constantly monitoring all nonmasked interrupt control leads so that, when the state changes (by a new incoming interrupt or the completion of a previous interrupt), this logic is performed all over again.

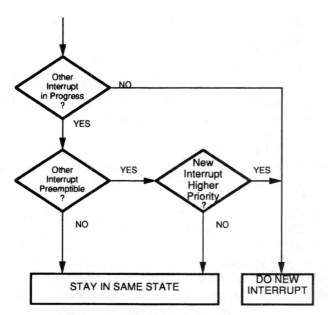

Figure 11.6 Interrupt controller logic example.

Calculation of the vector. Once the interrupt controller has determined that the incoming interrupt should be handled, it will calculate
a vector index based on the interrupt lead number. This, in turn, is
changed into an interrupt vector location which is used with a vector
base address to give us an interrupt processing entry point. In short,
the interrupt lead is mapped to an address which then is placed in
the instruction address register. (The current instruction address register contents are saved in a particular register or in a agreed-upon
location within the current process task.)

Preparation of the system

When we enter the interrupt processing routine, all of the registers (except for the instruction address register) are the same as within the
software routine that was interrupted. They will also need to be in that
same condition when the interrupt finishes and the original routine continues to execute. This means that all of the processor contents must be
saved. Some processors have a special instruction that saves all of the
registers by pushing them onto the local stack (this stack area can be
the same as the interrupted routine—or it may be changed to a special
interrupt stack). At any rate, the current contents must be saved.

After the previous context has been saved, it is next necessary to set
up the context for the interrupt routine. This may entail some initialization of registers or satisfying other constraints imposed by the language

and compiler. All of this may be accomplished within an assembly language entry *preface* to the interrupt processor. Some language extensions also add special keywords (such as *interrupt*) that indicate to the compiler that this special prefacing (and postprocessing) software needs to be automatically added to the beginning of the function.

Interrupt processing

The interrupt processing routine is now running. It examines the status registers to determine the incoming hardware event. It then processes the event and initiates any necessary responses. At some point in this process, the device interrupt will be canceled. This may be done as a side effect to reading the status register (in which case, the interrupt routine must be prepared to process *all* of the status bits that are currently set when it reads the register since, otherwise, interrupts will be lost). It may also be done explicitly by setting a command register acknowledgment bit. It depends on the design of the hardware device.

Once the processing is complete, the interrupt controller is acknowledged as the last thing before exiting. Sometimes the interrupt controller acknowledgment is implicit when the interrupt routine finishes. It will then restore the previous context and replace the previous instruction address register contents. The next instruction to be processed is that of the interrupted routine. Interrupt processing is complete.

Summary of Chapter

A real-time system requires the ability to react to events in accordance with chronological time. This means that it will need special programming to handle interrupts and other real-time asynchronous events. It will also need to be efficient, particularly in the interrupt processing routines. System resources will include buffers, timers, and data space use. These are not necessarily unique to a real-time system, but special requirements will be associated with real-time needs.

Real-time programming requires the use of LLDs and interrupt processors. It will usually need to use an operating system to provide asynchronous task handling.

Many aspects of the system design are common to those of other types of systems. These include profiling, to determine the efficiency of the system, and debugging, to locate and correct problems within the system. There are some special restraints in the use of these functions within a real-time system.

The next, and final, chapter will discuss the use of the coordinating and management entities. These provide the "glue" for the system and let the system route data primitives appropriately when the system is designed for multiple use (multiple links, multiple interfaces, multiple service types, etc.).

12

Coordination and Management Entities

The coordination and management functions form a set that acts as a method of controlling and coordinating the various layers into a working system. The management functions are primarily concerned with the maintenance of global data throughout the system. The coordination functions act to interpret primitives, and route them, on the basis of their contents to the appropriate layer entities. They also act as arbitrators of state conditions—not allowing an N_DATA_RQ primitive to be sent over a connection that has not yet been established, for example.

Management functions are described in greater detail in the various ITU-T recommendations than are coordination entities. In ITU-T Recommendation I.320, for example, the layers are marked as parallel to the local control plane, global control plane, and plane management function as can be seen in Fig. 12.1. In the B-ISDN protocol reference model (as seen in Chap. 8), the documents refer to the user, control, and management planes. Other documents sometimes refer to the supervisory functions.

Some of these apparent discrepancies are due to the gradual evolution of the architecture of ISDN. They can also be looked at as different "slices" of necessary system functionality. Sometimes they are referred to in terms of state, such as the distinction between control and user planes—where the user plane incorporates functions needed on an active connection and the control plane provides functions needed on a connection that is not yet established (or in the process of being torn down). In such a situation, some of the functions may overlap between the two planes. Sometimes the architectural areas are described in terms of functions. Thus, a supervisory and management plane may be used.

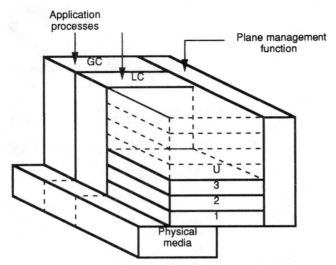

Figure 12.1   ITU-T generic protocol block. (*From ITU-T Recommendation I.320.*)

In any case, no matter how the particular document may choose to refer to them, these functions are used to provide coordination and global resource management. As such, they provide the means to create a coherent system. Coordination functions can be almost nonexistent in a system that supports only one service type and one protocol stack (an ISDN system that supports one voice channel, for example). It is when multiple logical links, physical interfaces, different service types, or different higher-layer interfaces need to be supported that the coordination functions become sufficiently complex to mandate the creation of a separate coordination entity (existing, perhaps, as a set of distinct tasks under an OS).

This chapter will present various aspects of the coordination and management functions as used in an ISDN system. It is meant to form an architectural foundation for the basic concepts used within ISDN as a means to integrate the system. A discussion of the basic differences between user and control plane use will be presented first. Then a more detailed analysis of management functions. Finally, the supervisory and coordination functions for a complex, varied, system will be discussed.

Control and User Planes

The Control plane (C-plane), and User plane (U-plane) are not fully separate areas. The C-plane is further divided into the local C-plane (LC) and global C-plane (GC). These distinctions are to be used for architectural methodology. The exact application processes that are specific to each plane are left by ITU-T Recommendation I.320 "for further study."

So, what kinds of information are associated with a C-plane? A CRV is associated with the control of an individual connection. It does not have any significance to the remote end (the remote end's CRV is completely independent from what the network has assigned to the local end) and, thus, is part of what can be called the LC functionality. On the other hand, the service type that has been negotiated (out-of-band or in-band) *does* have significance from end to end. It is something that can be considered part of the GC functionality.

C-plane data (and functions) are part of the necessary information to set up and tear down calls. A TEI is negotiated locally and is part of the LC plane. The process of negotiating the TEI is part of the management plane (MDL_AS_IN and MDL_AS_RQ primitives) but can also be treated as part of the LC plane (DL_U_DA_RQ and DL_U_DA_IN primitives on the broadcast logical link) since these procedures furnish control information that is of local significance. The channel identifier is also of only local significance. However, *user rate* information applies to the global, end-to-end, scenario.

If the C-plane pertains to connection information, to what does the U-plane relate? This is associated with information that the user, or application layers, may need to perform their services. A TEI and the TEI management procedures are irrelevant to a higher-layer entity. The same is normally true about channel identification information, but this is a gray area because it depends on whether the application is allowed to request a specific bearer channel when placing a call. If so, the channel identification is significant on the U-plane.

The service type will be significant on both the C- and U-planes. It is important on the C-plane because of the use of the IEs to support the negotiation and because of hardware interactions that will result. It is important on the U-plane because the user (and the application that is being used) needs to know whether voice is being provided or Group 4 fax or whatever. Caller ID information may be desired by the high-layer application.

These examples start to point out just why the various planes are so difficult to define precisely. Functions, and locations, depend on the use of the network entity. In some cases, there may be no C-plane necessary at all (fixed TEI and one semipermanent X.25 link). In others, information may be useful on the U-plane, and if so, it should be considered to be located there for planning purposes. However, if it is not useful, it still may be part of necessary C-plane data. In one piece of equipment, information x has only local significance and, in another, has global significance.

Management Plane

The U- and C-planes do not provide crisp boundaries or definitions to the system. The management plane provides a greater solidity because part of the management functions are based on the OSI model. Therefore, a subset of management functions are defined in various ITU-T protocol interface documents. An M primitive (MDL_, MPH_, MN_, etc.) passes between a layer entity and the management plane. As we have seen in the discussions about interfaces, most management primitives are able to be implemented in the form of a synchronous call.

The management plane functions can, thus, be seen as a library of function calls that are available to all of the layers. However, some functions are relevant to only one layer, and in most instances, the data that are passed along with the management request will vary even if the actual function remains the same.

What types of management plane interactions are defined in the various ITU-T recommendations? One large category is that of errors. Every layer has an ERR_IN primitive that is sent to the management plane. In some instances, such as Q.921, there are also recommendations as to the desired response from the management plane. Specific errors will be listed in the documentation, and these should be implemented. Each implementation, however, will have error possibilities that are specific to that equipment. These should also fall into the category of a management plane Mxx_ERR_IN primitive.

Another category is general information. The physical layer provides information to the management plane about the state of the physical interface. (It is even optionally possible for the management plane to send a specific request to deactivate the physical interface.) General status information is not included in the primitive definitions in the recommendations, but such would fall into this category.

One possibile extension would be to have Mxx_STAT_IN and Mxx_STAT_RQ primitives. This type of primitive grouping can provide automatic, or polled, statistics concerning the general operation of the system. It can also be used to obtain specific internal information that is not directly relevant to other layers but may be needed for maintenance. Information of this sort might include the current state of a logical link and how many data frames are waiting to be acknowledged.

In the discussion about the C-plane, we mentioned the situation of TEI assignment. TEIs are associated with the data link layer. However other layers also have identifiers. A CES is used in intertask communication. The network layer needs a unique identifier to identify the specific connection (if more than one connection is supported). All of these primitive types fall into *management assignment* categories.

Timers are mentioned in the various ITU-T recommendations, but only the use of, and response to, these events are specified. There is no mention of primitives that support this required event. Yet, to provide proper layer boundaries, it is necessary for the entry (and exit) points to be firmly delimited. It would be possible to use a generic management function to directly request a timer [start_timer( ), for example]. The problem arises in the opposite direction. What happens when a timer expires? Where will this event be intercepted? The most reasonable place is at the incoming management processing function [an mdl_in( ) function for the data link layer, for example]. If an Mxx_TM_RS primitive is needed, it may be cleaner, and more balanced, to also use a parallel primitive such as Mxx_TM_IN to initiate timers.

There are also functions that are needed in the system but which are, inherently, specialized to a particular piece of equipment or hardware. The main category of these is controlling hardware devices. An MPH_CONTROL_RQ primitive can be used as an umbrella, but the actual primitive would then be specified by some parameter to the primitive. Is it better to maintain the nomenclature even if it is not directly useful? This is a design decision to be made by the system architect and implementors.

Finally, there will be a category of system routines that are independent of the layers and which should probably be initiated only as function calls. For example, buffers differ from the timer situation since all of the interaction is initiated by the local entity. Thus a call of buffer_allocate( ) may be used to obtain a buffer, and a call of buffer_deallocate( ) could be used to free the buffer. Such functions may be used equally from all layers. Other functions may do things such obtaining the system clock time (this function might be provided by the operating system).

Sublayers and management primitives

The use of sublayers brings with it the concept of vertical management primitives. The management plane is normally presented in the form of an equally accessible plane that is parallel to the seven layers (the upper layers of which may not be present on the local node). When sublayers are involved, such as in SVCs for Frame Relay, management primitives may be presented directly between management subsections. This is to provide the proper layer boundaries. Otherwise, one sublayer would have to present a management primitive that is actually appropriate for another sublayer.

Sublayer A may send primitives G and H to the management plane and sublayer B may send primitives J and K. Normally, a layer will only send and receive primitives that are directly associated with the functions of that layer. In the case of sublayering, it is possible that sublayer A may want to send primitive J (for example, to transfer a negotiated DLCI). If A sent the primitive J directly, it would violate layer boundaries. However, if it is allowed to send primitive I to the management plane, and the management plane, in response to receiving primitive $I,$ sends a primitive $IJ,$ which is interpreted as the same as receiving J from sublayer B—layer boundaries have not been violated.

In some ways, this is a diplomatic issue. To the implementor, the most important task is to provide a robust and reliable system. To the architect, the most important task is to provide a framework that will allow for expansion and change. Both are right within the framework of the task they have. The final implementation decision depends on many factors. One of the primary aspects of such factors has to do with the reusability of the modules and the various layers. The architectural framework is designed primarily to allow ease of change. If change is not needed, the added overhead may be deemed counterproductive.

Errors

Error indications are the most widely specified type of management primitive. Errors can be sent from any of the various layers. They will often have parameters associated with them to indicate the identification of the logical entity that is involved in addition to specific information that is associated with the particular error.

Some errors call for specific responses. On this basis, some of the recommendations (ITU-T Q.921, for example) delineate errors based on the action that is to be taken. In Q.921, the errors are identified as error codes labeled A through O. These are subclassified into types of error such as receiving an unsolicited response, unsuccessful retransmission, peer initiated reestablishment, or a general error type.

After classification of errors (with a description of the condition which causes this error), ITU-T Recommendation Q.921 states the actions to be taken. Many of the errors call upon basic logging. Some will affect the status the TEI. These are error conditions that are sufficiently serious to call into doubt the basic reliability of the logical link. In many situations, the action is left as implementation dependent in the case of a user-side management situation.

The physical layer does not directly specify the appropriate management entity responses to errors. This is because many errors will be specific to the hardware and, thus, not applicable to a general architecture document. Other layers do not indicate the precise indications or the specific actions to be taken. It may be expected that, as the standards fully stabilize (if ever), the management interactions will be more clearly defined.

Hardware interactions

Hardware interactions fall into two categories—the generic and the specific. The generic management primitives deal with conditions that may be expected in multiple physical devices that provide the same general functions. The MPH_DI (Deactivation Indication) primitive is such a primitive. This primitive may be generated by many different LLDs that control many different hardware devices. The primitive is associated with all LLDs that provide a specific function. This function is that of giving access to a physical interface. Physical device LLDs that do *not* provide physical interface layer access will not incorporate this primitive. An example might be that of accessing a video processing unit that is directly controlled as a side effect of information relayed over the signalling channel.

Hardware-specific primitives.   Most hardware interactive management primitives fall into the "specific" category. They are specific to the hardware device, and they are specific in response based on the use of the entire system. These are not easily categorized and are not normally specified in any of the standards or specifications.

Hardware-specific management primitives are determined according to the needs of the system and must be part of the overall system design. A test interface may need access to the *loopback* functionality of the hardware device. An application that is associated with a voice service may need a command that tells the CODEC to turn up the gain on the audio output for a receiver. These commands make use of the available functions supported by a hardware device but are not inherently part of the ISDN protocol.

B-channel connection.   Other hardware-specific primitives *are* part of the ISDN protocol requirements. One such primitive is a command to route a demultiplexed TDM channel to another device. This is done at a point in the protocol when the bearer channel has been established from end to end. In a BRI, there are three channels multiplexed over the same interface and are designated as the D-channel and B-channels (one and two). The D-channel should be automatically demultiplexed, and routed, to an HDLC device as soon as line activation has occurred. This allows the signalling access to the network (or from the network to the user, depending on the equipment's location).

The B-channels, however, contain no valid information until the bearer channel has been routed to another peer endpoint by means of the network. Once end-to-end connection has been established, the allocated B-channel (or B-channels) will contain valid data. However, the data must be processed by something. In the case of unrestricted data, this may be an HDLC device. It may also be a V.110 frame-supporting device. Whatever the exact purpose, there will be a coordination between the Q.931 signalling protocol and the hardware to set up the B-channel appropriately.

The setting up (or connection) of the B-channel is mainly a situation where the data are now known to be valid, and it is important to route them to a device that can handle them. For this purpose, the interface device may be told to demultiplex the B-channel. It may then be told to route the demultiplexed channel to a specific device.

On the other hand, the data may always be available over a common data bus. They are just ignored until some device is told to service the data. The routing of the primitive will depend on the devices and the function of the system.

Statistics and system information

Statistical information is not usually specified in the ITU-T recommendations. Yet, it may be a vital part of the overall system requirements. An application cannot make use of any statistics if the data are not available. Additionally, an application is extremely limited in its ability to monitor the condition of a local system without the means to collect information.

These primitives are needed for the final system. They may also be useful during profiling or debugging periods. Thus, there will be two categories—one of which is a subset of the other. The large category is that of all statistical information to be gathered from the system during development. The subset is concerned with the data that are considered to be useful during the run-time of the system. The fact that these extra primitives add overhead must be taken into account and the overall number of the final primitives strictly limited.

If possible, these primitives should involve pure logging of data without any associated algorithm for suggested response. This is to reduce the overhead as much as possible.

Why should such a function be implemented as a primitive? Wouldn't it be better, and faster, to implement it as a system library function? That is certainly possible. Part of this design decision depends on whether polled information can be solicited. If so, a parallel balancing of primitives is architecturally easier to maintain. If not, a special log_data( ) function is more appropriate.

System-coordinated assignments

Many systems require data that are used to communicate between entities. Sometimes it is a specific identifier that is attached to the system as a whole. Other times it is negotiated by one entity, communicated to another, and then used as an identifier to indicate a specific use. An ISDN system has a number of these. Between the physical and the data link layers (or other upper-layer function which needs access to the physical device), this primarily consists of an interface and TDM (or other type of multiplexed data entity) channel. If only one interface is supported, the first is not needed.

The TEI is not actually used as an identifier between layers. It is used, rather, as an identifier for use with the network. Chapter 4 covers much of the negotiation needed for automatic TEI assignment. However, this negotiation is done by the management layer (though specified in ITU-T Recommendation Q.921) and must be passed, after negotiation, from the management plane (or entity) to the data link layer.

The network layers use a combination of TDM channel, CES, and SAPI to identify a specific type of logical link. These will be combined with the TEI to form the logical link address.

The network layers normally need a type of identifier within them. For ITU-T Recommendation Q.931, this is the CRV. If a call is placed remotely, it will be supplied by the switch. A call originated locally, however, must allocate a CRV to use in its dealings with the network. An X.25 network layer requires an LCI which is allocated based on LCIs in use and the available ranges.

Finally, in a system that supports multiple connections, the higher layers need a method of identifying the specific connection. In a pure Q.931 model, the CRV could be used for this function. In a system that supports in-band call management in addition to the Q.931 signalling, this will not be sufficient.

Physical layer identifier. The physical layer identifier consists of an interface and a physical channel identifier. If only one interface is supported, there is no need for an explicit interface component. As mentioned above, the signalling channel should be automatically initiated upon line activation. There will need to be a method of identifying this signalling channel (or channels, in some PRI and B-ISDN situations).

Other channels will come into use after call setup has been achieved. The bearer channel is connected to appropriate hardware. At this point, interactions with the physical layer, or with entities that can support multiple physical channels (such as layer 2, in many cases), need to have an identifier associated with this particular established physical channel. For an N-ISDN system, this can consist of the CHAN_ID field. Thus, a 00 can indicate the use of the D-channel on interface zero, and a 01 can indicate the use of B-channel number 1 on interface 0. These identifiers are implicitly defined according to the out-of-band call setup parameters *or* via the parameters as give by the network at subscription time for semipermanent links.

TEI assignment. TEI assignment results from the actions taken by the data link layer and the management entity which deals with the network. If a local request for use of a logical link appears, and the data link layer does not have a TEI assigned, it can respond with an MDL_AS_IN primitive. The response of the management entity is then to give the data link layer a TEI, if it is nonautomatic, or to begin the negotiation with the network.

TEI negotiation is covered in detail in Chap. 4. The important aspects for this chapter are the interactions with the management entity. Either after negotiation, or immediately upon request for nonautomatic TEIs, the management layer will initiate an MDL_AS_RQ primitive. This primitive then associates a TEI and a CES with a logical link. If requested, it may then begin establishment of multiple frame mode.

CES and SAPI usage.   The CES is assigned by the system as a purely local identifier to identify the local (or peer-to-peer) use of a connection. One identifier might be used for signalling use. Another might be used for X.31 B-channel use. A third might be used for use of D-channel packet layer handling. The use of, and choice thereof if used, is up to the system designer and has no network significance.

The SAPI, on the other hand, is specified by the ITU-T recommendations. Some values are currently assigned, others are in the process of being assigned, and others are still in reserve for future expansion. These are assigned, and used, according to function. Signalling has a specific SAPI and so do management functions (such as TEI negotiation). A D-channel packet link has a separate SAPI and, with switched Frame Relay, other SAPIs may be dedicated after signalling call setup.

The CES and SAPI are both used as identifiers. Unlike some other identifiers, these are fixed (according to system design or ITU-T recommendations). They are used for specific purposes to indicate a certain need.

Network layer identifier assignment.   Layer 3 often needs its own identifier—normally to interact with the peer rather than for interentity communication on the local side. This is called the CRV for the Q.931 signalling protocol, the LCI for X.25, the DLCI for V.120 (and for Frame Relay), and so forth.

When a network layer entity needs an identifier, it could use local data structures and functions to come up with a proper value. However, this will interact with the use of values for other channels, links, and identifiers. It is also a common function, even if the data involved may vary in form. Routing the request as a management entity primitive may be a very reasonable choice. This may not be required, depending on the variety of services to be supported by the system.

System higher-layer identifiers.   The layers above the chained layers (higher layers, upper layers, application layers, etc.) need a form to identify a logical connection. This is used in various primitives (N_DATA_RQ, N_RESET_RQ, N_CONN_IN, etc.) to communicate about a specific connection between the chained layers and the upper layers. CRVs are unique to Q.931 established connections. LCIs are unique to X.25 connections, and so forth. Each type of network layer protocol will have its own unique identifier (perhaps in conjunction with other identifiers) that can be used with a particular logical, or physical, connection.

If only one service type is used, the implementor may decide to use the CRV (for example) between the network layer and the higher layers. If not, it would be possible to use a CRV for a Q.931 channel, LCI for an X.25 channel, and so on. This means, however, that the primitive interface would not be service independent between the network layer and the higher layers. For these reasons, allocating a systemwide (but service independent) connection identifier may prove to be a good system design choice.

Timers

The use of timers is specified in many ITU-T recommendations. However, the actual format of these events is not specified. ITU-T Recommendation I.430, for example, lists actions that are to be taken when timer T3 expires. It also lists when the timer is to be activated. Nothing is mentioned about the form of the events.

In some ways, this is to be expected. Timers are a function that is based on the complete system design and implementation. As discussed in the previous chapter, timers may be provided as a specific hardware/software interaction, or they may be used within a generic fashion. Thus, a standard is not directly relevant.

However, these events must be supported in some way—particularly for the timer responses. If the management entity provides a common timer support mechanism to be used by the various layers of the system, timers become a management function. In this light, using Mxx_TM_IN and Mxx_TM_RS primitives becomes very reasonable. If a specific timer is exclusively used, such primitives are unnecessary overhead. (A single function would exist to operate the hardware timer, and it would be serviced as part of the interrupt processing function assigned to the interrupt.)

Some of the parameters to be associated with such management primitives were given in the previous chapter. These fall into the classes of identification and length. The length is used by the general timer mechanism to set up the data structures. The identification is used by a timer support task and within the layer that is expecting the timer response.

Layer-independent functions

A system will have a variety of functions that need to be used by various layers. In some systems, these functions are considered to be part of a common system library. Thus, they would be compiled into the protocol system by including the manage_func.lib as part of the compilation. This is an aspect of organization. It does not affect the use of, or need for, the functions.

Some functions may be associated with buffer use. Others may give access to logging, profiling, or debugging functions. Some may be generic routines that are often (but not always) provided as part of the compiler's standard libraries. These include string manipulation functions, conversion routines, and so forth. Whether these should be considered management or utility functions is an individual choice. Generally, management functions should deal with globally coordinated data (such as buffers, or system clock times).

Summary of use

Management plane functions provide for data coordination functions and synchronous actions that are common for a system and rely on globally available data. Sometimes they are directly specified in recommendations, standards, or specifications. Often they are left as functional requirements without formats being given. Sometimes they are needed to provide assignment coordination for identifiers or system timer use.

Coordination Entity

The coordination entity is needed in a multipurpose system. This may involve the coordination of multiple physical interfaces, connections, service types, functionality (TA versus coprocessor use, for example), or any other diverse set of functions.

A basic coordination entity will exist in fact, if not in name, in all systems. This is because the bringing up of a system, initializing the various data structures, and starting the needed tasks is part of the function of a coordination entity. Coordination implies sequencing and routing. In the start-up routine, it is important that sequences occur in order—that data structures are initialized before interrupts which may make use of the data structures are enabled.

The coordination entity has a large advantage for a complex system. It provides commonality. It lets the higher layers interact with the system in a common way—independent of the exact state, or use, of the system. It also filters and translates primitives coming up from the lower layers into the common interface primitives expected by the higher layers.

The duties of the coordination entity depend on the state of each of the layers of the system. It does not always need to be aware of the internal state of each of the layers. It *does* need to know information such as the fact that a B-channel is in use and it is being used for a particular bearer service type. It needs to know this information to be able to determine the legality of primitives between the lower and higher layers. It also needs to know this information to route primitives to the right entity if multiple service types are supported.

The coordinating entity is involved at the invocation of the system. It knows, implicitly or explicitly, that initialization is to occur and coordinates the sequence of events such that the system is brought into action in a safe and dependable manner. It may also wait for a particular stable system state to exist before notifying the higher layers that the system is available for use.

Later, the coordinating entity splits out primitives based on whether they are meant for the system as a whole or a specific connection. If the primitive is meant for a specific connection, it uses its locally stored information about the connection to determine where the primitive should be sent. An N_DATA_RQ primitive for a bearer channel that is supporting X.25 over an unrestricted data link should be routed to the X.25 network layer. However, if it is a V.120 link, it will need to be routed according to the layer 2 and 3 protocols as given in the Bearer Capability or LLC IE.

Sometimes the coordinating entity will block primitives that are obviously illegal. A data request on a voice channel is illegal. Is it better for the coordinating entity to block the primitive or to route it to a layer 3 entity for it to process in the error-handling method specified by its protocol? If a known entity exists, it is probably best to pass the primitive along, but what about that N_DATA_RQ primitive that is being used on a voice bearer channel? To where should the primitive be routed for the error handling?

For some conditions, no "right" place exists for the routing. In such cases, the coordinating entity can act to protect the rest of the system from possible confusion. The responsibility for error detection lies in either the layer protocols, coordination entity, or higher layers. Redundancy adds to overhead, so it is useful to be able to coordinate the actual functions provided by the different entities. The coordination functions are often incorporated into separate documents (such as X.31 for usage of X.25 over ISDN).

Supervisory functions

Supervisory functions include initialization, testing, and higher-layer hardware access. Initialization has been discussed already. Testing involves having access to protocol aspects that are not normally used in a typical call scenario. Higher-layer hardware access gives the possibility of an application being able to control hardware before, during, or after a connection has been established. This is useful in both gateways and for integrated stations that need to coordinate multiple interfaces and equipment.

Testing.   During the testing of a system, the developer may need access to special functions that are not useful in the final system. Some of these accesses may be imposed by testing or certification requirements. Others are useful during debugging to provide special mechanisms for automated testing.

What types of things might be useful in a testing environment? For testing of the layer 2, for example, it would be useful to be able to bypass the other layers—have the application layer request the establishment, or release, of a logical link without using the regular protocol stack. Using the regular protocol stack would result in higher-layer processes being invoked (since a data link layer has no normal need of being established without a higher-layer protocol needing to use it). The supervisory test function, thus, provides access to a special use of a layer.

It might also be useful to have the lower layers generate data requests without having the data passed from the higher layers. For example, say that there are throughput problems on the system. Do they lie in the disk I/O of a host system or in the protocol overhead in the lower layers? If the data are generated by the coordinating entity, one variable is eliminated.

The application may also want to cause the system to go into *local busy* state. This is potentially useful for an active system but can also cause software to be tested that might not normally be invoked in a running system. TEI assignments may also be invoked to test the TEI management procedures.

Higher-layer hardware access.   This type of supervisory function is similar to that of invoking lower-layer events without the intermediary layers. However, in this case, the request translates into a specific hardware configuration change—or hardware information. A primitive can be used via the coordinating entity to inquire as to the physical interface framing state.

Let's say that a voice bearer service connection has been set up. The system has automatically connected the bearer channel to a CODEC based on the service type. This CODEC will operate as a specialized digital-to-analog (and vice versa) device for voice-data translation. Under certain line circumstances (lots of static) or personal preference situations, the hardware may need to be changed. The volume (or gain) may need to be manipulated. A tone may need to be generated. These are examples of application layer access to the hardware.

Higher-layer data access. It is possible to access hardware devices via the application layer by use of the coordinating entity in a supervisory role. It is also possible to gain access to data (or to set it, although this would probably be useful only under testing, or debugging, conditions) in the same way.

The user may want to know how many buffers are in use and may also want to verify information that is available locally. This would be data such as how many connections are active, how many on hold, what are their service types, and so forth.

The use of the coordinating entity in its supervisory role allows network administration functions to be performed. It can verify the data that it holds versus what the system reports, can request periodic error statistics, and can check on activity levels at the different layers. How much data is being transported? What are the retransmission rates? How many line errors are being detected by the hardware?

A gateway application may, in particular, need to gain access to data of the system. In such a case, there is no human user to visually, or otherwise, verify the sanity of the system. If a user receives a call (assuming the optional incoming call timeout timer is not implemented), there will be audible, or visual, reminders that an incoming call is pending and is awaiting action. For an automated application, it will depend on the integrity of the application's database and the reliability of its auditing sections to make sure that pending events are eventually noticed.

Service types

One of the most useful functions of the coordinating entity in a real-life functioning system (as opposed to testing scenarios) is that of being able to support multiple service types within the same system.

Let's take a fairly complex system as an example. An ISDN system is able to support one (but only one) voice channel; it can also support two unrestricted data channels (if one is not occupied with voice). One of the bearer channels can be used for X.31/X.25 data traffic. The other bearer channel is connected to a serial interface that supports asynchronous command and data access.

A voice call arrives. Interlayer primitives are sent up and down the stack. To the coordinating entity (and application), the most important primitives are N_CONN_IN and the corresponding N_CONN_RS primitive. (Other primitives may also pass through the coordinating entity to detail progress information, INFO data segments, or other general information.)

It is possible that the system may automatically connect the B-channel to the local CODEC. However, it will also be necessary to inform the application that a voice connection is being made. The application will provide different stimuli and will be restricted in its primitive use, depending on the service type.

As a second example involved with service type (but also associated with connection management), let us consider a Case A X.31 connection. Two connections will need to be made. The first N_CONN_RQ is routed to the Q.931 out-of-band signalling protocol—invoking a service type of unrestricted data. The second N_CONN_RQ is routed to the X.25 call management section of its layer 3 protocol.

It would be possible for a single N_CONN_RQ primitive to cause both of these actions. All that would be needed would be both addresses and the appropriate mixed service type (one for the Q.931 and one for the X.25). At this point, the coordinating entity could send the N_CONN_RQ primitive to the Q.931 software and then, after receiving the corresponding N_CONN_CF, invoke a new N_CONN_RQ to be sent to X.25. Only after the N_CONN_CF primitive was received from X.25 would the confirmation be sent back to the application.

In many systems, this would not be the best scenario. This is because of how the standard application layer interfaces are configured. It would be better to have the application send two sequential N_CONN_RQ primitives, each with its own service type and address. The channel identification would need to pass back to the application. This channel identification can then be used by the coordinating entity to make the association between the two otherwise unrelated N_CONN_RQ primitives. Each N_CONN_RQ primitive would be routed according to the needs of the service type.

Connection management

Connection management means the actual routing of a connection's primitive stream based on the way that the system is being used. Above, we were most concerned with the connection establishment. Now, we will be concerned with the general data, and primitive, flow for the system.

As another example scenario based on the above complex system, say that an application wants to start a call via the R interface reference point. This information will arrive through the LLD that handles the serial interface device. It will then pass it to the TA functions. These functions, in turn, pass it along to the coordinating entity. Why does it do this? Why doesn't it send the request directly to the network layer?

The main reason is that this would decentralize the coordinating role. The secondary reason is that it would entail duplication of software, and general logic, needed by the system.

If all primitives pass between the lower and higher layers via the co-ordination entity, it alone needs to maintain, and use, the information collected about each connection. An N_CONN_RQ primitive message arrives from the TA function. The coordinating entity remembers this routing and associates it with a global connection identifier that it manages.

It will then proceed to establish the connection based on the service type as described in the above section. Once the connection is estab-lished (assuming that it was successful), the N_CONN_RS primitive will arrive back at the coordinating entity. The coordinating entity knows that this connection identifier is associated with the TA func-tion. Thus, it can route the primitive there. The TA then performs its necessary functions (possibly invoking data transfer via the LLD across the R interface reference point).

One connection is now established between the TA, network, and TE2. Any further primitives can continue to be routed based on the connection identifier (in the case of hardware supported V.110 there may not be any primitives until the connection is ready to be re-leased). Now another connection request arrives—a Case A X.31/X.25 connection just as mentioned in the earlier example. With the common coordination, one bearer channel may be used for the TA function and the other used on a shared memory function—or a separate (and dif-ferent) TA that interacts with a different type of user equipment.

Implementation Concerns

The U- and C-plane functions are incorporated as part of the layers, the management entity, and the coordinating entity. The manage-ment entity is primarily made up of synchronous functions that coor-dinate global data use. (Some, like the timers, may have greater complexity, but this has already been discussed in detail.) The coordi-nating entity is the most complex of these functional areas.

If the coordinating entity acts as an intermediary situation between the lower and higher layers, what will this mean in terms of architec-ture and implementation? From the higher layer, it will need to com-municate with any type of supported interface (TA-initiated, shared mailbox, or integrated application primitives, for example). Between it and the lower layers, it will need to know the primitive interface for any entity with which it will need to communicate.

Since the time needed within the coordinating entity is unknown, it should have a "software break" between the higher and lower layers. This could be done on either "side" of the coordinating entity, but the methods of primitive interfaces toward the higher layers tends to sup-port the break on the side which interacts with the lower layers.

So, a primitive arrives at the coordinating entity from "above" (via shared memory, TA primitives, or whatever). It then analyzes what is needed based on the current state and the parameters of the event. Toward the layer 3 or layer 2 entities, it then makes use of intertask message queues. (With the hardware, it may use a synchronous call if the task is brief.)

In the other direction, there will be message queues from the lower layers toward the coordinating entity. The coordinating entity can then analyze the primitive and pass it along to the higher layers. (In general, less analysis is needed for primitives coming up since there is little that the coordinating entity can do in most circumstances.)

These architectural aspects are of the most value in complex systems. In a simple system, the value is greatly decreased, and the coordinating entity may be simplified to a simple set of function calls and translating routines. The knowledge of how to work with a complex system is still useful for something simpler.

Summary of Chapter

The coordination and management entities act in a fashion to allow separate modules, services, and interfaces to work together in a system. Many of the tasks of these entities are not directly specified in architectural documents or specifications. The ITU-T recommendations mention local control planes, global control planes, management planes, supervisory functions, and even coordination functions. The names change according to how the system is to be used and how the functions are divided.

Coordination functions are primarily needed in a complex, multi-purpose, system. In a simple system, they may be reduced to simple functions and translation procedures. As the system complexity increases, the value of a central coordination function also increases. Together, the coordinating and management entities present a uniform and versatile system.

References and Selected Bibliography

The documents listed in the first two sections were used as reference material for this book. The first section refers to ITU-T recommendations. The next lists other technical reference specifications. The final section lists various books on ISDN, some of which are referred to in this book. This listing should not be considered to be a complete guide to all ISDN literature.

ITU-T Recommendations

F.310	Broadband Videotex Services
F.732	Broadband Videoconference Services
F.821	Broadband TV Distribution Services
F.822	Broadband HDTV Distribution Services
G.703	Physical/Electrical Characteristics of Hierarchical Digital Interfaces
G.704	Synchronous Frame Structures Used at Primary and Secondary Hierarchical Levels
G.711	Pulse Code Modulation (PCM) of Voice Frequencies
G.721	32 kbit/s Adaptive Differential Pulse Code Modulation (ADPCM)
G.722	7 kHz Audio Coding Within 64 kbit/s
I.112	Vocabulary of Terms for ISDNs
I.113	Vocabulary of Terms for Broadband Aspects of ISDN
I.120	Integrated Services Digital Networks
I.121	Broadband Aspects of ISDN
I.122	Framework for Providing Additional Packet-Mode Bearer Services
I.140	Attribute Technique for the Characterization of Telecommunication Services Supported by an ISDN and Network Capabilities of an ISDN

I.150	B-ISDN ATM Functional Characteristics
I.211	B-ISDN Service Aspects
I.230	Definition of Bearer Service Categories
I.231	Circuit-Mode Bearer Service Categories
I.233	Frame Mode Bearer Services
I.311	B-ISDN General Network Aspects
I.320	ISDN Protocol Reference Model
I.321	B-ISDN Protocol Reference Model and Its Application
I.325	Reference Configurations for ISDN Connection Types
I.327	B-ISDN Functional Architecture
I.361	B-ISDN ATM Layer Specification
I.362	B-ISDN ATM Adaptation Layer (AAL) Functional Description
I.363	B-ISDN ATM Adaptation Layer (AAL) Specification
I.411	ISDN User-Network Interfaces—Reference Configurations
I.413	B-ISDN User-Network Interface
I.430	Basic User-Network Interface—Layer 1 Specification
I.431	Primary Rate User-Network Interface—Layer 1 Specification
I.432	B-ISDN User-Network Interface—Physical Layer Specification
I.440	See Q.920
I.441	See Q.921
I.450	See Q.930
I.451	See Q.931
I.452	See Q.932
I.460	Multiplexing, Rate Adaptation, and Support of Existing Interfaces
I.461	See X.30
I.462	See X.31
I.463	See V.110
I.465	See V.120
I.610	OAM Principles of B-ISDN Access
Q.920	ISDN User-Network Interface Data Link Layer—General Aspects
Q.921	ISDN User-Network Interface Data Link Layer Specification
Q.922	ISDN Data Link Layer Specification for Frame Mode Bearer Services
Q.930	ISDN User-Network Interface Layer 3—General Aspects

Q.931   ISDN User-Network Interface Layer 3 Specification for
             Basic Call Control

Q.932   Generic Procedures for the Control of ISDN Supplementary
             Services

Q.933   Digital Subscriber Systems No. 1 (DSS 1) Signalling
             Specification for Frame Mode Basic Call Control

T.50   International Reference Alphabet (IRA) (Formerly
             International Alphabet No. 5 or IA5) Information Technology—
             7-bit Coded Character Set for Information Interchange

V.21   300 Bits per Second Duplex Modem Standardized for Use in
             the General Switched Telephone Network

V.22   1200 Bits per Second Duplex Modem Standardized for Use
             in the General Switched Telephone Network and on Point-
             to-point 2-wire Leased Telephone-type circuits

V.23   2400 Bits per Second Duplex Modem Using Frequency
             Division Technique Standardized for Use in the General
             Switched Telephone Network and on Point-to-point 2-wire
             Leased Telephone-type circuits

V.24   List of Definitions for Interchange Circuits Between Data
             Terminal Equipment (DTE) and Data Circuit-terminating
             Equipment (DCE)

V.25 bis   Automatic Calling and/or Answering Equipment on the
             General Switched Telephone Network (GSTN) Using the
             100-series Interchange Circuits

V.28   Electrical Characteristics for Unbalanced Double-current
             Interchange Circuits

V.35   Data Transmission at 48 Kilobits Per Second Using 60-108
             kHz Group Band Circuits

V.42   Error-correcting Procedures for DCEs Using Asynchronous-
             to-synchronous Conversion

V.110   Support of DTEs with V-Series Type Interfaces by an ISDN

V.120   Support by an ISDN of DTEs with V-Series Type Interfaces
             with Provisions for Statistical Multiplexing

X.3   Packet Assembly Disassembly Facility (PAD) in a Public
             Data Network

X.20 bis   Use on Public Data Networks of Data Terminal Equipment
             (DTE) Which is Designed for Interfacing to Asynchronous
             Duplex V-series Modems

X.21   Interface Between Data Terminal Equipment and Data
             Circuit-Terminating Equipment for Synchronous Operation
             on Public Data Networks

X.21 bis   Use on Public Data Networks of Data Terminal Equipment
             (DTE) Which is Designed for Interfacing to Synchronous V-
             series Modems

X.25	Interface Between Data Terminal Equipment (DTE) and Data Circuit-terminating Equipment (DCE) for Terminals Operating in the Packet Mode and Connected to Public Data Networks by Dedicated Circuit
X.28	DTE/DCE Interface for a Start-stop Mode Data Terminal Equipment Accessing the Packet Assembly/Disassembly (PAD) Facility and a Packet Mode DTE or Another PAD
X.29	Procedures for the Exchange of Control Information and User Data Between a Packet Assembly/Disassembly (PAD) Facility and a Packet Mode DTE or Another PAD
X.30	Support of X.21-, X.21 bis-, and X.20 bis-Based DTEs by an ISDN
X.31	Support of Packet Mode Terminal Equipment by an ISDN
X.32	Interface Between Data Terminal Equipment (DTE) and Data Circuit-terminating Equipment (DCE) for Terminals Operating in the Packet Mode and Accessing a Packet Switched Public Data Network Through a Public Switched Telephone Network or an Integrated Services Digital Network or a Circuit Switched Public Data Network
X.75	Packet Switched Signalling System Between Public Networks Providing Data Transmission Services

Other Technical References

ANSI T1.606	Integrated Services Digital Network (ISDN)—Architectural Framework and Service Description for Frame-Relaying Bearer Service
ANSI T1.617	Integrated Services Digital Network (ISDN)—Signaling Specification for Frame Relay Bearer Service for Digital Subscriber Signaling System Number 1 (DSS1)
ANSI T1.618	Integrated Services Digital Network (ISDN)—Core Aspects of Frame Protocol for Use with Frame Relay Bearer Service
ISO 2110	Information Technology—Data Communication—25-Pole DTE/DCE Interface Connector and Contact Number Assignments
ISO 4335	Information Technology—Telecommunications and Information Exchange Between Systems—High-Level Data Link Control (HDLC) Procedures
ISO 8208	Information Technology—Data Communication—X.25 Packet Layer Protocol for Data Terminal Equipment

Selected Bibliography of ISDN-Related Books

Black, Uyless. *Data Link Protocols.* Englewood Cliffs, NJ.: PTR Prentice-Hall, Inc., 1993.

Brewster, Ronald L. *ISDN Technology*. London; New York: Chapman & Hall, 1993.
Fritz, Jeffrey Neil. *Sensible ISDN Data Applications*. Morgantown, WV.: West Virginia University Press, 1992.
Goldstein, Fred R. *ISDN in Perspective*. Reading, Mass.: Addison-Wesley, 1992.
Griffiths, John M., ed. *ISDN Explained: Worldwide Network and Applications Technology.*
Händel, Rainer, and Manfred N. Huber. *Integrated Broadband Networks: An Introduction to ATM-Based Networks*. Reading, Mass.: Addison-Wesley, 1991.
Hardwick, Steve. *ISDN Design: A Practical Approach*. San Diego: Academic Press, 1989.
Kessler, Gary C. *ISDN: Concepts, Facilities, and Services*. 2d ed. New York: McGraw-Hill, 1993.
Smouts, M. *Packet Switching Evolution from Narrowband to Broadband ISDN*. Boston: Artech House, 1992.
Spohn, Darren L. *Data Network Design*. New York: McGraw-Hill, 1993.
Stallings, William. *Advances in Integrated Services Digital Networks (ISDN) and Broadband ISDN*. Los Alamitos, Calif.: IEEE Computer Society Press, 1992.
———.*Handbook of Computer-Communications Standards*. vol. 1. New York: Macmillan, 1987.
———.*ISDN: an Introduction*. New York: Macmillan, 1989.
———.*ISDN and Broadband ISDN*. 2d ed. New York: Macmillian, 1992.
———.*Networking Standards: A Guide to OSI, ISDN, LAN, and MAN Standards*. Reading, Mass.: Addison-Wesley, 1993.

Index

ABOUT THE AUTHOR

Charles K. Summers is Vice-President of Engineering at TeleSoft International, Inc., where he is involved with ongoing software design and development, customer support, and marketing technical support. He was formerly a member of the technical staff at AT&T Bell Laboratories in Denver and currently lives in Seattle, Washington.